中国文化遗产研究院
吴哥古迹保护研究系列第一种

茶胶寺庙山建筑研究

Architectural Research on Ta Keo
Temple-Mountain at Angkor Site

中国文化遗产研究院　温玉清　著

文物出版社

责任编辑：李　睿
责任印制：张道奇

图书在版编目（CIP）数据

茶胶寺庙山建筑研究／温玉清著．—北京：文物出版社，2013.9
ISBN 978－7－5010－3807－7

Ⅰ.①茶…　Ⅱ.①温…　Ⅲ.①寺庙－宗教建筑－建筑艺术－研究－柬埔寨　Ⅳ.①TU－098.3

中国版本图书馆CIP数据核字（2013）第199217号

茶胶寺庙山建筑研究

温玉清　著

*

文 物 出 版 社 出 版 发 行
（北京市东城区东直门内北小街2号楼）
http：//www.wenwu.com
E-mail：web@wenwu.com
北京京都六环印刷厂印刷
新 华 书 店 经 销
889×1194　1/16　印张：32.5
2013年9月第1版　2013年9月第1次印刷
ISBN 978－7－5010－3807－7　定价：360.00元

序

柬埔寨吴哥古迹是公元九世纪至十五世纪古代高棉帝国繁盛时期都城与寺庙建筑的遗迹，由于吴哥在东南亚文明史中举足轻重的地位，使之成为东南亚地区乃至全世界最为著名的古代史迹之一。

援助柬埔寨吴哥古迹保护的国际行动始于上世纪九十年代。在联合国教科文组织统一协调下，先后有包括中国在内的十多个国家的工作队或学术机构，在现场开展吴哥古迹的保护与研究工作，吴哥古迹成为文化遗产保护的国际舞台与合作典范。1998 年，中国政府援助柬埔寨吴哥古迹周萨神庙（Chausay Tevoda）保护修复工程正式启动，这是中国政府首次大规模参与文化遗产保护的国际合作项目。2008 年 12 月，由中国文化遗产研究院主持完成的周萨神庙保护修复工程竣工，恢复了寺庙原有建筑格局与艺术风貌，赢得了柬埔寨政府、国际组织以及各国同行的赞誉。2009 年 12 月，中柬两国政府签署换文，正式确认由中国政府提供援助经费，用于开展吴哥古迹茶胶寺庙山（Ta Keo temple-mountain）的保护修复工程。

作为中国政府援助柬埔寨吴哥古迹保护修复工程的执行机构，中国文化遗产研究院结合国际文化遗产保护思潮与我国文化遗产保护实践，按照"科学研究贯穿保护修复工程全过程，抢险加固，排除险情，局部维修与全面修复相结合"总体思路的要求，在茶胶寺开展了详细的前期勘察工作，主要涉及建筑、考古、结构工程、岩土工程、保护修复技术等诸多领域，取得了重要的阶段性成果。本书作者温玉清同志即是我院援柬团队中青年学者的代表，他的《茶胶寺庙山建筑研究》也是上述阶段性研究成果中第一部付梓出版的著作。

茶胶寺（Ta Keo Temple-Mountain），亦称茶胶寺庙山，是吴哥古迹中最为雄伟且具有鲜明特色的庙山建筑之一。这座未曾完工的庙山遗构，为人们探寻和了解古代高棉石构建筑意匠提供了弥足珍贵的实例。温玉清博士的著作正是基于吴哥学术史，从茶胶寺庙山的建筑本体入手，从更为微观的庙山建筑形制与设计尺度入手，并将其置于"印度文明 - 东南亚文化 - 中华文明"的范围内进行考察和剖析。我认为，从某种意义上说，吴哥时代所呈现出的文化面貌，显而易见是多种文化融合的结果；是文化的多元，成就了吴哥时代的辉煌。这是我拜读温著之后的一种真切的感受。

谈及本书关于柬埔寨吴哥古迹研究与保护的学术史的论述，恕我再赘言几句。

众所周知，回溯学术史，在探讨前辈足迹及功过得失之时，其实也是在选择某种传统和规范，并确定自己的学术方向。只有通过了解前人研究的经历和当前研究的现状，以已有的基础作为工作的起点，认清自己所处的地位。只有弄清学术渊源与发展沿革，调整自身的知识结构，才可避免炒前人的冷饭。对于任何置身吴哥古迹保护与研究中国学者而言，吴哥古迹的研究与保护的历史进程如何？如何评析吴哥古迹保护研究的方法和理路？如何客观地评价不同研究方法的特点及其局限性？都是需要面对和解决的问题。拜读温玉清博士的这部著作，处处皆能体会到他密切关注学术史的自觉和"知行合一"的情怀，严谨扎实的学术训练基础由此可见一斑。我更觉得，这也应是本书最为难能可贵之处。

如此想来，书中汇总梳理近百年来吴哥古迹考古清理、调查测绘与学术研究等方面的相关史料，并围绕吴哥古迹的保护修复史进行了概括与论述，力求了解其研究理路的来龙去脉，想必对于中国学者参与吴哥古迹保护与研究局面的改观将是不无裨益的。

今年欣逢中柬建交五十五周年，是中柬友好年，也是中国文化遗产研究院参与柬埔寨吴哥古迹保护十五周年。作为院长，我将温玉清博士这部《茶胶寺庙山建筑研究》的出版付梓看做是具有纪念和庆贺意义的一份礼物；对于温博士，这将是他在吴哥历史与文化学习、研究的新起点之一。记得他曾经谦虚地向我表示，在"吴哥学"当中，他还是个小学生。我把他的这本著作视为他的"吴哥小学"毕业证书，相信他今后还能在吴哥研究的学术殿堂里上中学，上大学，也能拿到博士学位。

2012年，中国国家主席胡锦涛、国务委员刘延东都曾经专程到茶胶寺工地慰问我院的援柬吴哥古迹保护工作队，温玉清博士是他们在暹粒参观吴哥古迹的主要讲解员。两位领导人都勉励温博士继续深入学习、研究吴哥的历史与文化，多出成果。我衷心希望温博士以更多的成果、更好的业绩，回报国家领导人的亲切关怀，回报吴哥古迹对他的熏陶与滋润。

是为序。

刘曙光
2013年3月26日
于北京高原街

前　言

作为东南亚地区最为重要的古代史迹，举世闻名的吴哥（Angkor）是公元九世纪至十五世纪古代高棉帝国繁盛时期都城与寺庙建筑的遗迹。以大吴哥城（Angkor Thom）与吴哥窟（Angkor Wat，亦称小吴哥）为代表的九十余处建筑组群及其数百座单体建筑遗构，散布在今之柬埔寨北部暹粒省大约四百余平方公里的热带丛林之中。

古代吴哥璀璨绚烂的文化在东南亚文明的历史进程中扮演着举足轻重的角色，尤其对历史上湄公河及湄南河流域的众多王权国家和民族文化都产生了深远的影响。以印度教与佛教为主流的宗教信仰深深地铭刻在吴哥时代的文化脉络之中，并由此构成了古代高棉建筑与艺术的主题。作为集中体现古代高棉建筑艺术与技术的巅峰之作，诸多遗存至今的吴哥史迹仍以其撼人心魄的空间布局与象征冠绝古今。尤其是那些雄伟壮观的庙山（temple-mountain）建筑成为吴哥时代建筑特征的显著标志，它们不仅以缜致精微、美轮美奂的雕刻与装饰代表着古代高棉石刻艺术的高超水平，还以其雄浑精丽的建筑风格展示出古代高棉建筑技术的无穷魅力，堪称吴哥时代建筑的代表作。

吴哥古迹还包括公元九世纪至十二世纪修建的罗莱池（Indratataka Baray）、东池（Eastern Baray）、西池（Western Baray）、北池（Jayatataka Baray）和皇家浴池（Srah Srang）等大型古代水利工程设施，以及连通各地的道路、人工运河、桥梁等交通设施，时至今日，其中的某些古代水利设施与道桥系统仍在发挥着重要的现代功用。

茶胶寺（Ta Keo temple-mountain），亦称茶胶寺庙山，是吴哥古迹中最为雄伟且具有鲜明特色的庙山建筑之一。茶胶寺庙山坐落于今之柬埔寨西北部暹粒省（Siem Reap Province）首府暹粒市北郊的吴哥考古遗址公园（Angkor Archaeological Park）之内，位于大吴哥城胜利门（Victory Gate）外东约1公里处。茶胶寺庙山西侧的环壕距暹粒河（Stung SiemReap）约280米，庙山东侧环壕之外以长度约为500米的神道与东池相接。在茶胶寺庙山的四周，分布着诸如达布隆寺（Ta Prohm）、班蒂克黛寺（Banteay Kdei）、达内寺（Ta Nei）、蟠龙水池（Neak Pean）、周萨神庙（Chausay Tovada）、托马侬神庙（Thommanon）等多座吴哥时代的重要庙宇。

作为早期吴哥庙山建筑的重要遗构，茶胶寺庙山依然保留着千年前创建之初尚未完工的状态，并以其完全以砂岩砌石构筑庙山中央五塔的做法，或因其巨型砌石的使用、十字形平面且四面开敞皆出抱厦的塔殿，以及环绕须弥台回廊平面格局的出现，开创了吴哥时代庙山建筑风气之先的形制；加之古代高棉宗教与文化赋予茶胶寺庙山独特的建筑风格和艺术魅力，使之成为见证古代高棉历史变迁的重要的吴哥古迹之一。

援助柬埔寨吴哥古迹保护的国际行动始于二十世纪九十年代初。历经多年战火的摧残，柬埔寨吴哥古迹列入世界文化遗产名录伊始，即被列为濒危遗产[1]。在联合国教科文组织（UNESCO）的统一协调下，先后有包括中国在内的来自十几个国家的工作队或学术机构，共同开启了援助柬埔寨吴哥古迹保护国际行动的序幕。各国的专家和学者在吴哥古迹保护领域密切合作，相互借鉴，使吴哥古迹保护事业成为展示文化遗产保护与研究能力的国际舞台，吴哥古迹保护国际行动的模式亦成为国际社会共同推动世界文化遗产保护事业发展的合作典范。

1998年，中国政府援助柬埔寨吴哥古迹保护周萨神庙保护修复工程正式启动，这是中国政府首次大规模参与文化遗产保护的国际合作项目。2008年12月，由中国文化遗产研究院主持勘察、设计及施工的周萨神庙保护修复工程竣工，基本恢复了寺庙原有建筑格局与艺术风貌，赢得了柬埔寨政府、国际组织以及各国同行的赞誉。在完成周萨神庙保护修复工程后，经中柬两国政府友好协商，选定茶胶寺庙山作为中国政府援助柬埔寨吴哥古迹保护二期项目，中国国家文物局与柬埔寨有关方面共同签署了合作协议。2009年12月，中柬两国政府签署换文，正式确认由中国政府提供4000万元人民币援助经费，用于吴哥古迹茶胶寺庙山的保护修复工程。根据中柬两国政府换文的规定，茶胶寺保护修复工程主要包括"建筑结构加固、建筑材料修复、考古研究"等内容，全部工程计划将于2018年完成。

2007年以来，在中国国家文物局、柬埔寨吴哥古迹保护与发展管理局（APSARA Authority）、联合国教科文组织吴哥古迹保护与发展协调委员会（ICC - Angkor）的具体指导下，作为中国政府援助柬埔寨吴哥古迹保护修复工程的执行机构，中国文化遗产研究院及其援柬吴哥古迹保护工作队（Chinese Government Team for Safeguarding Angkor - CSA）以国际文化遗产保护领域相关理念、法规、宪章、准则等为参照，以我国文化遗产保护实践作为借鉴，通过全面深入的科学研究制订具体保护修复方案，按照"科学研究贯穿保护修复工程全过程，抢险加固，排除险情，局部维修与全面修复相结合"总体思路的要求，组织力量对茶胶寺庙山建筑本体及其周围环境进行了较为详细的调查、测绘与勘察工作，主要涉及建筑、考古、结构工程、岩土工程、保护修复技术等诸多领域，先后完成各类研究报告与技术文件二十余项，取得了重要的阶段性成果，这是迄今为止围绕茶胶寺庙山保护修复工程所开展的最大规模的保护与研究工作。其中，中国文化遗产研究院编制完成的《柬埔寨吴哥古迹茶胶寺保护修复工程总体研究报告》、《援柬茶胶寺保护修复总体设计方案》、《援柬茶胶寺修复项目第一、二阶段施工项目工程设计》等技术文件均已获得中国国家文物局、中国商务部、柬埔寨吴哥古迹保护与发展管理局、联合国教科文组织ICC - Angkor的正式批复。2010年11月27日，茶胶寺保护修复工程举行了开工典礼。目前，茶胶寺保护修复工程第一阶段施工项目正在实施过程之中。

为配合茶胶寺保护修复工程的顺利开展，在中国文化遗产研究院副院长、总工程师侯卫东研究员主持的《茶胶寺保护修复工程综合研究》整体框架内，作为子项课题负责人，笔者先后承担了《茶胶寺庙山建筑调查与测绘》、《茶胶寺庙山建筑形制复原研究》、《柬埔寨吴哥古迹茶胶寺建筑保护修复史研究》等多项专题研究工作，对柬埔寨境内分布的吴哥时代庙山建筑遗存进行了较为全面的田野调查和踏勘；作为设计项目负责人，笔者还主持完成了《援柬茶胶寺保护修复工程总体设计方案》、《援柬茶胶寺保护修复工程第一阶段实施项目施工图设计》、《援柬茶胶寺保护修复工程第二阶段实施项目施工图设计》等多项工程设计任务。另外，在数年的工作过程中，笔者通过与法国、日本、意大利、德

---

[1] 正是得益于援助柬埔寨吴哥古迹保护的国际合作，2004年在中国苏州举行的第28届世界遗产大会上，吴哥古迹从濒危世界遗产名录中删除，重新回归世界遗产的大家庭。（著者注）

国、印度等从事吴哥古迹保护的国际同行进行交流，还较为系统地梳理了援助吴哥古迹保护国际行动的整体面貌及进展情况。

正是基于上述研究项目及工程设计工作中所取得的阶段性成果，初步奠定了本书的学术框架和研究基础，本书的编撰付梓也主要是在整理、修订、汇总、完善相关阶段性研究成果的基础上展开的。围绕茶胶寺庙山建筑研究的主题，本书的正文部分可以大致厘定为六章，每章内容各有侧重，现将本书的主要内容概述如兹，以供读者参考。

作为茶胶寺庙山建筑研究的历史背景介绍，本书第一章为《古代高棉史略》，其内容旨在通过追溯公元一至六世纪中叶的扶南时代、公元六世纪中叶至九世纪初的真腊时代、公元九世纪初至十五世纪中期吴哥时代以及公元十五世纪中叶以降的后吴哥时代的史料及史迹概要，简略地梳理了柬埔寨古代历史的渊源与文化脉络，参照以古代高棉社会在政治、经济、文化、宗教等领域的"印度化（Indianization）"的深刻影响，借此或可深入探赜柬埔寨古代建筑历史的渊薮，尤其是庙山建筑形制的源流与变迁。如果把吴哥时代的建筑放在"印度文明－东南亚文化－中华文明"的范围内考察，那么古代高棉帝国所呈现出的文化面貌，显而易见，是多种文化融合的结果。就其文化内涵的多样性而言，从某种意义上说，文化的多元成就了吴哥时代的辉煌。因此，本书第一章还概括述及吴哥文明兴起、演进及其衰微的历史文化进程，并初步探讨了"水与水利"在吴哥都城变迁与其地理区位、自然气候之间在物质生活和思想观念等不同层次上的交互影响。

关于柬埔寨吴哥古迹研究与保护的学术史，亦是笔者感兴趣且至为关心的课题。本书第二章为《吴哥古迹研究保护史概略》，主要对近百余年来的柬埔寨吴哥古迹保护研究历程进行了简略的回顾与综述。吴哥古迹的研究与保护的历史进程如何？如何评析吴哥古迹保护研究的方法和理论？如何客观地评价不同研究方法的特点及其局限性？管窥法国学者研究吴哥古迹的规模、气象及门径，我们不时被当时学者治学的专注和敬业的虔诚所震撼。近百年来柬埔寨吴哥古迹保护修复的历史中，以法国远东学院（École Française d'Extrême-Orient，亦简称为 EFEO）为代表的法国学者扮演着极其重要的角色。法国远东学院以其建立和积累的学术理论与研究方法，开创并奠定了东南亚历史研究，尤其是古代高棉历史、吴哥建筑历史与艺术、吴哥古迹保护修复等领域的基石。时至今日，法国远东学院仍是在古代高棉历史与考古、吴哥古迹保护研究领域蜚声世界的学术重镇。回溯学术史，在探讨前辈足迹及功过得失之时，其实也是在选择某种传统和规范，并确定自己的学术方向。只有通过了解前人研究的经历和当前研究的现状，以已有的基础作为工作的起点，才能认清自己所处的地位；只有弄清学术渊源与发展沿革，调整自身的知识结构，才可避免炒前人的冷饭。因此，本书第二章的内容主要是初步汇总梳理近百年来吴哥古迹考古清理、调查测绘与学术研究等方面的相关史料，围绕吴哥古迹的保护修复史进行概括与论述，力求了解廓清其研究理论的来龙去脉。这对于中国学者参与吴哥古迹保护与研究局面的改观或进步将不无裨益。

二十世纪二十年代以降，围绕茶胶寺庙山建筑源流及其变迁的探讨，诸多法国学者尝试从平面布局、建筑装饰以及碑铭内容释读等方面入手，针对茶胶寺庙山的建筑形制及其始建年代开展了较为详缜的梳理和考证，以期追溯茶胶寺庙山的历史源流与变迁历程。其中，尤以菲利普·斯特恩（Philippe Stern）、柯瑞尔·雷慕沙（Coral-Rémusat）、维克多·格罗布维（Victor Goloubew）、乔治·赛代斯（Georges Cœdès）等人的研究考证成果最为显著。本书第三章为《茶胶寺庙山源流考》，其主要内容即是基于上述学者的研究成果及其学术框架，对茶胶寺庙山的历史沿革与建造背景进行了详细的梳理和考证。另外，第三章的内容还对茶胶寺庙山保护研究的历史进行了简要的回顾与概述。

本书第四章为《茶胶寺庙山的建筑形制》，主要基于利用三维激光扫描技术完成茶胶寺庙山建筑测绘所取得的阶段性成果，对茶胶寺庙山的整体布局与建筑形制进行了详细的记录与描述，以期能够更为全面和精确地反映茶胶寺庙山建筑形制特征及其维修前的现状情况。值得说明的是，虽然运用了最为先进的测绘技术，但是所取得的阶段性测绘成果在深度和广度上仍有待继续提升且表达形式较为单一，未能系统地形成高质量正射影像及三维数字模型成果；另一方面，阶段性测绘成果对散落构件亦未能进行全面系统的测量记录，尚不能充分满足研究与工程的实际需要。鉴于此，该章内容较为侧重于茶胶寺庙山建筑形制的描述及其源流的探讨。

本书第五章为《茶胶寺庙山建筑复原初探》，主要通过公元十世纪末以降吴哥时代庙山建筑源流与变迁的梳理，较为深入地剖析了古代高棉庙山建筑形制特征、风格演变、宗教象征之所在，并主要围绕吴哥时代庙山建筑的形制特征及其艺术特色，针对茶胶寺庙山的整体布局构成、中央五塔、塔门、各种类型（藏经阁、长厅、回廊、角楼）的附属建筑以及建筑装饰等内容进行了初步的建筑复原研究。毋庸讳言，作为茶胶寺庙山建筑复原研究的阶段性成果，该章所论述的茶胶寺庙山建筑复原，仅仅涉及庙山建筑的宏观体量及其建筑空间布局，而对茶胶寺庙山砌石方式（特征描述及分类）、尺度规律（设计基准）、构造做法、工艺流程等方面的诸多细节，皆未能进行系统的阐明，尚不能厘清茶胶寺庙山建筑形制与空间布局在历史上可能出现过的某些重大改变。

本书第六章为《茶胶寺庙山的营造尺度与设计方法探微》，其内容主要针对茶胶寺庙山的设计尺度及其规律的研究所存在的诸多缺环和空白，在参考法国及日本学者相关研究成果的基础上，选择茶胶寺庙山作为分析和研究的对象，主要针对庙山平面布局、立面设计、建筑单体布局、细部构造等方面所涉及的营造尺度与设计方法进行初步的剖析和讨论，借此或可管窥古代高棉庙山建筑的独特构造及其设计意匠。

过去数年间，我们的研究工作始终得到了外交部、财政部、商务部、文化部、国家文物局和我驻柬埔寨大使馆等单位各级领导的热情关怀与大力支持，并且先后有来自天津大学、北京特种工程设计研究院、湖南大学、西安建筑科技大学、北京大学等合作单位的数十人参加了茶胶寺保护修复工程的前期调查和相关研究工作，本书的所附实测图、复原设计图等皆是集体智慧和团队协作的结晶。鉴于笔者亲历了以上整个过程，现将相关的合作情况简述如下。

2008年6月—2010年6月，中国文化遗产研究院与天津大学建筑学院合作开展了茶胶寺庙山建筑的调查与测绘工作。吴葱教授作为合作方的项目负责人，承担了茶胶寺庙山建筑三维激光扫描测绘的艰巨任务，并根据三维激光扫描点云数据绘制完成了茶胶寺庙山的CAD实测图，成为茶胶寺保护修复工程设计的主要基础资料。中国文化遗产研究院还与天津大学建筑学院共同开展了《茶胶寺庙山建筑形制与复原研究》、《建筑信息模型技术BIM在茶胶寺保护修复工程的应用》等专项研究工作。另外，天津大学建筑学院的张春彦博士、张宇博士还承担了本书征引法文参考文献的翻译和审校工作。

2008年11月-12月，2009年10月-11月，以及2011年12月-2013年1月，中国文化遗产研究院还与北京特种工程设计研究院合作，开展茶胶寺庙山工程地质勘察、部分构件石材的检测和检验，先后完成了茶胶寺庙山基台稳定性评估报告和材料加固试验，以及茶胶寺庙山场地排水方案的初步设计；另外，中国文化遗产研究院还先后与湖南大学、西安建筑科技大学、北京大学、中国社会科学院考古研究所等单位合作开展专项研究；其中，湖南大学霍静思副教授在对所选择的茶胶寺庙山两处具有代表性的单体建筑的结构稳定性验算方面提供了帮助；西安建筑科技大学陈平教授赴现场对结构加固方案提供了咨询意见；北京大学段晴教授实地考察茶胶寺古代碑铭遗存并提供了相关的指导与咨询；中国社会科学院考古研究所李裕群研究员对茶胶寺庙山的考古调查与发掘工作进行了现场指导。国家文物局古建筑专家组的张之平高级工程师（中国文化遗产研究院）、吕舟教授（清华大学）、李永革研究员（故宫博物院）、张克贵研究员（故宫博物院）、张立方研究员（河北省文物局）、黄滋研究员（浙江古建筑设计研究院）、张宪文研究员（广西壮族自治区文物保护研究设计中心）以及联合国教科文组织 ICC-Angkor 专家组的 Mounir Bouchenaki 教授（ICCROM）、Giorgio、CROCI 教授（罗马大学）、Pierre-André LABLAUDE 教授（法国凡尔赛宫博物馆）、日高健一郎教授（筑波大学）等中外专家曾多次对茶胶寺保护修复工程设计方案提供技术咨询意见。

上述专项研究取得的学术成果或专家咨询意见，皆为本书的编撰提供了重要的参考资料，在此谨向合作单位的诸位老师、各位专家及同仁们致以最诚挚的谢意。

另外，法国远东学院（EFEO）为茶胶寺庙山建筑研究提供了数十帧关于茶胶寺庙山的珍贵历史照片，以及雅克·杜马西先生关于茶胶寺庙山建筑的研究著作，柬埔寨吴哥古迹保护与发展管理局（APSARA Authority）资料中心（Angkor Documentation Center 及 GIS UNIT）为本书提供了数帧关于吴哥古迹及茶胶寺庙山的重要地图，在此一并致谢。

令人特别感念的是，国家文物局为本书出版提供了专门的经费支持，诸位局领导及相关司处领导始终关注着研究的进展和本书的出版情况。文物出版社为此也费尽心力，作为本书出版工作的统筹者，李睿先生承担了大量的编辑工作，在此深表谢意。

毋庸讳言，任何学术研究的成果都是阶段性的，本书也仅仅是体现了围绕茶胶寺庙山建筑研究而进行的某些初步探索而已，这些所谓的阶段性成果和探索可能是非常粗浅，甚至是片面的。作为一部中国学者研究柬埔寨吴哥建筑的学术著作，囿于笔者学识浅陋，各方资料挂漏、错讹在所难免，敬希方家赐教指正。

目　录

序 1

前言 1

第一章　古代高棉史略 1
　　一、扶南（公元一世纪初 – 约公元 550 年） 1
　　二、真腊（约公元 550 年 – 公元 802 年） 12
　　三、吴哥：古代高棉历史的黄金时代（公元 802 年 – 公元 1432 年） 17
　　四、善水之城：吴哥时代都城的选址与经营 24
　　五、吴哥时代的历史终结（公元 1432 年） 35

第二章　吴哥古迹研究保护史概略 48
　　一、法国远东学院溯往 48
　　二、吴哥古迹研究史概述 53
　　三、吴哥古迹保护修复史概略 59

第三章　茶胶寺庙山源流考 73
　　一、概述 73
　　二、茶胶寺庙山建造年代的考证 75
　　三、茶胶寺庙山碑铭研究述略 83
　　四、茶胶寺庙山历史沿革 86
　　五、茶胶寺庙山研究保护简史 88

第四章　茶胶寺庙山的建筑形制 103
　　一、茶胶寺庙山的整体布局 103
　　二、茶胶寺庙山建筑形制概略 110
　　三、茶胶寺庙山的保存现状及建筑勘察简述 130

第五章　茶胶寺庙山建筑复原初探 141
　　一、茶胶寺庙山总体格局的复原 141

1

二、茶胶寺庙山建筑的初步复原 …………………………………………………… 148

第六章　茶胶寺庙山的营造尺度与设计方法探微 …………………………………… 183
　　一、概述 ………………………………………………………………………………… 183
　　二、茶胶寺庙山平面布局的尺度分析 ………………………………………………… 186
　　三、茶胶寺庙山建筑立面设计的尺度分析 …………………………………………… 192
　　四、茶胶寺庙山建筑单体设计尺度分析 ……………………………………………… 196
　　五、关于茶胶寺庙山营造尺度与设计方法的初步结论 ……………………………… 200

主要参考文献 …………………………………………………………………………… 207

附录Ⅰ　柬埔寨吴哥古迹保护处茶胶寺庙山清理发掘月报摘录 …………………… 209
附录Ⅱ　茶胶寺庙山历史照片 ………………………………………………………… 215
附录Ⅲ　吴哥时代国王世系及其主要建筑遗迹 ……………………………………… 225

实测图 …………………………………………………………………………………… 227

图版 ……………………………………………………………………………………… 319

补记 ……………………………………………………………………………………… 481

后记 ……………………………………………………………………………………… 483

插图目录

第一章

图 1-1　公元 6 世纪的亚洲贸易及文化交流线路示意图（图片来源：James C. M. Khoo, *Art&Archaeloogy of Fu Nan—Pre-Khmer Kingdom of the Lower Mekong Valley*, 2003）

图 1-2　柬埔寨古迹分布图（图片来源：柬埔寨文化艺术部与法国远东学院绘制完成，法国远东学院提供）

图 1-3　柬埔寨吴哥古迹的地理区位

图 1-4　柬埔寨主要考古遗迹分布图（图片来源：Claude Jacques, *The Khmer Empire: Cities and Sanctuaries from 5th to the 13th Century*, 2007）

图 1-5　吴哥古迹卫星影像图之一（图片来源：柬埔寨吴哥古迹保护与发展管理局 APSARA Authority 资料中心提供）

图 1-6　吴哥古迹卫星影像图之二（图片来源：© Google Earth）

图 1-7　湿婆神的终极象征摩诃林伽（图片来源：Vo Si Khai, *The Kingdom Of FU-NAN and the Culture of Oc Eo*, 2003）

图 1-8　扶南时期的毗湿奴像（图片来源：私人收藏）

图 1-9　公元 6 世纪的毗湿奴造像及其发掘出土的情形（图片来源：Vo Si Khai, *The Kingdom Of FU-NAN and the Culture of Oc Eo*, 2003）

图 1-10　公元 6 世纪扶南时期的诃里诃罗洛造像（图片来源：柬埔寨国立博物馆/National Museum of Cambodia）

图 1-11　公元 5 世纪扶南时期的观音造像（图片来源：James C. M. Khoo, *Art&Archaeloogy of Fu Nan—Pre-Khmer Kingdom of the Lower Mekong Valley*, 2003）

图 1-12　湄公河下游扶南时代的遗址分布图（图片来源：引自 Claude Jacques, *The Khmer Empire: Cities and Sanctuaries from 5th to the 13th Century*, 2007）

图 1-13　真腊时期的建筑遗址桑坡布瑞窟之一（图片来源：温玉清 摄）

图 1-14　真腊时期的建筑遗址桑坡布瑞窟之二（图片来源：温玉清 摄）

图 1-15　真腊时期的建筑遗址桑坡布瑞窟之三（图片来源：温玉清 摄）

图 1-16　桑坡布瑞窟南组群中央圣殿轴测图（图片来源：引自 Dumarçay Jacques & Royère Pascal, *Cambodian Architecture: Eighth to thirteenth centuries.* 2001）

图 1-17　桑坡布瑞窟南组群中央圣殿内景（图片来源：温玉清 摄）

图 1-18　桑坡布瑞窟南组群小型神龛立面图（左）及八边形平面的砖塔轴测剖面图（右）（图片来

源：引自 Dumarçay Jacques & Royère Pascal，*Cambodian Architecture：Eighth to thirteenth centuries*. 2001）

图1-19　吴哥都城的选址与洞里萨湖及湄公河的关系（图片来源：温玉清 绘制）

图1-20　吴哥地区考古遗址分布图（图片来源：法国远东学院、澳大利亚悉尼大学 Great Angkor Project 研究项目组、柬埔寨吴哥古迹保护与发展管理局 APSARA Authority 共同完成，法国远东学院 Christophe Pottier 博士提供）

图1-21　吴哥时代都城经营区位变迁示意图（图片来源：Cunin Olivier，*DE TA PROHM AU BAYON：Analyse comparative de l'histoire architecturale des principaux monuments du style du Bayon*，Doctorat de l'Institut National Polytechnique de Lorraine，2004）

图1-22　吴哥时代的毗湿奴造像（图片来源：法国巴黎吉美博物馆/Musée Guimet）

图1-23　吴哥时代的湿婆造像（图片来源：柬埔寨国家博物馆/National Museum of Cambodia）

图1-24　吴哥时代的建筑雕刻（图片来源：温玉清 摄）

图1-25　吴哥时代建筑门楣雕饰之一（图片来源：法国巴黎吉美博物馆/Musée Guimet）

图1-26　吴哥时代建筑门楣雕饰之二（图片来源：法国巴黎吉美博物馆/Musée Guimet）

图1-27　今之东池现状（图片来源：王元林 摄）

图1-28　今之西池现状（图片来源：温玉清 摄）

图1-29　吴哥都城的选址与古伦山（荔枝山）及洞里萨湖的关系（图片来源：Cunin Olivier，*DE TA PROHM AU BAYON：Analyse comparative de l'histoire architecturale des principaux monuments du style du Bayon*，Doctorat de l'Institut National Polytechnique de Lorraine，2004）

图1-30　吴哥都城选址与经营的源流之一：高布思滨的千林伽河（图片来源：温玉清 摄）

图1-31　吴哥都城选址与经营的源流之二：古伦山（荔枝山）的千林伽河（图片来源：温玉清 摄）

图1-32　吴哥窟鸟瞰（图片来源：柬埔寨吴哥古迹保护与发展管理局 APSARA Authority 提供）

图1-33　吴哥窟远眺（图片来源：陈鹤 摄）

图1-34　吴哥窟中央五塔（图片来源：陈鹤 摄）

图1-35　吴哥通王城（大吴哥城）南门（图片来源：温玉清 摄）

图1-36　巴戎寺（图片来源：刘锦标 摄）

图1-37　阇耶跋摩七世（Jayavarman VII）雕像（图片来源：柬埔寨国家博物馆/National Museum of Cambodia）

图1-38　巴戎寺脸型塔之局部（图片来源：温玉清 摄）

图1-39　巴戎寺立面图（图片来源：Maurice Glaize，*The Monuments of the Angkor Group*，the 4th edition，1993）

图1-40　巴戎寺平面图（图片来源：Maurice Glaize，*The Monuments of the Angkor Group*，the 4th edition，1993）

图1-41　达布隆寺内景（图片来源：刘锦标 摄）

图1-42　吴哥时代主要君主及其城市与寺庙的建设之一（图片来源：EFEO&Musée Cernuschi，*Archaeologists in Angkor*，2011）

图1-43　吴哥时代主要君主及其城市与寺庙的建设之二（图片来源：EFEO&Musée Cernuschi，*Archae-

图1-44　阇耶跋摩七世时期的都城与寺庙的建设分布图（图片来源：EFEO&Musée Cernuschi, *Archaeologists in Angkor*, 2011）

图1-45　大吴哥城的格局及考古遗址分布图（图片来源：法国远东学院Jacques Gaucher教授提供）

图1-46　吴哥时代末期战争形势示意图（图片来源：温玉清 绘制）

图1-47　吴哥时代主要国王世系简表（图片来源：王巍 绘制）

第二章

图2-1　以"重新发现吴哥"而闻名的法国博物学家亨利·穆奥（Henri Mouhot, 1816-1861）（图片来源：Henri Mouhot, *Travels in the Central Parts of Indo-China, Siam, Cambodia, and Laos During the Years* 1858, 1859, *and* 1860, 1868）

图2-2　1863年法国政府驻柬埔寨代表杜达德特·拉格雷组建的湄公河考察团（图片来源：Francis Garnier, *Travels in Cambodia and Part of Laos, The Mekong Exploration Commission Report* 1866-1868, *Volume I* , 1996）

图2-3　法国远东学院的首任院长路易斯·芬诺（Louis Finot, 1864-1935）（图片来源：EFEO&Musée Cernuschi, *Archaeologists in Angkor*, 2011）

图2-4　1900年成立于越南河内的法国远东学院旧影（摄于1951年）（图片来源：Catherine Clémentin-Ojha&Pierre-YvesManguin, *A Century in Asia：The History of Ecole Française d'Extrême-Orient*, 2007）

图2-5　1900年法国远东学院拉云魁尔（Lunet de Lajonquière）与路易斯·芬诺（Louis Finot）及其考察队（图片来源：Catherine Clémentin-Ojha & Pierre-Yves Manguin, *A Century in Asia：The History of Ecole Française d'Extrême-Orient*, 2007）

图2-6　法国远东学院高棉历史学家乔治·赛代斯（George Cœdès, 1886-1969）（图片来源：法国远东学院网站 www.efeo.fr）

图2-7　法国远东学院考古学家亨利·帕尔芒捷（Henri Parmentier, 1871-1949）（图片来源：法国远东学院网站 www.efeo.fr）

图2-8　法国远东学院早期学者合影（图片来源：Catherine Clémentin-Ojha & Pierre-Yves Manguin, *A Century in Asia：The History of Ecole Française d'Extrême-Orient*, 2007）

图2-9　亨利·帕蒙蒂埃绘制的瓦普寺测绘图（图片来源：Catherine Clémentin-Ojha&Pierre-YvesManguin, *A Century in Asia：The History of Ecole Française d'Extrême-Orient*, 2007）

图2-10　1909年拉云魁尔主持调查并绘制的柬埔寨古迹分布图（图片来源：Lunet de Lajonquère: *Inventaire Descriptif des Monuments du Cambodge*, 1911）

图2-11　1909年拉云魁尔主持调查并绘制的吴哥地区的古迹分布图（图片来源：Lunet de Lajonquère, *Inventaire Descriptif des Monuments du Cambodge*, 1911）

图2-12　德拉蓬特绘制的今之柏威夏省圣剑寺测绘图（图片来源：L. Delaporte, *Les Monuments du Cambodge：Etudes d'Architecture Khmére*, 1920）

图2-13　德拉蓬特绘制的吴哥圣剑寺测绘图（图片来源：L. Delaporte, *Les Monuments du Cambodge：*

Etudes d'Architecture Khmére, 1920）

图 2-14　德拉蓬特绘制的皇家浴池及其雕刻测绘图（图片来源：L. Delaporte, *Les Monuments du Cambodge：Etudes d'Architecture Khmére*, 1920）

图 2-15　吴哥古迹保护处的首任负责人简·康迈勒（Jean Commaille, 1868-1916）（图片来源：EFEO&Musée Cernuschi, *Archaeologists in Angkor*, 2011）

图 2-16　吴哥古迹保护修复的先驱者亨利·马绍尔（Henri Marchal, 1876-1970）（图片来源：Catherine Clémentin-Ojha & Pierre-Yves Manguin, *A Century in Asia：The History of Ecole Française d'Extrême-Orient*, 2007ⓒ EFEO）

图 2-17　吴哥古迹保护修复"原物归位法"（Anastylosis）的最早工程实例：女王宫（1934 年）（图片来源：EFEO&Musée Cernuschi, *Archaeologists in Angkor*, 2011ⓒ EFEO）

图 2-18　吴哥古迹保护修复工程实例：大吴哥城南门神道（1961 年）（图片来源：EFEO&Musée Cernuschi, *Archaeologists in Angkor*, 2011ⓒ EFEO）

图 2-19　吴哥古迹保护修复工程实例：龙蟠水池（1938 年）（图片来源：EFEO&Musée Cernuschi, *Archaeologists in Angkor*, 2011ⓒ EFEO）

图 2-20　吴哥古迹保护修复工程实例：圣剑寺（1930 年代）（图片来源：EFEO&Musée Cernuschi, *Archaeologists in Angkor*, 2011ⓒ EFEO）

第三章

图 3-1　茶胶寺庙山西南侧外观（图片来源：温玉清 摄）

图 3-2　茶胶寺庙山的区位

图 3-3　巴空寺庙山（BaKong）平面图（图片来源：Maurice Glaize, *The Monuments of the Angkor Group*, the 4th edition, 1993）

图 3-4　巴肯山（Bakheng）平面图（图片来源：Maurice Glaize, *The Monuments of the Angkor Group*, the 4th edition, 1993）

图 3-5　比粒寺庙山（Pre Rup）平面图（图片来源：Coral-Rémusat, V. Goloubew, G. Cœdès, *LA DATE DU TÀKÈV, Bulletin de l'Ecole française d'Extrême-Orient-BEFEO*, Tome XXXIV, 1934）

图 3-6　比粒寺庙山（Pre Rup）西北侧外观（图片来源：温玉清 摄）

图 3-7　东梅奔寺（Eastern Mebon）平面图（图片来源：Coral-Rémusat, V. Goloubew, G. Cœdès, *LA DATE DU TÀKÈV, Bulletin de l'Ecole française d'Extrême-Orient-BEFEO*, Tome XXXIV, 1934）

图 3-8　东梅奔寺（Eastern Mebon）中央五塔外观（图片来源：温玉清 摄）

图 3-9　女王宫（Banteay Srei）轴测图（图片来源：Dumarçay Jacques & Royère Pascal, *Cambodian Architecture：Eighth to thirteenth centuries.* 2001）

图 3-10　空中宫殿（Phimanakas）平面图（图片来源：Coral-Rémusat, V. Goloubew, G. Cœdès, *LA DATE DU TÀKÈV, Bulletin de l'Ecole française d'Extrême-Orient-BEFEO*, Tome XXXIV, 1934）

图 3-11　空中宫殿（Phimanakas）外观（图片来源：永昕群 摄）

图 3-12　巴塞增空寺（Baksei Chamkrong）立面图（图片来源：Dumarçay Jacques & Royère Pascal, *Cambodian Architecture：Eighth to thirteenth centuries.* 2001）

图 3-13　巴方寺（Baphuon）平面图（图片来源：Dumarçay Jacques & Royère Pascal, *Cambodian Architecture: Eighth to thirteenth centuries.* 2001）

图 3-14　推测茶胶寺庙山年代的"花萼"纹样（图片来源：温玉清 摄；Coral-Rémusat, V. Goloubew, G. Cœdès, *LA DATE DU TÀKÈV*, Bulletin de l'Ecole française d'Extrême-Orient - BEFEO, Tome XXXIV, 1934）

图 3-15　茶胶寺庙山碑铭遗存分布位置（图片来源：Coral-Rémusat, V. Goloubew, G. Cœdès, *LA DATE DU TÀKÈV*, Bulletin de l'Ecole française d'Extrême-Orient - BEFEO, Tome XXXIV, 1934）

图 3-16　茶胶寺庙山碑铭（编号 K.275）局部（图片来源：永昕群 摄）

图 3-17　茶胶寺庙山碑铭（编号 K.278）局部（图片来源：永昕群 摄）

图 3-18　艾莫涅尔（É. Aymonier）绘制的吴哥地区古迹分布图（1904 年）（图片来源：É. Aymonier, *Le Cambdoge III Le Groupe D'Angkor*, 1904）

图 3-19　艾莫涅尔（É. Aymonier）绘制的茶胶寺庙山平面图（1904 年）（图片来源：É. Aymonier, *Le Cambdoge III Le Groupe D'Angkor*, 1904）

图 3-20　拉云魁尔（L. Lajonquière）绘制的茶胶寺庙山平面图（1911 年）（图片来源：Lunet de Lajonquère, *Inventaire Descriptif des Monuments du Cambodge*, 1911）

图 3-21　德拉蓬特（L. Delaporte）完成的茶胶寺庙山平面图及其复原图（1920 年）（图片来源：L. Delaporte, *Les Monuments du Cambodge Etudes d'Architecture Khmére*, 1920）

图 3-22　1920-1923 年法国远东学院茶胶寺庙山清理发掘历史照片以及当时的调查记录手稿（图片来源：法国远东学院提供© EFEO）

图 3-23　1920 年初法国远东学院吴哥古迹保护处茶胶寺清理发掘现场历史照片之一（图片来源：法国远东学院提供© EFEO）

图 3-24　1920 年初法国远东学院吴哥古迹保护处茶胶寺庙山清理发掘现场历史照片之二（图片来源：法国远东学院提供© EFEO）

图 3-25　雅克·杜马西（Jacques Dumarçay）测绘的茶胶寺庙山立面图（图片来源：Jacques Dumarçay. *TA KEV: ETUDE ARCHITECTURALE DU TEMPLE*, l'Ecole française d'Extrême-Orient EFEO：1971）

第四章

图 4-1　茶胶寺庙山整体格局平面图之一（图片来源：柬埔寨吴哥古迹保护与发展管理局资料信息中心 APSARA Authority GIS Unit 提供）

图 4-2　茶胶寺庙山整体格局平面图之二（图片来源：柬埔寨吴哥古迹保护与发展管理局资料信息中心 APSARA Authority GIS Unit 提供）

图 4-3　利用激光雷达（LiDAR）获取的茶胶寺庙山整体格局影像图（图片来源：柬埔寨吴哥古迹保护与发展管理局资料信息中心 APSARA Authority GIS Unit，ROS Rorath 先生提供）

图 4-4　东池西侧的建筑遗址及实测图（图片来源：Jacques Dumarçay. *TA KEV: ETUDE ARCHITECTURALE DU TEMPLE*, l'Ecole française d'Extrême-Orient EFEO, 1971）

图 4-5　茶胶寺庙山东南角的建筑遗迹现状实测图（图片来源：Jacques Dumarçay. *TA KEV: ETUDE*

ARCHITECTURALE DU TEMPLE，l'Ecole française d'Extrême‑Orient EFEO，1971）

图 4－6　茶胶寺庙山东南角的建筑遗迹现状（图片来源：温玉清 摄）
图 4－7　茶胶寺庙山西北侧阇耶跋摩七世时期的建筑遗迹（图片来源：温玉清 摄）
图 4－8　1923 年法国远东学院发掘清理茶胶寺庙山神道情况（图片来源：法国远东学院提供© EFEO）
图 4－9　2012 年茶胶寺庙山考古发掘所揭示的环壕驳岸构造情况之一（图片来源：温玉清 摄）
图 4－10　2012 年茶胶寺庙山考古发掘所揭示的环壕驳岸构造情况之二（图片来源：温玉清 摄）
图 4－11　2012 年茶胶寺庙山考古发掘所揭示的环壕淤积土层情况（图片来源：王元林 摄）
图 4－12　2012 年茶胶寺庙山考古发掘所揭示的环壕底部构造情况（图片来源：王元林 摄）
图 4－13　茶胶寺庙山整体布局示意（图片来源：中国文化遗产研究院《援柬二期茶胶寺保护修复工程综合研究》© CACH&CSA）
图 4－14　茶胶寺庙山建筑构成（图片来源：Jacques Dumarçay. TA KEV：ETUDE ARCHITECTURALE DU TEMPLE，l'Ecole française d'Extrême‑Orient EFEO：1971）
图 4－15　茶胶寺庙山一层台东北角的"建筑基址"（图片来源：温玉清 摄）
图 4－16　茶胶寺庙山一层台东南角的"建筑基址"（图片来源：温玉清 摄）
图 4－17　茶胶寺庙山须弥台第一层雕刻纹样分布示意（图片来源：中国文化遗产研究院《援柬二期茶胶寺保护修复工程综合研究》陈筱 绘制）
图 4－18　茶胶寺庙山须弥台砌石工艺调查（图片来源：中国文化遗产研究院《援柬二期茶胶寺保护修复工程综合研究》陈筱 绘制）
图 4－19　茶胶寺庙山须弥台砌石工艺流程示意（图片来源：中国文化遗产研究院《援柬二期茶胶寺保护修复工程综合研究》陈筱 绘制）
图 4－20　茶胶寺庙山内发现的尚未完成雕刻的构件（图片来源：温玉清 摄）
图 4－21　茶胶寺庙山中央主塔形制构成示意（图片来源：Jacques Dumarçay. TA KEV：ETUDE ARCHITECTURALE DU TEMPLE，l'Ecole française d'Extrême‑Orient EFEO：1971）
图 4－22　茶胶寺庙山中央五塔散落构件分布（图片来源：中国文化遗产研究院《援柬二期茶胶寺保护修复工程综合研究》© CACH&CSA）
图 4－23　茶胶寺中央五塔石切石石材类型示意（图片来源：温玉清 摄）
图 4－24　茶胶寺庙山中央五塔顶部塌落的构件（图片来源：温玉清 摄）

第五章[1]

图 5－1　茶胶寺庙山整体格局的复原
图 5－2　茶胶寺庙山东立面及剖面现状图
图 5－3　茶胶寺庙山立面及剖面复原图
图 5－4　茶胶寺庙山复原图之一
图 5－5　茶胶寺庙山复原图之二

---

[1]　第五章插图除注明出处者外，图片来源皆应注明："（中国文化遗产研究院、天津大学建筑学院：《茶胶寺庙山建筑形制与复原研究》，项目负责人：温玉清、吴葱，参加者：伍沙、张宇、马庆阳、王祥、任思捷、闫金强，2009 年 © CACH&CSA）"

图 5-6　巴方寺庙山复原图（图片来源：法国远东学院 Pascal Royére 博士提供）
图 5-7　茶胶寺庙山塔殿平面形式示意
图 5-8　吴哥时代庙山中央塔殿布局形式的演变
图 5-9　茶胶寺庙山中央五塔散落构件的分布
图 5-10　茶胶寺庙山中央主塔及角塔的建筑形制
图 5-11　茶胶寺庙山中央主塔复原图
图 5-12　茶胶寺庙山角塔复原图
图 5-13　吴哥时代庙山塔门布局形式的演变
图 5-14　茶胶寺庙山塔门平面组合形式示意
图 5-15　茶胶寺庙山塔门的建筑形制
图 5-16　茶胶寺庙山塔门类型及其散落构件分布
图 5-17　茶胶寺庙山东外塔门建筑形制复原
图 5-18　茶胶寺庙山南内塔门建筑形制复原
图 5-19　茶胶寺庙山南外塔门建筑形制复原
图 5-20　茶胶寺庙山东内塔门建筑形制复原
图 5-21　茶胶寺庙山各座塔门保存现状
图 5-22　茶胶寺庙山塔门山花构件的分布及砌石构件形制分析
图 5-23　吴哥时代庙山建筑长厅平面布局的演变
图 5-24　吴哥时代庙山中长厅平面布局的组合方式
图 5-25　茶胶寺庙山长厅遗存现状及其建筑形制
图 5-26　茶胶寺庙山长厅建筑形制的复原
图 5-27　茶胶寺庙山长厅建筑形制复原
图 5-28　茶胶寺庙山长厅形制复原之构成示意
图 5-29　吴哥时代庙山建筑藏经阁平面布局的演变
图 5-30　吴哥时代庙山建筑藏经阁平面组合方式
图 5-31　茶胶寺庙山藏经阁建筑形制复原之一
图 5-32　茶胶寺庙山藏经阁建筑形制复原之二
图 5-33　茶胶寺庙山藏经阁建筑形制复原之三
图 5-34　吴哥时代主要寺庙回廊的平面格局
图 5-35　茶胶寺庙山回廊形制复原之一
图 5-36　茶胶寺庙山回廊形制复原之二
图 5-37　茶胶寺庙山回廊角塔建筑形制复原之一
图 5-38　茶胶寺庙山回廊角塔建筑形制复原之二
图 5-39　茶胶寺庙山塔门山花构件分布及其拼对之一
图 5-40　茶胶寺庙山塔门山花构件分布及其拼对之二
图 5-41　茶胶寺庙山塔门山花构件形制分析
图 5-42　茶胶寺庙山塔门山花构件雕饰纹样的复原

图 5-43　茶胶寺庙山塔门花柱构件分布
图 5-44　茶胶寺庙山塔门花柱复原
图 5-45　茶胶寺庙山须弥台雕刻纹饰组合方式
图 5-46　茶胶寺庙山须弥台雕刻纹饰复原之一
图 5-47　茶胶寺庙山须弥台雕刻纹饰复原之二
图 5-48　茶胶寺庙山须弥台雕刻纹饰复原之三
图 5-49　茶胶寺庙山须弥台雕刻纹饰复原之四
图 5-50　茶胶寺庙山须弥台雕刻纹饰复原之五

第六章

图 6-1　茶胶寺庙山平面基准尺度的推测（图片来源：温玉清 绘制）
图 6-2　茶胶寺庙山东西方向轴线的偏移（图片来源：温玉清 绘制）
图 6-3　茶胶寺庙山平面设计尺度基准（图片来源：温玉清 绘制）
图 6-4　茶胶寺庙山立面设计尺度分析（图片来源：温玉清 绘制）
图 6-5　茶胶寺庙山立面设计尺度分析-须弥台底面（图片来源：温玉清 绘制）
图 6-6　茶胶寺庙山立面设计尺度分析-须弥台顶面（图片来源：温玉清 绘制）
图 6-7　茶胶寺庙山立面设计尺度分析-林伽高度（图片来源：温玉清 绘制）

实测图目录

1. 茶胶寺庙山总平面图
2. 茶胶寺庙山东立面图
3. 茶胶寺庙山南立面图
4. 茶胶寺庙山北立面图
5. 茶胶寺庙山西立面图
6. 茶胶寺庙山 1–1 剖面图
7. 茶胶寺庙山 2–2 剖面图
8. 茶胶寺庙山散落构件分布示意图
9. 须弥台西南角平面图
10. 须弥台东北角平面图
11. 须弥台西北角平面图
12. 须弥台东南角平面图
13. 须弥台东立面图
14. 须弥台东立面影像图
15. 须弥台南立面图
16. 须弥台南立面影像图
17. 须弥台西立面图
18. 须弥台西立面影像图
19. 须弥台北立面图
20. 须弥台北立面影像图
21. 须弥台 1–1 剖面图
22. 须弥台 2–2 剖面图
23. 须弥台 3–3 剖面图
24. 须弥台 4–4 剖面图
25. 须弥台 5–5 剖面图
26. 须弥台 6–6 剖面图
27. 须弥台 7–7 剖面图
28. 东外塔门平面图
29. 东外塔门东立面图
30. 东外塔门西立面图

31. 东外塔门南立面图
32. 东外塔门北立面图
33. 东外塔门1-1剖面图
34. 东外塔门2-2剖面图
35. 西外塔门平面图
36. 西外塔门西立面图
37. 西外塔门东立面图
38. 西外塔门北立面图
39. 西外塔门南立面图
40. 西外塔门1-1剖面图
41. 西外塔门2-2剖面图
42. 南外塔门平面图
43. 南外塔门南立面图
44. 南外塔门北立面图
45. 南外塔门东立面图
46. 南外塔门西立面图
47. 南外塔门1-1剖面图
48. 南外塔门2-2剖面图
49. 南内塔门平面图
50. 南内塔门南立面图
51. 南内塔门北立面图
52. 南内塔门东立面图
53. 南内塔门西立面图
54. 南内塔门1-1剖面图
55. 南内塔门2-2剖面图
56. 北藏经阁平面图
57. 北藏经阁北立面图
58. 北藏经阁东立面图
59. 北藏经阁南立面图
60. 北藏经阁西立面图
61. 北藏经阁1-1剖面图
62. 北藏经阁2-2剖面图
63. 北藏经阁3-3剖面图
64. 北藏经阁4-4剖面图
65. 南藏经阁平面图
66. 南藏经阁南立面图
67. 南藏经阁北立面图

68. 南藏经阁东立面图
69. 南藏经阁西立面图
70. 南藏经阁1-1剖面图
71. 南藏经阁2-2剖面图
72. 南藏经阁3-3剖面图
73. 南藏经阁4-4剖面图
74. 南内长厅平面图
80. 南内长厅2-2剖面图
81. 北内长厅平面图
82. 北内长厅东立面图
83. 北内长厅西立面图
84. 北内长厅北立面图
85. 北内长厅南立面图
86. 北内长厅1-1剖面图
87. 北内长厅2-2剖面图
88. 南外长厅平面图
89. 南外长厅东立面图
90. 南外长厅西立面图
91. 南外长厅1-1剖面图
92. 北外长厅平面图
93. 北外长厅东立面图
94. 北外长厅西立面图
95. 北外长厅1-1剖面图

图版目录[*]

1. 茶胶寺庙山鸟瞰（引自 Claude Jacques/Philippe Lafond，*The Khmer Empire：Cities and Sanctuaries from the 5th to the 13th Century*，© River Books，2007）
2. 茶胶寺庙山东侧及神道局部
3. 茶胶寺庙山西南侧全景
4. 茶胶寺庙山东北侧
5. 茶胶寺庙山西北侧
6. 热带暴雨中的茶胶寺庙山东侧（永昕群 摄）
7. 茶胶寺庙山东北侧（永昕群 摄）
8. 茶胶寺庙山西侧
9. 茶胶寺庙山南侧
10. 中央主塔东侧外观（刘锦标 摄）
11. 中央主塔中厅仰视
12. 中央主塔西侧过厅内景
13. 中央主塔中厅内景（刘锦标 摄）
14. 中央主塔西侧外观
15. 中央主塔顶部假层的砌筑方式（陈艺文 摄）
16. 中央主塔东侧抱厦
17. 中央主塔北抱厦的抱框石
18. 中央主塔南侧外观
19. 中央主塔塔顶砌石的构造方式之一（陈艺文 摄）
20. 中央主塔塔顶砌石的构造方式之二（陈艺文 摄）
21. 中央主塔塔顶俯瞰抱厦顶部
22. 东北角塔东侧外观（刘锦标 摄）
23. 东北角塔北侧外观（刘锦标 摄）
24. 东北角塔西侧外观
25. 东北角塔南侧外观
26. 东北角塔东侧抱厦内景
27. 东北角塔西侧抱厦内景

---

* 图版照片除注明拍摄者外，皆为著者拍摄。

28. 东北角塔中厅内景之一
29. 东北角塔中厅内景之二
30. 东北角塔俯瞰（陈艺文　摄）
31. 东北角塔中厅内景之三
32. 东南角塔西侧外观
33. 东南角塔东侧外观
34. 东南角塔北侧外观
35. 东南角塔俯瞰（陈艺文　摄）
36. 东南角塔中厅内景之一
37. 西北角塔仰视
38. 西北角塔北侧踏道
39. 西北角塔俯瞰（陈艺文　摄）
40. 西南角塔东侧外观（刘锦标　摄）
41. 西南角塔北侧外观
42. 西南角塔顶部砌石细部之一（陈艺文　摄）
43. 西南角塔顶部砌石细部之二（陈艺文　摄）
44. 西南角塔俯瞰（陈艺文　摄）
45. 须弥台东北角顶部现状
46. 须弥台东侧顶部现状（自北向南拍摄）
47. 须弥台西北侧顶部现状（自东向西拍摄）
48. 须弥台西南侧顶部现状（自北向南拍摄）
49. 须弥台东北角部现状之一
50. 须弥台东北角部现状之二
51. 须弥台之西北侧
52. 须弥台东侧踏道
53. 须弥台东侧踏道之二
54. 须弥台东侧踏道局部
55. 须弥台东侧踏道垛台之一
56. 须弥台东侧踏道垛台之二
57. 须弥台北侧踏道
58. 须弥台北侧踏道垛台
59. 须弥台西侧踏道
60. 须弥台西侧踏道细部
61. 须弥台东侧的砂岩雕刻遗存
62. 须弥台东侧的砂岩雕刻（仰莲）
63. 须弥台东侧的砂岩雕刻纹样之一
64. 须弥台东侧的砂岩雕刻纹样之二

65. 须弥台东侧的砂岩雕刻纹样之三
66. 须弥台东侧的砂岩雕刻风化现状之一
67. 须弥台东侧的砂岩雕刻风化现状之二
68. 须弥台东侧的砂岩雕刻风化现状之三
69. 须弥台东侧的砂岩雕刻风化现状之四
70. 须弥台东侧的砂岩雕刻风化现状之五
71. 须弥台东侧的砂岩雕刻风化现状之六
72. 东外塔门东侧立面
73. 东外塔门西侧立面
74. 东外塔门北侧室内景
75. 东外塔门主室内景
76. 东外塔门东北侧
77. 东外塔门及东内塔门西侧俯瞰
78. 北外塔门北侧立面
79. 北外塔门南侧俯瞰
80. 北外塔门南侧立面
81. 北外塔门东侧外观
82. 北外塔门西侧俯瞰
83. 西外塔门东侧立面
84. 西外塔门内北侧外观
85. 西外塔门南抱厦
86. 西外塔门西侧
87. 南外塔门西侧
88. 南外塔门东侧
89. 南外塔门东侧室内景
90. 南外塔门北侧
91. 南外塔门西侧室内景
92. 南外塔门南侧
93. 东内塔门东侧立面
94. 东内塔门西侧立面
95. 东内塔门内北侧
96. 东内塔门北侧室砖叠涩屋顶遗迹（永昕群　摄）
97. 东内塔门北侧砖叠涩遗迹（永昕群　摄）
98. 东内塔门外南侧
99. 北内塔门南侧
100. 北内塔门西侧
101. 北内塔门北侧

102. 北内塔门俯瞰
103. 北内塔门主室仰视
104. 西内塔门东侧
105. 西内塔门东侧
106. 西内塔门东侧俯瞰
107. 西内塔门西侧
108. 西内塔门内侧的假层山花
109. 南内塔门内东侧
110. 南内塔门南侧
111. 南内塔门内侧
112. 南内塔门内西侧
113. 南内塔门内侧俯瞰（刘锦标 摄）
114. 南外长厅西侧明窗之局部
115. 南外长厅西北侧
116. 南外长厅西侧的明窗
117. 南外长厅后室的明窗
118. 南外长厅内景
119. 南外长厅北侧抱厦
120. 南外长厅东北侧俯瞰
121. 南外长厅西侧
122. 北外长厅后室明窗细部
123. 北外长厅西南侧俯瞰
124. 北外长厅南侧抱厦遗迹之一
125. 北外长厅抱厦遗迹之二
126. 北外长厅后室西侧
127. 南内长厅西北侧俯瞰（闫明 摄）
128. 南内长厅西侧窗棱细部
129. 南内长厅北侧抱厦（闫明 摄）
130. 南内长厅西侧
131. 南内长厅东侧（闫明 摄）
132. 南内长厅主室内景
133. 南内长厅主室地面铺砌遗迹
134. 北内长厅西南侧俯瞰
135. 北内长厅南侧抱厦入口
136. 北内长厅抱厦西侧
137. 北内长厅西侧
138. 北内长厅主室西侧立面

139. 北内长厅主室内景（闫明　摄）

140. 北内长厅抱厦内景之一（闫明　摄）

141. 北内长厅抱厦内景之二（闫明　摄）

142. 北内长厅山花遗迹

143. 北藏经阁西侧俯瞰

144. 北藏经阁东南侧

145. 北藏经阁东北侧

146. 北藏经阁南侧

147. 北藏经阁西侧抱厦局部

148. 北藏经阁东北侧

149. 北藏经阁西北侧

150. 北藏经阁俯瞰

151. 北藏经阁内景

152. 北藏经阁东侧

153. 北藏经阁西侧抱厦

154. 北藏经阁抱厦门楣

156. 北藏经阁西南侧之一

157. 北藏经阁西南侧之二

158. 南藏经阁西北侧俯瞰（于志飞　摄）

159. 南藏经阁西南侧（于志飞　摄）

160. 南藏经阁南侧

161. 南藏经阁西南侧俯瞰

162. 南藏经阁东侧（于志飞　摄）

163. 南藏经阁东北侧（于志飞　摄）

164. 南藏经阁西北侧俯瞰（于志飞　摄）

165. 暴雨中的南藏经阁

166. 烈日下的南藏经阁

167. 第二层台回廊西北侧

168. 第二层台回廊西北侧俯瞰

169. 第二层台回廊东南侧俯瞰

170. 第二层台回廊东南角俯瞰

171. 第二层台回廊西北侧俯瞰

172. 庙山内院回廊北侧东段

173. 庙山内院回廊东侧南段

174. 庙山内院回廊西侧局部

175. 庙山内院回廊西侧南段

176. 庙山内院回廊西南角

177. 庙山内院回廊东侧南段
178. 庙山内院回廊
179. 庙山内院回廊西侧南段
180. 庙山内院（二层台）东北角楼俯瞰
181. 庙山内院（二层台）东北角及角楼
182. 庙山内院东南角及角楼内侧
183. 庙山内院（二层台）东南角及角楼
184. 庙山内院（二层台）西南角及角楼俯瞰
185. 庙山内院（二层台）西南角及角楼内侧
186. 庙山内院（二层台）西南角楼俯瞰
187. 庙山内院（二层台）西南角楼内景（自北向南拍摄）
188. 庙山内院东北侧　（陈鹤　摄）
189. 庙山内院西侧（陈鹤　摄）
190. 庙山内院南侧（陈鹤　摄）
191. 庙山内院的北藏经阁与北内长厅（陈鹤　摄）
192. 庙山内院北侧（陈鹤　摄）
193. 庙山内院北侧俯瞰（陈鹤　摄）
194. 庙山内院之东南角
195. 庙山内院之东北角
196. 庙山内院东北角及北藏经阁（陈鹤　摄）
197. 庙山外院东南侧
198. 庙山外院东北侧
199. 庙山外院北侧俯瞰
200. 庙山外院围墙西侧局部
201. 庙山外院围墙北侧
202. 庙山外院围墙西北角
203. 庙山外院围墙东北角北侧
204. 庙山外院围墙东北侧内部
205. 庙山外院围墙南侧
206. 北环濠西侧（自东向西）
207. 北环濠通道西侧（自西向东）
208. 东门外南池
209. 东门外神道
210. 东门外神道上的建筑基址（自东向西）
211. 西侧环濠（自北向南）
212. 东内塔门编号K278古代高棉文碑铭（永昕群　摄）
213. 茶胶寺庙山东内塔门编号K278古代高棉文碑铭之局部（永昕群　摄）

214. 茶胶寺庙山东内塔门编号 K275 古代高棉文碑铭之局部（永昕群 摄）
215. 建筑装饰构件之南外长厅明窗窗棱局部
216. 建筑装饰构件之各种类型的窗棱
217. 建筑装饰构件转角部位雕刻的蛇神（NAGA）纹样
218. 建筑装饰构件表面雕刻的植物纹样之一
219. 建筑装饰构件表面雕刻的植物纹样之二
220. 建筑装饰构件之东外塔门主室的山花
221. 建筑装饰构件之花柱
222. 建筑装饰构件之花柱细部
223. 建筑装饰构件之花柱未完成的细部雕饰
224. 建筑装饰构件细部雕刻（摩羯 Makara）之一
225. 建筑装饰构件细部雕刻（摩羯 Makara）之二
226. 建筑装饰构件细部之三
227. 建筑装饰构件细部之四
228. 建筑装饰构件细部之五
229. 施工流程遗迹之一
230. 施工流程遗迹之二
231. 施工流程遗迹之三
232. 施工流程遗迹之四
233. 第二层基台西南角内部角砾岩的砌筑方式（陈艺文 摄）
234. 西南角楼砌石间的金属拉结之一（陈艺文 摄）
235. 西南角楼砌石间的金属拉结之二（陈艺文 摄）
236. 北内长厅山花背面木构架插隼遗迹之一
237. 北内长厅山花背面木构架插隼遗迹之二
238. 南内长厅附近的散落挡头瓦构件
239. 砂岩砌石表面的圆孔及封石（北内塔门）
240. 墙身角部特殊的砌石方式
241. 回廊抱框石的连接方式
242. 塔门窗棱的固定方式
243. 须弥台内部的角砾岩砌石
244. 庙山内院的建筑遗迹之一
245. 庙山内院的建筑遗迹之二
246. 庙山内院的建筑遗迹之三
247. 庙山内院的建筑遗迹之四
248. 庙山内院的建筑遗迹之五
249. 茶胶寺庙山神道全景之一
250. 茶胶寺庙山神道全景之二

251. 位于东池西堤上与茶胶寺庙山相接的建筑基址之东北侧
252. 位于东池西堤上与茶胶寺庙山相接的建筑基址之南侧
253. 位于东池西堤上与茶胶寺庙山相接的建筑基址之西侧
254. 位于东池西堤上与茶胶寺庙山相接的建筑基址之东南侧局部（王元林 摄）
255. 位于东池西堤上与茶胶寺庙山相接的建筑基址之北侧（王元林 摄）
256. 位于东池西堤上另外一座与茶胶寺庙山相接的建筑基址之南侧局部（王元林 摄）
257. 位于东池西堤上另外一座与茶胶寺庙山相接的建筑基址之西侧（王元林 摄）
258. 位于东池西堤上另外一座与茶胶寺庙山相接的建筑基址东侧（王元林 摄）
259. 位于东池西堤上另外一座与茶胶寺庙山相接的建筑基址东侧石狮子雕像遗迹
260. 位于东池西堤上另外一座与茶胶寺庙山相接的建筑基址东侧石狮子雕像底部的铭文
261. 石狮子基座底部铭文拓片（王元林 摄）
262. 位于茶胶寺庙山西侧耶跋摩七世时代所建的小型寺庙（HOSPITAL）遗址
263. 从茶胶寺东遗址北至1号遗址（王元林 摄）
264. 神道东端与东池西堤相接的建筑基址北部（自南向北）
265. 茶胶寺庙山东南角的建筑遗址

历史照片目录

1. 茶胶寺庙山二层台回廊东南侧（法国远东学院摄影档案馆 CAM01667© EFEO）
2. 茶胶寺庙山二层台北内长厅解体维修情况（法国远东学院摄影档案馆 CAM08693© EFEO）
3. 茶胶寺庙山西北侧全景（法国远东学院摄影档案馆 CAM08698© EFEO）
4. 茶胶寺庙山神道南侧之发掘情况（法国远东学院摄影档案馆 CAM08713© EFEO）
5. 茶胶寺庙山二层台（法国远东学院摄影档案馆 CAM08723© EFEO）
6. 茶胶寺庙山西侧两座塔门俯瞰（法国远东学院摄影档案馆 CAM08725© EFEO）
7. 茶胶寺庙山东侧全景（法国远东学院摄影档案馆 CAM08757© EFEO）
8. 茶胶寺庙山神道发掘情况（法国远东学院摄影档案馆 CAM08780© EFEO）
9. 茶胶寺庙山西侧全景（法国远东学院摄影档案馆 CAM08784© EFEO）
10. 茶胶寺庙山西侧阇耶跋摩七世时期的建筑遗址（法国远东学院摄影档案馆 CAM08761© EFEO）

第一章　古代高棉史略

一、扶南（公元一世纪初 – 约公元 550 年）

柬埔寨是东南亚地区历史悠久的文明古国[1]，仅见于中国古代史籍中关于"扶南"的记载[2]，是迄今所知关于柬埔寨古代历史最早的文献记录[3]，其历史可以追溯至公元一世纪前后[4]。一般的解释认为，"扶南"一词源自高棉语"phnom"古音"bnam"的汉译[5]，其意为山，扶南国王则被称为"Kurung Bnam"，[6]其意大致相当于梵文中的"Parvatabhūpâla"或"Çailarâja"，意为"山王"或"山的守护神"。[7]对于扶南之名源流的考证，或许能够反映出古代高棉人对于山岳原始崇拜的深刻影响。

---

[1] 今之柬埔寨（Cambodia, Kampuchea）位于东南亚中南半岛的东南部，与越南中部为界，西临暹罗湾及泰国东部，南毗越南南部，北部与泰国东北及老挝西部接壤，其疆域位于北纬10°20′–14°32′、东经102°18′–107°37′，国土总面积约为18.1万平方公里。中国史籍中，汉时将今之柬埔寨称为"扶南"，隋及唐初称"真腊"，中唐以降则称"吉蔑"，元时改称"吉孛智"（或甘孛智），明万历以后则称"柬埔寨"，而柬埔寨人始终自称为"吉蔑"或"柬埔寨"。居住在柬埔寨的华人则将其译为"高棉"（Khmer），应是"吉蔑"的转译，西方则称之为"柬埔寨"（Cambodia, Cambodge）。"吉蔑"、"高棉"或指种族之名，"柬埔寨"则为国家的名称。一般认为，古代高棉人与湄南河流域的孟族（Mons）在种族关系上较为密切，高棉族与孟族的祖先，皆共同生活在中南半岛中西部地区，后来由于泰族的南徙渗进，高棉人居于中南半岛中部，而孟族则主要居于中南半岛之西。还有学者认为，东南亚的早期国家首先出现在大河下游及滨海港口区域：一是湄公河下游及其三角洲地带，即高棉人的古国——扶南；二是今之越南中部地区，即占婆人的古国——林邑；三是马来半岛的北部滨海地带，有狼牙修等古国；四是中南半岛的孟人地区，有金邻（金地、素旺那普米、直通）、堕罗钵底、罗斛；今之缅甸地区有骠国、太公、前期蒲甘、维萨里等国；五是海岛地区，有古戴、多罗磨、诃罗单、阁婆、婆利、室利佛逝等王国。（著者注）
[2] 中国史籍中关于扶南源流的记载，或可参见《三国志》卷四七《吴书·吴主传》："（赤乌）六年，扶南王范旃遣使献乐人及方物。"《三国志》卷六〇《吴书·吕岱传》："岱既定交州，复进讨九真，斩获以万数。又遣从事南宣国化，既缴外扶南、林邑、堂明诸王，各遣使奉贡。"《梁书》卷五四《海南诸国传总叙》："及吴孙权时，遣宣化从事朱应、中郎康泰通焉。其所经及传闻则有百数十国，因立记传。"据向达先生的考证：朱应撰有《扶南异物志》（早佚），康泰也曾撰有《吴时外国传》、《扶南土俗》、《扶南传》、《（康泰）扶南记》等，恐为同一书籍惟其称呼不同，皆早佚；另外，康泰或为康居人。参见向达：《汉唐间西域及南海诸国古地理书》，载于《国立北平图书馆馆刊》第四卷第六号。
[3] 今之越南南部芽庄地区出土的梵文碑铭武康碑（Vo–Canh），是迄今所知东南亚地区最早的梵文碑铭之一，立于公元三世纪。
[4] 《后汉书》卷七六《南蛮传》："肃宗元和元年，日南徼外蛮夷究不事人邑豪献生犀、白雉。"有些学者认为"究不事"殆为今之柬埔寨的古音，应是柬埔寨国名最早见于中国史籍的记载。但也有学者以为"究不事"与今之柬埔寨的对音中，除"不"字外，其他对音并不符合，其所产犀雉亦为印支半岛及苏门答腊各地所共有，认为此说依据尚不充分，似难作为定论。参见陆峻岭、周绍泉编注：《中国古籍中有关柬埔寨资料汇编》，北京：中华书局1986年版，第1页。
[5] G. I. Gerini, *Research in Ptolemy's geography of Eastern Asia*, Asiatic Society Monograph, I. London, 1909, p. 275.
[6] 此处 Kurung 乃高棉语 Kurun，暹罗语 Krun，占婆语 Klun 之对音，其发音却近似于中文的"昆仑"。《通典》卷一八八："隋时其国王姓古龙，诸国多姓古龙，讯耆老言，昆仑无姓氏，乃昆仑之讹";《太平御览》卷七八六："扶南国在林邑西三千余里，自立为王，诸属皆有官长，及王之左右大臣，皆号为昆仑。"其中的"昆仑"皆为古时东南亚诸国的国王尊号，在中国史籍中还有歌伦、故伦、昆仑等多种不同译称。（著者注）
[7] 梵文"Çailarâja"与南印度跋罗婆（Pallava）王朝诸王的称号类似，而在中国史籍中，扶南国王的名字多被冠以范姓，如范旃、范师蔓、范寻，推测应与源自南印度地区王号称谓梵文"Varman"的略译有关，并含有"保护"或"保护者"之意。另外，公元四世纪中叶扶南国王竺旃檀之名，可能源自印度贵霜王朝迦腻色迦王世系王号 Chandan 的汉译，他亦有可能是贵霜王朝覆亡时逃亡扶南的原王室成员。由此皆可见印度文化影响之一斑。（著者注）

公元一世纪以降，中国与印度之间海上贸易渐趋繁荣，庞大的海上贸易网络与先前途经中亚的陆路贸易南北呼应。从印度东海岸出发的商船，横跨孟加拉湾，到达马来半岛的西海岸。陆路运输货物经过克拉地峡，到达暹罗湾的西海岸。然后，他们再沿着到达中国南部，中南半岛逐渐发展成为这条商路沿线的重要贸易地点（图1-1）。公元四世纪中叶，印度笈多王朝在沙摩陀罗·笈多（Samudragupta，公元335－380年在位）统治时期，展开对恒河流域和南印度地区的征服，并对马来半岛、苏门答腊、爪哇等地实行扩张政策。与此同时，特别是以南印度地区跋罗婆王朝（Pallava）婆罗门僧侣为代表的知识精英也随之流动、迁徙，从而开启了东南亚地区第一批"印度化"[1]王国的历史序幕（图1-2，图1-3，图1-4，图1-5，图1-6）。[2]

从东南亚地区所发现的古代碑铭和宗教造像等文物皆反映出与南印度地区的密切关系。毋庸置疑，印度对东南亚的影响是多方面的，但最为显著的表现为宗教的影响，这是由古代印度的历史文化特征所决定的。古代印度宗教文化的繁兴与传播，使其对外扩张中也往往带有深厚的宗教色彩。因此，无论在政治、经济、文化、宗教各领域[3]，包括扶南在内的中南半岛诸王国皆受到来自印度文化的深刻影响，[4]或可将此描述为一种"在当地基础上印度化的上层建筑"[5]，中国史籍记载的扶南创建者混填[6]据信也是起源于印度北方的憍陈如（Kaundinya）婆罗门氏族[7]。在全面系统地吸收印度多元复

---

[1] "印度本土与外印度之间的关系可以追溯至史前时代，只是从某个时期起，这些关系才在印度支那半岛和群岛上导致了一批印度化王国的建立。这些国家给我们留下的最古老的考古遗存，不一定就是第一次印度文化浪潮的证据。很可能最初已经有一些航海者、商人或移民先于那些为第一批婆罗门教或佛教寺庙举行仪式以使之神圣化的僧侣，以及撰写最早梵文碑铭的文人到那里，他们是第一批印度化殖民地的创建者。印度化过程在本质上应当理解为一种系统的文化传播过程。这种文化建立在印度的王权观念上，其特征表现在婆罗门教和佛教的崇拜、《往世书》里的神话和遵守印度教神圣法典，特别是《摩奴法典》等方面，并且用梵文作为表达工具。这种梵文或印度的文化确实是自然而然地就移植到了东南亚。它与孟加拉和达罗毗荼人地区的梵文文化之间的唯一差别，可能即在于这一事实：前者是由海路传播，而后者则从陆路传播，也可以说是通过相互渗透传播的。东南亚的这种印度文化，我们根据地区，分别称之为印度—高棉文化、印度—爪哇文化。"参见 G. 赛代斯著，蔡华等译：《东南亚的印度化国家》（G. Cœdès, *LES ÉTATS HINDOUISÉS D'INDOCHINE ET D'INDONÉSIE*），北京：商务印书馆2008年版，第32－34页。

[2] "印度化"过程在本质上应当理解为一种系统的文化传播过程。这种文化建立在印度的王权观念上，其特征表现在婆罗门教和佛教的崇拜、《往世书》里的神话和遵守印度教神圣法典，特别是《摩奴法典》等方面，并且用梵文作为表达工具。这种梵文或印度的文化确实是自然而然地就移植到了东南亚。它与孟加拉和达罗毗荼人地区的"梵文文化"之间的唯一差别，可能即在于这一事实：前者是由海路传播，而后者则从陆路传播，也可以说是通过相互渗透传播的。东南亚的这种印度文化，我们根据地区，分别称之为"印度—高棉文化"、"印度—爪哇文化"等。参见 G. 赛代斯著，蔡华等译：《东南亚的印度化国家》（G. Cœdès, *LES ÉTATS HINDOUISÉS D'INDOCHINE ET D'INDONÉSIE*），北京：商务印书馆2008年版，第33页。

[3] 最近三十余年的考古调查与发掘已经证明：东南亚地区的印度化过程是在经历了大约一千多年相互交流的基础上开始建立的，在印度化的过程中，整个东南亚地区的政治体系日趋复杂，尤其是以黄金与锡为主的航海贸易网络起到了决定性的作用。参见 Jacques Claude, *Évolution des pratiques religieuses au Bayon*, in Sixth Symposium on the Bayon – Final Report, UNESCO, Paris, 2002, pp. 70－79.

[4] 近世以来吴哥等地考古发掘出土的扶南时期印度教及佛教造像，古朴典雅，多具有强烈的印度笈多王朝风格。关于东南亚印度艺术风格之影响，参见 A. K. 库马拉斯瓦米：《印度和印度尼西亚的艺术史》，伦敦，1927年版；《印度艺术的影响》，伦敦，1927年版；R. 格鲁塞：《在外印度的巴拉和森纳的艺术》，载于《东方学研究》卷 I, pp. 277－285；A. J. 伯内·肯帕斯：《那烂陀的青铜制品和印度化爪哇的艺术》，莱顿，1933年版。

[5] "东南亚诸印度化国家从来不是印度的政治藩属，而是它的文化殖民地。"参见 G. 赛代斯著，蔡华等译：《东南亚的印度化国家》（G. Cœdès, *LES ÉTATS HINDOUISÉS D'INDOCHINE ET D'INDONÉSIE*），北京：商务印书馆2008年版，第419页。

[6] 《梁书》卷五十四《海南诸国·扶南国传》："其后王憍陈如，本天竺婆罗门也，有神语曰，应王扶南，憍陈如心悦，南至盘盘，扶南人闻之，举国欣戴，迎而立焉，复改制度，用天竺法。"

[7] 根据中国史籍《晋书》及《梁书》的记载，伯希和在其撰写的《扶南考》中，推定混填至扶南的时代，最晚不迟于公元一世纪。伯希和推测混填一名，即是梵文憍陈如（Kaundinya）的对音，此名出于印度婆罗门种姓。按公元一世纪前后，印度人已渐东移，定居东南亚，其后东南亚各地便有了一些印度化国家。混填（或憍陈如）至扶南为王，是为印度统治东南亚及其文化影响最深远之事。参见伯希和：《扶南考》（Paul Pelliot, *Le Fou－nan*），载于《法国远东学院学报》第3卷（*Bulletin de l'Ecole française d'Extrême－Orient－BEFEO*, Volume III），1903年版，第248－303页。

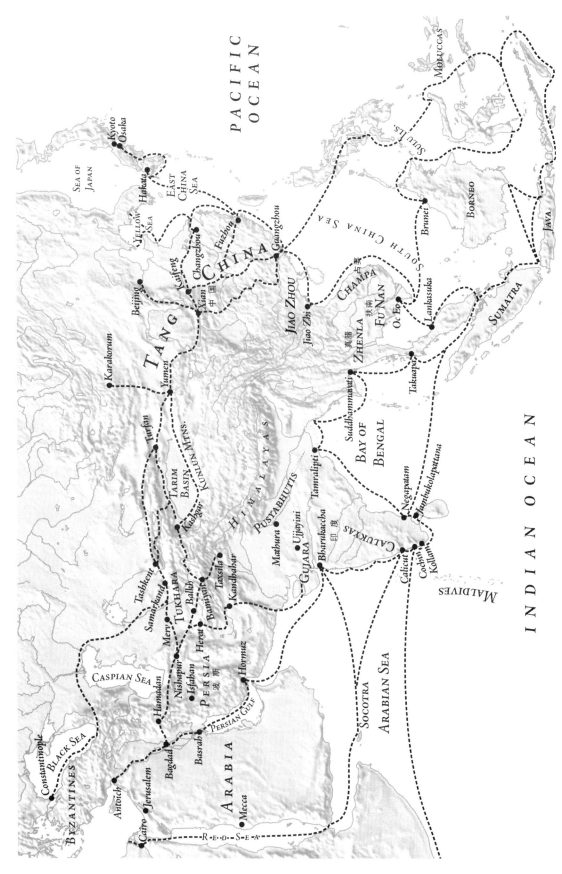

图 1-1 公元 6 世纪的亚洲贸易及文化交流线路示意图

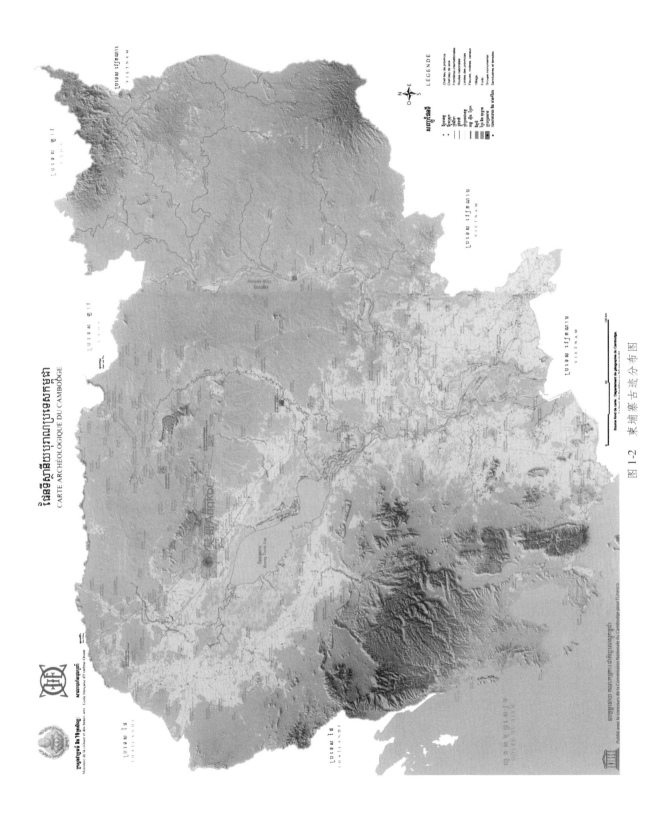

图 1-2 柬埔寨古迹分布图

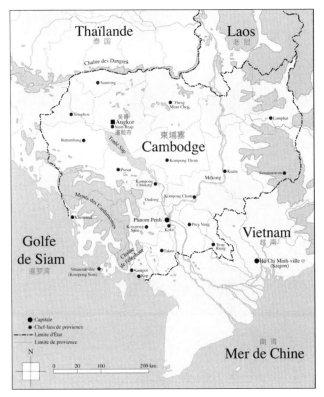

图1-3 柬埔寨吴哥古迹的地理区位

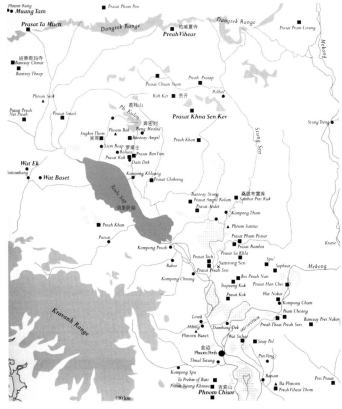

图1-4 柬埔寨主要考古遗迹分布图

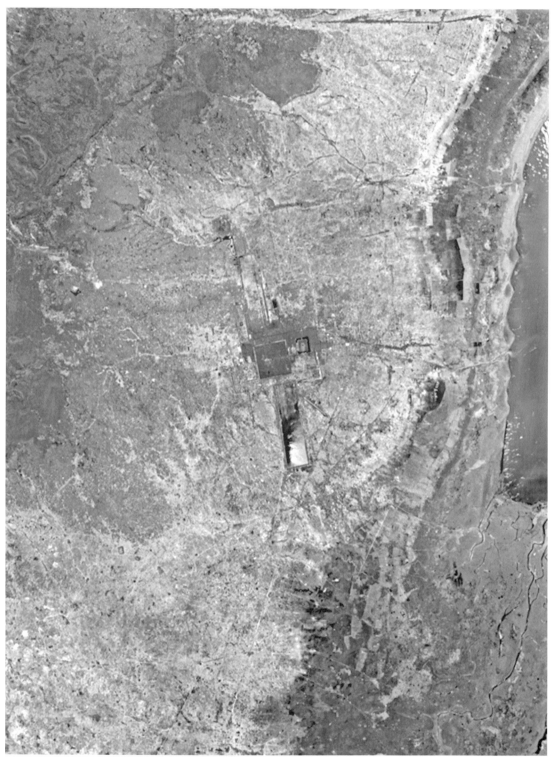

图 1-5 吴哥古迹卫星影像图之一

图 1-6　吴哥古迹卫星影像图之二

杂的宗教信仰体系基础上[1]，扶南逐渐确立了以"湿婆－林伽（Maheśvara－Linga，Maheśvara 其意为伟大的湿婆神）"[2]为核心的国家信仰与王权统治。[3] 摩诃林伽作为创造之神湿婆神（Shiva）的终极象征和最高表现形式，凭借婆罗门阶层对于国家信仰所施加的巨大影响，将其视为国家的守护神以及国王的化身，而由婆罗门主持的国家祭祀仪式（state cults）更被赋予了极其崇高的尊严和地位。[4] 因此，"湿婆－林伽"崇拜成为一种把国王神化为精神力量的手段。[5] 另外，由于受到尊奉一个人格神为世界的主宰而融合不同信仰将湿婆（Shiva）、毗湿奴（Vishu）这两位神祇和梵天（Brahma）结合起来而形成的印度教"三位一体观（Brahmanism Trimurti）"之影响[6]，扶南对诃里诃罗（Harihara，毗湿奴和湿婆合为一体的神祇）崇祀也曾盛行一时，亦是当时宗教信仰的显著特征之一。[7]（图 1-7，图 1-8，图 1-9，图 1-10，图 1-11）

---

[1] 扶南时期的佛教，应为印度贵霜王朝迦腻色迦时期尊崇佛教的影响所及。根据中国史籍记载，扶南名僧伽婆罗、曼陀罗和须菩提等，皆曾造访中国进行佛经翻译。（著者注）

[2] "大部分在外印度组成的王国，很快就采用了湿婆教的王权观念，这种王权是建立在婆罗门和刹帝利这两个种姓上的，表现为对国王林伽的崇拜。" R. 海因－格尔登：《东南亚的国家观念和王权观念》，载于《远东季刊》第 II 卷（Robert Heine－Geldern, *Conceptions of State and Kingship in Southeast Asia*, *Far Eastern Quarterly*, Vol. II）1942 年版，第 15－30 页。

[3] "公元初的几个世纪，在绝大多数王国，湿婆崇拜以及毗湿奴崇拜、大乘佛教、小乘佛教信仰和平共处，虽然随着时间的推移发生了分化，逐渐改变了不同宗教派别的分布状况。"引自尼古拉斯·塔林主编，贺圣达等译：《剑桥东南亚史》，昆明：云南人民出版社 2003 年版，第 237 页。

[4] 《梁书》卷五十四《海南诸国·扶南国传》："俗事天神，天神以铜为像，二面者四手，四面者八手，手各有所持，或小儿，或鸟兽，或日月。"

[5] 湿婆教的婆罗门，以仪式的方式将狂迷之道转化成林伽（Linga）崇拜，同时采用古老的古典吠陀救世论，而将人格性的世界主宰导入其体系。如此一来，湿婆教里便存在着最高度内在异质性的种种形态，一端是作为新正统派的贵族婆罗门阶层的信徒，另一端则是作为村落寺庙崇拜的农民大众的信徒。参见马克思·韦伯著，康乐等译：《印度的宗教：印度教与佛教》，桂林：广西师范大学出版社 2005 年版，第 428－430 页。

[6] "梵天本身为一个理论角色，实际上从属于湿婆与毗湿奴。对其崇拜仅限于唯一的贵族婆罗门的寺庙，除此其地位完全落居于湿婆和毗湿奴之后。"参见马克思·韦伯著，康乐等译：《印度的宗教：印度教与佛教》，桂林：广西师范大学出版社 2005 年版，第 420 页。

[7] 参见 D. G. E. 霍尔著，中山大学东南亚历史研究所译：《东南亚史》（D. G. E. Hall, *A History of Southeast Asia*），北京：商务印书馆 1982 年版，第 136 页。

图 1-7 湿婆神的终极象征摩诃林伽

图 1-8 扶南时期的毗湿奴像

图 1-9 公元 6 世纪的毗湿奴造像及其发掘出土的情形

图1-10 公元6世纪扶南时期的诃里诃罗造像

图1-11 公元5世纪扶南时期的观音造像

公元四世纪至六世纪中叶，扶南国势日渐强盛，属国众多，疆域广袤，称雄中南半岛。根据当时的中国史籍《梁书》"以扶南国王憍陈如阇耶跋摩，为安南将军"的记载[1]，阇耶跋摩或称憍陈如阇耶跋摩（Kaundinya Jayavarman）是扶南历史上最重要的国王之一，其在位统治时期亦是扶南历史的极盛时代。[2] 阇耶跋摩的继承者留陁跋摩（Rudravarman）所立的梵文碑铭，位于今之柬埔寨南部茶胶省（Ta Keo Province）的达布隆巴蒂寺内（Ta Prohm of Bati），为研究当时宗教和历史提供了宝贵资料[3]。"在整整五个世纪中，扶南都是中南半岛占统治地位的强大国家，衰落之后的很长一段时间，它在以后几代人的记忆中仍然享有盛誉。"[4]（图1-12）

由于年代久远，扶南时代的城市、宫殿、寺庙遗迹早已荡然无存，中国史籍的枝节记载则展示出迄今所知的关于扶南社会生活及风土人情的概况[5]。例如，关于扶南早期都城"治特牧城，俄为真

---

[1] 《梁书》卷二《武帝本纪》。另据《梁书》卷五四《扶南国传》载：天监二年（公元503年）阇耶跋摩复遣使送珊瑚佛像，并献方物。诏曰：扶南王憍陈如阇耶跋摩，介居海表，世纂南服，厥诚远著，重译献琛，宜蒙酬纳，班以荣号，可安南将军、扶南王。"

[2] 亦可参见本书第2页注[6]。此时亦为佛教流行中国并肇兴扶南之时，多有扶南名僧前往中国译经传法。《续高僧传》卷一《梁扬都正观寺扶南国沙门僧伽婆罗传》："僧伽婆罗，梁言僧养，亦云僧铠，扶南国人也。幼而颖悟，早附法津，学年出家，偏业《阿毗昙论》。声荣之盛，有誉海南。具足已后，广习律藏。勇意观方，乐崇开化。闻齐国弘法，随舶至都。住正观寺，为天竺沙门求那跋陀之弟子也。复从跋陀研精方等，未盈炎燠，博涉多通，乃解数国书语。值齐历亡坠，道教陵夷，婆罗静洁身心，外绝交故。拥室栖闲，养素资业。大梁御宇，搜访术能，以天监五年被敕征召于扬都寿光殿、华林园、正观寺、占云馆、扶南馆等五处传译，讫十七年，都合一十一部，四十八卷，即《大育王经》《解脱道论》是也。"另，"梁初又有扶南沙门曼陀罗者，梁言弘弱，大赍梵本，远来贡献。敕与婆罗共译《宝云》《法界体性》《文殊般若经》三部，合一十一卷。虽事传译，未善梁言，故所出经文多隐质。"

[3] Lawrence Palmer Briggs, *The Ancient Khmer Empire*, New Series Volume 41, Part 1, *The American Philosophical Society*, 1951, p. 31.

[4] 参见G. 赛代斯著，蔡华等译：《东南亚的印度化国家》（G. Cœdès, *LES ÉTATS HINDOUISÉS D'INDOCHINE ET D'INDONÉSIE*），北京：商务印书馆2008年版，第108页。

[5] 《梁书》卷五十四《海南诸国·扶南国传》："在日南群之南，海西大湾中，去日南可七千里，在林邑西南三千余里，城去海五百里，有大江广十里，西北流东入于海。其国轮广三千余里，土地洿下而平博。"

腊所并，益南徙那弗那城"的记载[1]，通过二十世纪初叶以来的考古学研究表明，特牧城之故址大致位于今之柬埔寨波罗勉省（Prey Veng province）之巴普农（Baphnom）与巴南（Banam）附近，而作为扶南末期统治中心的"那弗那城"，概为梵文"Navanagara"（新城）的译名，推测其位置大致位于湄公河下游及其三角洲地区，今之柬埔寨南部茶胶省的吴哥波雷（Angkor Borei）及其附近的扶农达山（Phnom Da）遗址则被认为是扶南末期的都城。全盛时期的扶南，其疆域还应当包括今之越南南部、湄公河中游以及湄南河流域和马来半岛之大部。

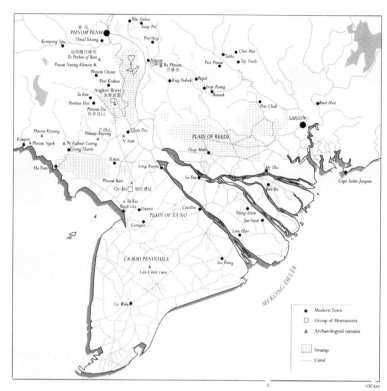

图 1-12　湄公河下游扶南时代的遗址分布图

论及扶南对于东南亚历史进程的地位及其影响，正如英国著名的东南亚史学家丹尼尔·霍尔在其著作《东南亚史》（D. G. E. Hall, *A History of Southeast Asia*）中的评述："扶南是东南亚历史上第一个大国，像欧洲历史上的罗马一样，其威望在亡国后还维持了很久。其风俗特别是对圣山和龙女（Soma）的崇拜，还被其后的高棉王国所延续。虽然它的建筑已经完全没有了，我们有理由相信它的某些特征仍然保存在遗存至今的前吴哥时代[2]的一些建筑遗迹中，而且那个时代带有浓郁的笈多艺术风格的佛教造像、头戴法冠的毗湿奴像和诃里诃罗像，可以使我们对扶南雕塑家塑造人像的方式获得某些概念。"[3]

由于扶南时代的历史文献与建筑实物早已荡然无存，唯有通过前吴哥时代的某些城市与建筑遗迹中可以约略了解当时的情况。根据帕蒙蒂埃（Henri Parmentier）的研究[4]，扶南时期的建筑平面多为等臂十字形，结构形式以单层木结构为主，在其墙身转角部位或屋顶多设有装饰。[5]而扶南时期寺庙建筑平面多呈方形或长方形，素平的砖砌墙体上设有垂直扶壁柱，墙面不设假门或"天宫"装饰（Flying Palace）；其屋顶形制与构造颇具特色，一般由逐层向内收分的多层密檐组成，出挑的密檐由多重叠涩线脚承托，檐口之上设有类似南印度地区跋罗婆风格壁龛（Pallava Kudus），这

---

[1]《新唐书》卷二二二《南蛮下·扶南传》。
[2] 学术界通常引用中国史籍中真腊的相关记载以界定公元六世纪中叶至公元九世纪初的古代高棉历史，又将其称为"前吴哥时代"（Pre-Angkorian）。（著者注）
[3] 参见 D. G. E. 霍尔著，中山大学东南亚历史研究所译：《东南亚史》（D. G. E. Hall, *A History of Southeast Asia*），北京：商务印书馆 1982 年版，第 57-58 页。
[4] H. Parmentier, *History of Khmer architecture*, Eastern Art, volume 3: 1931, pp. 141-179.
[5]《南齐书》卷五十八："（扶南）伐木起屋，国王居重阁，以木栅为城。海边生大箬叶，长八九尺编其叶以覆屋，人民亦为阁居。"

些雕饰壁龛大多以建筑物和瑞兽作为表现题材,这在某种程度上不仅体现出扶南时期建筑形制与构造的简略信息,而且这也应是探究公元七世纪以降真腊及吴哥时代砖塔原型及其构造方法的门径。另外,在塔殿入口两侧的花柱(colonette)通常是圆形的,某些砂岩门楣(Lintel)之上雕刻有摩羯形象(Makara)的雕饰。

总体而言,扶南时代的建筑风格更多地受到来自南印度跋罗婆王朝的影响,建筑形制简洁,其主要特征包括入口前出抱厦(Mandapa)、逐层向内收分的密檐、密檐上的壁龛雕饰、石质门楣的摩羯图案雕饰等。[1] 帕蒙蒂埃认为,湄公河下游和洞里萨湖之间形成"V"字形区域内的几座建筑遗址,应该属于扶南末期的风格,其中包括今之柬埔寨磅湛省的西埃特陶克(Prasat Preah Theat Toc)遗址、磅通省的库特邦窟(Kuk Trapeang Kuk)遗址等。[2]

---

[1] H. Parmentier, *L'art khmer primitif*. Vol 2. P. E. F 21 and 22, Paris, 1927. p. 126.
[2] Lunet de Lajonquière, *Inventaire Archeologique de l'Indochine*, 1902, 1907, 1912, 3 E. F. 4, 8 and 9, Paris Leroux.

二、真腊（约公元 550–802 年）

公元六世纪下半叶，扶南盛极而衰[1]。曾经作为扶南属国之一的真腊[2]，位于扶南疆土的北方，占据着今之柬埔寨上丁（Stung Treng）以北的湄公河中下游的广阔地区。伴随着拔婆跋摩一世（Bhavavarman I，约公元 550–600 年在位）及摩醯因陀罗跋摩（Mahendravarman，中国史籍亦称之质多斯那，约公元 600–615 年在位）统治时期的集权巩固和武力征伐，真腊的势力逐渐强盛起来。自此之后，真腊历经伊奢那跋摩一世（Isanavarman I，中国史籍称之为伊奢那先，约公元 615–635 年在位）、阇耶跋摩一世（Jayavarman I，约公元 657–681 年在位）统治期间的远征近伐，国势日盛，疆域广袤，并以"富贵真腊"之声名而威震宇内。尤其是伊奢那跋摩一世创建经营的伊奢那补罗城（Isanapura）更是曾经辉煌一时。时至今日，柬埔寨磅通省的桑坡布瑞窟（Sambor Prei Kuk）地区，尚遗存有诸多当时的城市、建筑、雕塑、碑铭等重要史迹[3]。及至公元七世纪末，扶南最终为真腊取代而覆亡。我国的史籍《隋书·真腊传》中对此亦有所记载[4]，这也是追溯真腊取代扶南史实最为弥足珍贵的文献记录，历来为学界所倚重[5]。

---

[1] 根据中国史籍的记载，扶南的最终覆亡大约在唐朝贞观初年，即公元七世纪二三十年代。（著者注）

[2] 真腊一名之来源，曾有种种推测。施古德（G. Schlegel）曾将之还原为 Tchanda。伯希和初注之绪言中谓此种还原方法固有其可能，然此种假定，今尚未见有反证或实证也。该利尼（. E. Gerini）则以为真腊古音当为 Chon-ra 或 Son-rai，而今日斯提安（Stieng，乃柬埔寨境内山居之少数民族）语中仍称柬埔寨为 Sorai，故认为真腊一名即由斯提安语 Sorai 而来。伯希和指出隋唐古音中腊字读 lap，不读 ra 或 rai，真字读 tsien，并不读 so。帕克（E. H. Parker）以为真腊古音为 Sienrap，其意似为与今日柬埔寨之暹粒（Siem-reap）有关。伯氏驳之曰：暹粒之原意为"被驯服之暹民"，其时代当在公元十三世纪暹民迁入其地而被驯服之后，故不能作为得名来源之依据。（伯希和《真腊风土记笺注》增订本，1951 年法文本，下简称伯氏新注，71 页。）汪大渊《岛夷志略》藤田丰八注真腊条云，真腊一名乃由暹粒得名。苏继庼不同意其说，而另创一说，谓"似为 Khmer 之讹转。"（苏继庼《岛夷志略校释》稿本）实则二者并不同音，Khmer 唐人译为吉蔑（新、旧《唐书》或阇蔑（唐沙门慧琳《一切经音义》卷 100 引慧超《往五天竺传》），其所谓"讹转"，实出臆测。伯希和谓中国载籍有真腊改国名曰占腊，实则并无其事，乃出于中国之臆测或误会。（新注 79–80 页）藤田丰八《岛夷志略校注》亦谓："占腊乃真腊之音讹，宋以来有之。或云，宋末其王灭占城，役属之，号占腊。殆明人附会之说耳。"（21 页）伯氏又云："占腊"二字有征服占城之意，此或宋绍熙间柬埔寨国王战胜占城后，于其尊号加入平定占城一辞，而致生臆测或误会欤？伯氏更提出新说，谓真腊之"真"，其古音可读为 cin，此即"支那"之译名。当时柬埔寨国王新兴，尊号中曾自夸战胜支那，故取真腊（Cin-rap）一名。（伯氏新注 81–82 页）今按伯氏之对音虽属可能，但其推论乃完全出于臆测。当时真腊新兴，并不与中国接壤，更无交兵之事。纵使自夸，恐亦不至于如此荒谬。伯氏新说，不足为凭也。至于真腊有时亦称占腊，当由占、真二者音近，占城、真腊又接壤相邻，用字易于混淆，并无深意。伯氏求之过深，又进一步推论真腊之字源，未免作丝益紊。总之，真腊一名之起源，迄今尚未能知之也。《郛》甲本将书名及《总叙》与第一则中之真腊皆写作真蜡。初疑为误字。但《元史》中亦有书真腊之腊为蜡者，如《元史·世祖本纪》（卷 13）至元二十二年九月丙子条。《说海》本明费信《星槎胜览》卷一真腊国条，目录作腊，而正文本条两处皆作蜡，知元、明时作为国名，此二字似通用。我国正史自《隋书》至《明史》，皆作真腊，则当以作腊字者较为通行。参见（元）周达观原著，夏鼐校注《真腊风土记校注》，北京：中华书局，1981 年版，第 16–17 页。

[3] Lawrence Palmer Briggs, *The Ancient Khmer Empire*, New Series Volume 41, Part 1, *The American Philosophical Society*, 1951, p. 51.

[4] "真腊"之名始见于《隋书》卷八二《南蛮·真腊传》："真腊国，在林邑西南，本扶南之属国也……其王姓刹利氏，名质多斯那。自其祖渐已强盛，至质多斯那，遂兼扶南而有之。死，子伊奢那先代立。居伊奢那城，郭下二万余家。城中有一大堂，是王听政之所。总大城三十，城有数千家，各有部帅，官名与林邑同。"同书卷四《炀帝纪》所记"大业十二年（公元 616 年）二月己未，真腊国遣使贡方物"，这是中国史籍中最早提及真腊的记载；又《明史·真腊传》载"其国自称甘孛智，后误为甘破蔗，万历后又改为柬埔寨"，由此可见"真腊"之名一直沿用到明万历年间。（著者注）

[5] 根据有关古代高棉碑铭研究，《隋书·真腊传》中记载的真腊王质多斯那，应为拔婆跋摩一世（Bhavavarman I）之弟 Śitrasena Mahendravarman 的译称。详见 *Les Inscription de Baksei Chamkron*, *Journal Asiatique*, Paris, 1909, ser. 10, 9 (13) pp. 467–515.

另外，根据中国史籍的记载亦可大致了解真腊时代宗教信仰的基本面貌。例如，《隋书》卷八二《真腊传》载："近都有陵伽钵婆山，上有神祠，每以兵二千人守卫之；城东有神名婆多利，祭用人肉，其王年别杀人，以夜祀祷，亦有守卫者千人，其敬鬼如此。多奉佛法，尤信道士；佛及道士，并立像于馆。"又如，《旧唐书》卷一九七《真腊传》载："国尚佛道及天神，天神为大，佛道次之。"由此可见，真腊时代的宗教信仰应以婆罗门教最为盛行，尤以奉祀湿婆或林伽为多[1]，或有毗湿奴与湿婆合体的诃里诃罗。此外，大乘佛教亦多有流布，甚至还包括诸多对祖先及自然神灵的崇拜等等。

伊奢那跋摩的继位者是拔婆跋摩二世（Bhavavarman II），其碑铭可以确定的时间为公元639年。拔婆跋摩二世之后是阇耶跋摩一世（Jayavarman I，约公元650－690年），阇耶跋摩一世继承了伊奢那跋摩时代的对外扩张政策，号称"众王中光荣之狮，胜利的阇耶跋摩"，遂征服中寮和上寮，向北直抵中国的南部边境。但是迅速扩张领土的真腊内部并不安宁，权力纷争埋下分裂的种子。至阇耶跋摩一世死后，真腊终于陷入混乱分裂的局面。

公元八世纪初，真腊分裂为南北两部，其北部地处山地丛林，称为陆真腊或上真腊；因其南部地处湄公河下游及三角洲地区的水乡泽国，河道湖汊纵横密集，称为水真腊或下真腊。[2] 水真腊据有扶南的旧境，都城称为毗耶德诃补罗（Vyadhapura），其位置大致位于今之柬埔寨南部茶胶省吴哥波雷（Angkor Borei）地区；陆真腊位于真腊的祖居之地，包括今日湄公河中游及丹克瑞克（Dangrek 亦译作扁担山）山脉以北的广大地区，其疆土向北延伸至中国边境，与南诏接壤，我国史籍又称其为"文单国"[3]。水真腊包括了伊奢那补罗城在内的真腊南部地区，是阇耶跋摩一世时代以降真腊的真正继承者。唯其内部有太阳王朝和太阴王朝的互相争夺，世系混乱，两部势力通过政治联姻也最终未能归于统一。

公元八世纪下半叶，水真腊陷入混乱。爪哇的岳帝王朝兴起，势力及于马来半岛及中印半岛沿岸，曾于公元774年及787年侵袭占城沿海之地，占据水真腊南部沿海很多地区，并且最终攻陷了水真腊的都城桑坡布瑞窟。公元八世纪末叶，随着苏门答腊室利佛逝（Śrivijaya，在我国史籍中亦称其为"三佛齐"）王国入侵，以及信奉大乘佛教的爪哇夏莲特拉王朝（Śailendra）[4] 的兴起和扩张，昔日声威煊赫的真腊逐渐式微，直至其后波澜壮阔的吴哥时代的来临取而代之。

纵观整个真腊时代，正如赛代斯在其名著《东南亚的印度化国家》（George Cœdès, *LES ÉTATS HINDOUISÉS D'INDOCHINE ET D'INDONÉSIE*）中所言："真腊时期的古代高棉文明，在水利设施方面，以及宗教和艺术方面都继承了扶南的成就；在建筑艺术方面则深受占婆（Champa）的影响[5]。这一文明在整个七世纪中获得了活力，而这种活力将使古代高棉在八世纪短暂地衰落一时之后，又能够长

---

[1] 陵伽钵婆意为"林伽之山"，今之老挝南端的湄公河西岸，山名占巴索（Cham Pasak），海拔高度为1397米，山顶之上有一天然巨石，形似奉祀之林伽。真腊祖居之地最初的都城，即建于此山麓。（著者注）
[2] 参见伯希和：《八世纪中国赴印度两道考》（Paul Pelliot, *Deux itinéraires de Chine en Inde à la fin du VIIIe siècle*），（*Bulletin de l'Ecole française d'Extrême-Orient - BEFEO*, Volume IV），第 IV 卷，1904 年，第 131－143 页。另据《新唐书·真腊传》记载："神龙（公元705－706年）后分为二半，北多山阜，号陆真腊；半南际海，饶陂泽，号水真腊。半水真腊地八百里，王居毗耶驮补罗。陆真腊或曰文单，曰婆镂，地七百里，王号笪屈。"
[3] 《旧唐书》卷十一《代宗本纪》："大历六年十一月己亥，文单国王婆弥来朝，献驯象十一……十二月……庚午，制以文单王婆弥为开府仪同三司，试殿中监。"
[4] 爪哇夏莲特拉王朝的势力范围涉及今之苏门答腊或马来半岛，以及印度尼西亚爪哇群岛之大部。（著者注）
[5] 占婆（Champa）是世界上由于受到外族征服而完全消亡的国家之一，历史上的占婆曾是一个相当繁荣的国家，占族人精通源自印度的文化和艺术，他们吃苦耐劳且骁勇善战，一直都是古代高棉人的宿敌。（著者注）

时间地统治着中南半岛的南部和中部。"[1]

真腊时期的建筑遗存较为丰富，主要分布在今之柬埔寨磅通省的桑坡布瑞窟（Sambor Prei Kuk）地区，多以砖石结构的寺庙塔殿为主。其中，桑坡布瑞窟建造时代约为公元七世纪，布雷克蒙（Prei Khmeng）遗址建造时代大约是公元八世纪上半叶，而布瑞贡磅（Preah Kompong）遗址的建造时代则是公元八世纪下半叶。[2]作为扶南建筑与艺术的延续和继承，真腊建筑形制与艺术风格整体上趋向秀丽挺拔，建筑尺度比例的控制更为精巧细腻，建筑材料多以砖石为主，墙身及屋顶以红砖砌筑，门楣、花柱、假门及窗棂等构件多以砂岩雕刻而成，而寺庙围墙以及建筑基础则以红色角砾岩砌筑。（图1-13，图1-14，图1-15）帕蒙蒂埃将真腊时期的建筑特征归纳如下：其一，寺庙院落平面格局呈方形，塔殿平面以长方形、六边形、八边形为主，通常各面墙体不辟假门，墙面上刻有砖雕精细的壁龛（天宫壁龛及雕刻表面原本覆以白色灰泥及色彩，现仅存大致轮廓和表面雕凿的痕迹）；其二，塔殿屋顶部分为多层叠涩拱构成，向上收分显著，立面自下而上逐层设有尺度较小的门或假门与之协调；其三，砂岩门楣、花柱、门窗等构件，雕刻精美，线条繁复，其中门楣的形制和雕刻题材或可分为两类，扶南型和真腊型，皆可从中看出扶南时期建筑风格的痕迹。[3]（图1-16，图1-17，图1-18）

关于扶南与真腊的历史，当时的文献资料早已散佚殆尽，唯余少量的碑铭存世。在迄今已经发现整理的一千二百余块古代高棉碑铭中，绝大多数雕刻梵文和古代高棉文。诸如芬诺（Louis Finot）、赛代斯为代表的高棉史学前辈已经将其大部分释读，编纂出版皇皇巨著《古代高棉碑铭全集》（*Inscriptions du Cambodge*）。虽然这些古代高棉碑铭的主要内容多为记述寺庙宗教仪式以及各种形式的祈祷祝词等，但通过诸多学者的研究和解

图1-13　真腊时期的建筑遗迹桑坡布瑞窟之一

图1-14　真腊时期的建筑遗迹桑坡布瑞窟之二

---

[1] 参见 G. 赛代斯著，蔡华等译：《东南亚的印度化国家》（G. Cœdès, *LES ÉTATS HINDOUISÉS D'INDOCHINE ET D'INDONÉSIE*），北京：商务印书馆2008年版，第131页。

[2] 1927年，法国艺术史家菲利普·斯特恩在其著作《吴哥的巴戎寺和高棉艺术的发展》（Philippe Stern, *Le Bayon d'Angkor et l'évolution de l'art khmer : étude et discussion de la chronologie des monuments khmers*）中，对古代高棉艺术风格进行了系统的分期与分类，其后柯瑞尔·雷慕沙（G. de Coral Rémusat）又将真腊时期的建筑风格进行了深入细致的研究。（著者注）

[3] 参见 H. 帕蒙蒂埃（Henri Parmentier）：《扶南时代的艺术》（*L'art présumé du FUNAN*），BEFEO, XXXII, 1932, pp. 183–189.

图 1-15 真腊时期的建筑遗迹桑坡布瑞窟之三

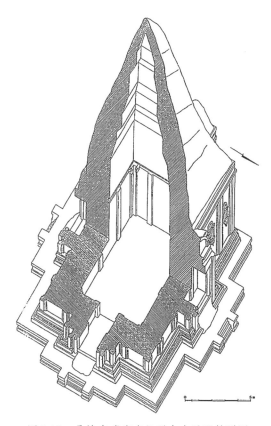

图 1-16 桑坡布瑞窟南组群中央圣殿轴测图

图 1-17 桑坡布瑞窟南组群中央圣殿内景

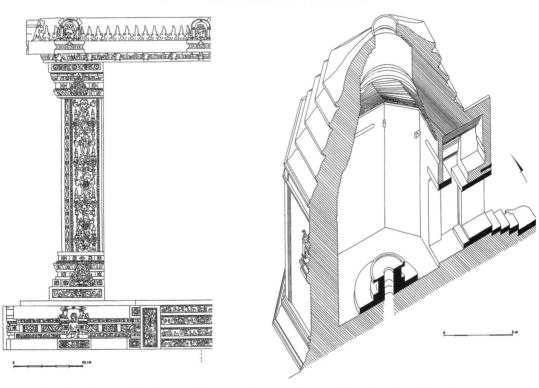

图 1-18 桑坡布瑞窟南组群小型神龛立面图（左）及八边形平面的砖塔轴测剖面图（右）

读，逐步从中梳理出吴哥王朝较为可信的世系年代，以及当时的社会、经济、政治等方面的信息。然而，自《汉书·地理志》、《三国志·吴志》以降，中国史籍如历代的官修正史、典籍，以及稗史、杂乘、地志、行纪等历史文献中，保留了许多有关扶南与真腊的文字记载，包括古代风物、社会习俗、政治经济、与中国的交往等诸多方面，因而成为研究柬埔寨古代历史最为珍贵的史料，近世以来的高棉古代历史研究也多以引用中国史籍中的记载作为信史依据。

三、吴哥：古代高棉历史的黄金时代（公元 802－1432 年）

古代高棉历史最为辉煌灿烂的黄金时代出现在公元九世纪至十五世纪。

作为东南亚地区最为重要的古代史迹，举世闻名的吴哥（Angkor）是公元九世纪至十五世纪古代高棉帝国繁盛时期都城与寺庙建筑的遗迹。以大吴哥城（Angkor Thom）与吴哥窟（Angkor Wat，亦称小吴哥）为代表的九十余处建筑组群及其数百座单体建筑遗构，散布在今之柬埔寨北部暹粒省大约四百余平方公里的热带丛林之中。古代吴哥璀璨绚烂的文化在东南亚文明的历史进程中扮演着举足轻重的角色，尤其对历史上湄公河及湄南河流域的众多王权国家和民族文化都产生了深远的影响。以印度教与佛教为主流的宗教信仰深深地铭刻在吴哥时代的文化脉络之中，并由此构成了古代高棉建筑与艺术的主题。作为集中体现古代高棉建筑艺术与技术的巅峰之作，诸多遗存至今的吴哥史迹仍以其撼人心魄的空间布局与象征冠绝古今。尤其是那些雄伟壮观的庙山（temple-mountain）建筑成为吴哥时代建筑特征的显著标志[1]，它们不仅以缜致精微、美轮美奂的雕刻与装饰代表着古代高棉石刻艺术的高超水平，还以其雄浑精丽的建筑风格展示出古代高棉建筑技术的无穷魅力，堪称吴哥时代的建筑代表作。

穿越前吴哥时代（Pre-Angkorian）的混沌与征伐，并得益于信奉佛教的夏莲特拉王朝在爪哇诸岛的衰落，公元 802 年，号称"众王之王"[2] 的阇耶跋摩二世（Jayavarman II，公元 802－850 年在位）统领高棉人逐步摆脱了爪哇的宗主权。阇耶跋摩二世在摩醯因陀罗山（Mahendraparvata）[3] 的容琛寺（Rong Chen）通过举行提婆罗阇仪式[4]（梵文 Devarâja，即为"神王合一"之意，其核心是通过对象征国王神性的林伽的崇拜仪式，使之成为古代高棉国王权力与统治的象征）自立为王（Charkravartin），[5] 从而结束了古代高棉民族持续一个多世纪的分裂局面，扶南及真腊时代的疆域和荣耀得

---

[1] 庙山建筑是印度教神话中须弥山的象征，主要形制特点表现为阶梯状的砌石须弥台（a stepped pyramid in a stupa）关于庙山建筑的形制源流，详见本书第五章相关叙述。
[2] 高棉语发音为 "Kamraten jagad ta Raja"。（著者注）
[3] E. 艾莫涅尔（Étienne Aymonier）将此山的位置定位于今之柬埔寨暹粒东北四十余公里的古伦山（Phnom Kulen，亦译作荔枝山）。二十世纪三十年代，菲利普·斯特恩和亨利·马绍尔在古伦山的考古发掘中，将其发现寺庙遗迹的建筑风格定义为过渡性的，即前吴哥时代与吴哥时代早期盛行的建筑艺术风格糅杂在一起，并且受到来自爪哇和占婆的影响。而实际上由于阇耶跋摩二世统治时期都城具体位置至今尚未固定，可能一直处于古伦山与大湖之间广袤的冲积平原地区。诃里诃拉洛耶（Hariharalaya）或许是其中地位最为显著的一座，位于今之柬埔寨暹粒市东南约 15 公里的罗莱士建筑群遗址（Roluos）中的 Svay Pream，Prasat Prei，Trapeang Phong，Prasat Olak 等，据考证皆属于公元八世纪末期的建筑遗迹。另外，根据帕蒙蒂埃的研究，位于柬埔寨马德望省丹克瑞克（Dangrek Mountain）山区的阿摩罗因陀罗补罗（Amarrendrapura）遗址，以及吴哥地区西池（Western Baray）附近的阿约寺（Ak Yum Temple）遗址，推测其时代为公元八世纪初，或为阇耶跋摩二世时期的都城遗址之一。参见 Lawrence Palmer Briggs，*The Ancient Khmer Empire*，New Series Volume 41，Part 1，*The American Philosophical Society*，1951，pp. 84－85.
[4] "与其说国王是作为行政管理者出现，不如说他更像一个活在世上的神。他那筑有城墙掘有壕沟的都城是宇宙的缩影。都城的中央以一座象征须弥山的庙山为标志，山顶是借助于婆罗门从湿婆神那里领受的国王之林伽——提婆罗阇（代表王权的林伽）。我们不清楚是否这座包含着王权本质的林伽，经过相继的这几位国王统治的时期，始终都是独一无二的。反过来说，国王在自己登基时举行仪式使之神圣化了的、刻有他们尊号的林伽，与提婆罗阇是否即为同一座林伽，抑或不是。每一位有足够时间和财力的国王，都在他们都城的中央修建自己的庙山。" 参见 G. 赛代斯著，蔡华等译：《东南亚的印度化国家》（G. Cœdès，*LES ÉTATS HINDOUISÉS D'INDOCHINE ET D'INDONÉSIE*），北京：商务印书馆 2008 年版，第 206－207 页。
[5] 因陀罗跋摩一世（Indravarman I 公元 877－889 年）时期的一块碑铭将阇耶跋摩二世称为"柬埔寨之王"（Sovereign of Kambuja），这是古代高棉历史上首次出现柬埔寨的名称。（著者注）

以光复。[1]

阇耶跋摩二世之后的继位者阇耶跋摩三世（Jayavarman III，公元850－877年在位）平庸守成，遂为其堂兄篡夺王位，称为因陀罗跋摩一世（Indravarman I，公元877－889年在位）。因陀罗跋摩一世在诃里诃罗洛耶（Hariharalaya，位于今之吴哥东南约15公里的罗莱士地区）创建巴空寺（Bakong）庙山作为国寺，并兴建神牛寺（Preah Ko）奉祀其父母及祖先。这两座寺庙标志着吴哥古典建筑艺术的开端，它们的建筑形制和艺术风格影响着此后的吴哥寺庙建筑。因陀罗跋摩一世还建造了庞大的水利灌溉工程，极大促进了农业的繁荣与发展，为吴哥王朝的兴盛奠定了丰厚的物质基础。

因陀罗跋摩的继位者是其子耶输跋摩一世（Yasovarman I，公元889－900年在位）。耶输跋摩一世将其都城从诃里诃罗洛耶迁移至耶输陀罗补罗（Yasodharapura）城，从而奠定今之吴哥城最初的都城格局。根据考古调查，耶输陀罗补罗城远大于今之所见大吴哥城的范围，面积接近四十平方公里。在这座新都城中心的巴肯山之巅，耶输跋摩一世创建了属于他自己的国寺巴肯寺（Phnom Bakheng），这座庙山的规模超越了其父建造的巴空寺。为保障都城建设和农业灌溉之需，他还在都城的东侧兴建了规模宏大的耶输陀罗达塔卡（Yasodharatataka），亦称为东池（Eastern Baray）的大型水库[2]，将暹粒河水引入东池调蓄利用。依靠日益充裕的国力，耶输跋摩一世时代将其统治下的疆土扩张至更为广阔的区域，其版图北抵中国，东接占婆，西濒印度洋，南至马来半岛北部，广袤的疆域面积可与扶南与真腊的极盛时期比肩。（图1-19，图1-20，图1-21）

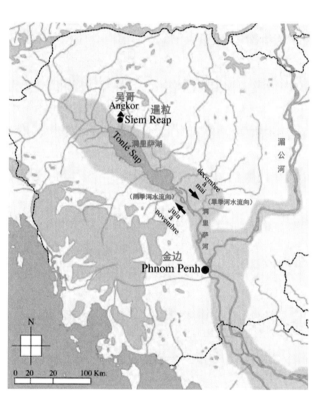

图1-19　吴哥都城的选址与洞里萨湖及湄公河的关系

---

[1]　阇耶跋摩二世统治时代及其Devarāja仪式，皆源自在斯多加通（Sodk Kak Thom）发现的一块古代高棉碑铭（1052年），1915年法国远东学院的首任院长路易·芬诺将此碑铭译出。（著者注）
[2]　东池坐落于大吴哥城胜利门东约1.5公里处，今之水池已经干涸且原有的水利调蓄功能皆已不存，唯其堤坝轮廓完整，旧时格局尚清晰可辨。东池呈1800米×7000米的矩形，其南北方向轴线大约向西偏移了1度。东池的建造历史或可追溯至因陀罗跋摩一世经营诃里诃罗洛耶时代，其后耶输跋摩一世迁都耶输陀罗补罗城可能考虑到都城建设濒临东池之便利，遂以巴肯山为中心进行选址。由于阇耶跋摩四世将都城搬迁到贡开的几十年间，东池可能被废弃，或许由于年久失修，阇耶跋摩四世及其继任者赫萨跋摩二世返回耶输陀罗补罗城伊始即对东池进行了修复，直至罗贞陀罗跋摩统治时期得以完成，主要的改变是将东池南边堤岸往南迁移，以至于水面从9平方公里达及14平方公里，而且堤岸也从2米高增加到大约5米左右。这不仅增大了东池的灌溉蓄水量，而且对防御来自暹粒河的洪水起到重要作用。公元952年，罗贞陀罗跋摩在完成扩建的东池中心修建了东梅奔寺。（著者注）

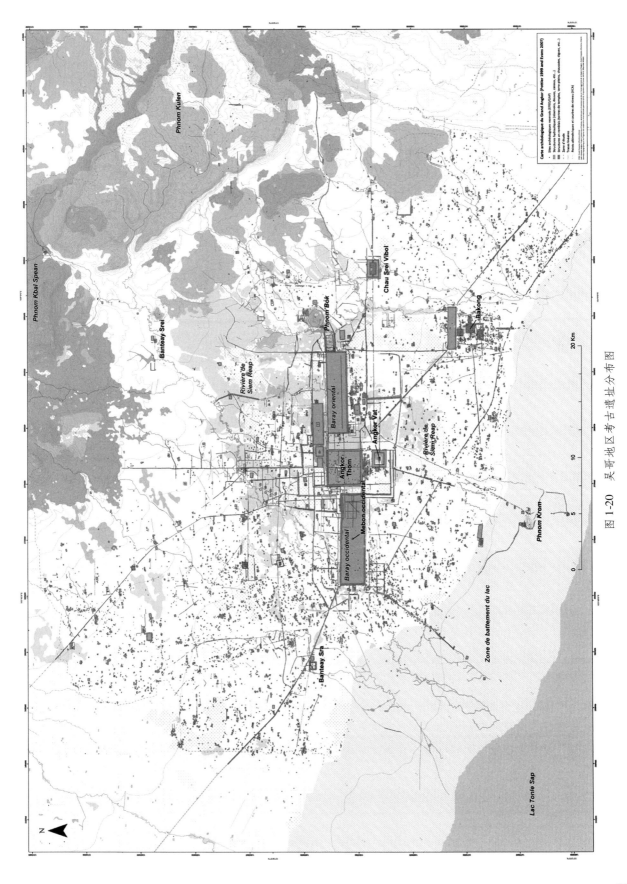

图 1-20 吴哥地区考古遗址分布图

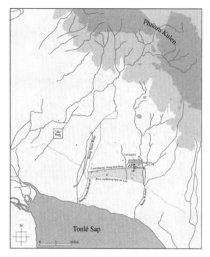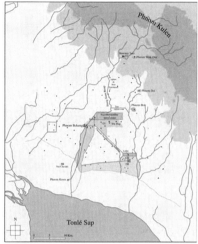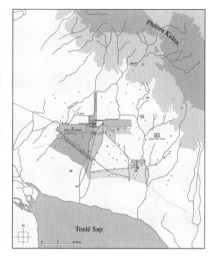
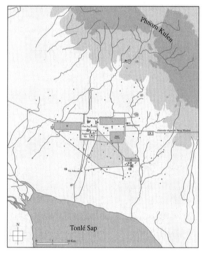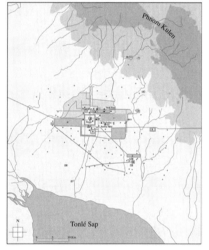

图 1-21　吴哥时代都城经营区位变迁示意图

迄自阇耶跋摩二世以降，直至公元 1432 年，高棉人被来自北方的暹罗人彻底击溃而迁移都城为止。在延续六百余年的吴哥时代，古代高棉人曾经征服过东南亚地区的大部疆域，国势鼎盛，文化灿烂，艺术辉煌，史称"高棉帝国"（The Khmer Empire）。

公元十二世纪上半叶，吴哥建筑艺术与技术日臻成熟，及至十二世纪末至十三世纪初达及顶峰，类型丰富，形制多样，艺术风格异彩纷呈。吴哥时期建筑与艺术的题材，虽然其原型来自印度，但在遵循印度模式的过程中，高棉匠师融入当时社会风俗以及对于自然风致的民族情感，形成了区别于印度本土风格新的变化与发展。

概括而言，吴哥时代初期的雕刻与装饰风格质朴古拙，更多地受到源自印度的影响，至吴哥时代盛期雕刻艺术形式多变，题材丰富，充满活力，至末期则处于衰退阶段，逐渐走向题材僵化与表现手法程式化。[1] 总之，吴哥时代建筑艺术的演变趋势主要表现为建筑艺术随着建造技术的发展日趋繁

---

[1] 吴哥时代建筑与艺术所反映出的风格演变和类型发展，包括题材、造型、纹饰、材质等因素，无不与当时的宗教信仰与社会生活紧密关联，并具有宗教礼仪功能和广泛的象征性。遗存至今的体积庞大，集建筑、雕塑、装饰于一身的宗教性和纪念性建筑，最为集中地反映出吴哥时期当时人们对于视觉形式的追求和为此付出的代价。今之吴哥地区遗存的建筑与艺术题材可以（转下页注）

复，并有着越来越强调其内容与题材的叙事性。例如，伟大的印度古代史诗《摩诃婆罗多》与《罗摩衍那》，广泛深入地渗透在南亚和东南亚文化艺术之中，深刻影响了吴哥时代的高棉艺术，成为古代高棉众多叙事性表现形式的原型[1]。因此，艺术史研究通常按历史时期及建筑形制特征将吴哥时代的建筑艺术风格分为：达山寺（Phnom Da）、古伦山（Phnom Kulen，亦译作荔枝山）、神牛寺（Preah Ko）、巴肯寺（Bakheng）、比粒寺（Pre Rup）、女王宫（Banteay Srei）、南北仓（Khleang）、巴方寺（Bphoun）、吴哥窟（Angor Wat）、巴戎寺（Bayon）等十种不同的类型[2]。（图1-22，图1-23，图1-24，图1-25，图1-26）

图1-22　吴哥时代的毗湿奴造像

图1-23　吴哥时代的湿婆造像

需特别指出的是，吴哥古迹中还包括了公元九世纪至十二世纪修建的罗莱池（Indratataka）、东池（Yaçodharatataka，Eastern Baray）、西池（Western Baray）、北池（Jayatataka）和皇家浴池（Srah Srang）等大型古代水利工程与灌溉设施，以及庞杂密集的水系网络，其中有些古代水利设施及其纵横交错的水网体系至今仍发挥着重要的功用。（图1-27，图1-28）

（接上页注[1]）大致归纳为两类：叙事型（narrative），描述历史或是宗教题材的富有故事性的装饰，或非叙事性的装饰纹样型（Heraldic – not narrative）；有的学者甚至认为：吴哥建筑雕刻与装饰艺术的发展历经了非叙事型（Heraldic）、简单叙事型（Simple narrative）、复合叙事型（Complex narrative programs）、终极装饰性（虚拟叙事）（finally Murals – visual narrative）等几个阶段。根据遗存至今的实物推断，六世纪至七世纪，古代高棉哲匠就已经开始运用高超的技巧表达各类建筑雕刻与装饰题材，但是若要达到今之吴哥古迹建筑雕刻与装饰所具备的艺术品第，其间必是经历了漫长的历史进程，但是具体细节尚不为人所详知。Vittorio Roveda, *Images of the Gods: Khmer Mythology in Cambodia, Laos & Thailand*, Floating World Editions, 2006, p. 243.

[1] Jacques Dumarçay and Pascal Royère, *Cambodian Architecture 8th to 13th centuries*, Paris, 2001, p. 27.

[2] J. Boisselier, *Le décor architectural in Asie du Sud – Est*, *Le Cambodge* Tome I, p. 26.

图 1-24 吴哥时代的建筑雕刻

图 1-25 吴哥时代建筑门楣雕饰之一

图 1-26 吴哥时代建筑门楣雕饰之二

图1-27　今之东池现状

图1-28　今之西池现状

另外，颇为值得一提的是，中国元朝的使者周达观，于元成宗元贞元年（公元1295年）奉命随使赴真腊，大德元年（公元1297年）六月得返，周氏返国后根据亲身见闻所撰《真腊风土记》，全文约八千五百字，较为翔实生动地记述了吴哥王朝盛世末期的城郭宫室、风土物产、语言文字、政治法律、经济贸易、宗教人伦、人文风俗以及平民生活等诸多方面的状况，所记与今之吴哥古迹遗存情况大多相符合，是现存同时代人所撰写的吴哥文化极盛时期的唯一记载，甚至在柬埔寨本国的文献之中，也没有像这样一部详述中古时代文物风俗生活的著作，所以历来研究柬埔寨历史的中外学者对其极为重视。[1]

---

[1] 公元十八世纪末期，《真腊风土记》逐渐引起欧洲学者的关注，这主要得益于耶稣会传教士所作的部分翻译，该译文于1789年在巴黎出版。1819年，阿贝尔·雷慕沙（A. Rémusat）则根据《古今图书集成》中的版本进行了更完整、更准确的翻译。1902年，伯希和（P. Pelliot）又根据《古今说海》中的版本译成新的法译本，并进行了注释（此本有冯承钧的中文译本，收入《西域南海史地考证译丛》）。伯希和曾计划增订译注本。1924年，此项计划因故中断，未及成书（注释仅写到第三条"服饰"）。伯希和去世后，由戴密微（P. Démiville）和赛代斯（G. Cœdès）加以整理，作为伯希和的遗著第三种于1951年出版。赛代斯本人对此书也曾两次作过补注；整理伯希和遗著时，赛代斯还曾在脚注中作过附注。此外，《真腊风土记》还有1936年的日译本（松枫居主人译），1967年吉尔曼（D. Paul Gilman）的英译本（系由伯希和1902年的法译本转译而成），而夏鼐所作《真腊风土记校注》以明刊本《古今逸史》为底本，对勘各本，择善而从，至于考证部分则参考伯希和的译注并加以注释。参见周达观原著、夏鼐校注：《真腊风土记校注》，北京：中华书局1981年版，第1－13页。

四、善水之城：吴哥时代都城的选址与经营

回溯百年以来关于吴哥古迹的学术研究史，迄自二十世纪五十年代起，以法国远东学院的小格罗斯利埃（Bernard Philippe Groslier）为代表的诸多学者，即已较早地开始关注并研究水利工程与吴哥文明兴衰之间的关系。他们通过对吴哥史迹中水利设施遗迹的调查研究，特别是分析吴哥都城变迁与其特殊的地理区位、自然气候之间的互动影响等，提出吴哥文明衰亡与其水利设施及水系网络的破坏息息相关。在吴哥文明的兴起、发展、演进以及走向衰微的过程之中，水在其间始终都在扮演着一个异常突出且非常关键的角色。[1]

作为东南亚最重要的国际河流，湄公河从老挝进入柬埔寨，自北向南贯穿柬埔寨全境直至今之越南南部的湄公河三角洲入海。湄公河水量丰沛，由于受到热带季风气候的影响，每年雨季的大量降水导致其水位猛涨，湄公河的洪水遂通过洞里萨河倒灌入东南亚最大的淡水湖洞里萨湖（Tonel Sap），亦称之为大湖（The Great Lake）。湄公河水的倒灌，导致洞里萨湖的水域面积会比枯水期扩展数倍，年复一年，周而复始，因而形成了东南亚地区最为丰饶肥沃的冲积平原。作为古代高棉历史的黄金时代，吴哥的历史进程与兴衰变迁，似乎都与大湖地区水域周而复始的退涨息息相关。自阇耶跋摩二世以降，高棉帝国的权力中心和都城选址大多都没有超出过古伦山（Phnom Kulen）至洞里萨湖之间冲积平原地区的范围。纵观吴哥时代都城的选址与经营始终游移在古伦山脉与大湖之间广阔的冲积

图 1-29 吴哥都城的选址与古伦山（荔枝山）及洞里萨湖的关系

平原上的原因，亦如人类历史上诸多文明形态的兴衰皆取决于人与水之间的互动关系一样，吴哥时代文明的兴衰历程也莫不如此。（图 1-29）

吴哥时代印度教和大乘佛教之间的冲突贯穿始终，更是一个此消彼长的漫长历史进程。但是，其间一以贯之的却是以水作为脉络的文化与艺术的传承。在以信奉湿婆教和毗湿奴教为主流的古代高棉社会中，水被赋予了极其丰富而深刻的象征意义和宗教文化内涵。在印度教的神话体系中，作为宇宙

---

[1] 菲利普·斯特恩对于吴哥时代的标志性进行如下三个方面的归纳：其一，通过对自然地理形势的人工改造而建成极为庞大复杂的水利设施，以示国王对臣民的关怀和对土地水神的敬意，并以此昭示国王权力和国王与神之间的关系，并将国王与神之居所须弥山（Meru mountain）联系起来；其二，国王建造寺庙以纪念其尊奉为神灵的父母和祖先；其三，国王为自己兴建一座庙山，以此作为宇宙中心须弥山的象征，并以此作为神之居所。参见菲利普·斯特恩：《柬埔寨王朝的更迭及其多样性》（Philippe Stern, *Diversité et rythme des fondations royales khmères*, Bulletin de l'Ecole française d'Extrême - Orient - BEFEO, Volume44 - 2, 1951, pp. 649 - 687.

中心的终极象征，须弥山（Meru Mountain）即是漂浮在无尽的宇宙海洋之中的五座金色山峰（Hemasringagri），不仅象征善恶的较量而获取的人生甘露源自这片神话的海洋，而且印度教中三大主神梵天、毗湿奴、湿婆之间象征着宇宙万物间"创造－保护－破坏"的辩证与轮回，亦是以这片无尽的神话海洋为背景而彰显出来。印度教中的湿婆教，信奉湿婆神，而湿婆神最高境界的化身是林伽（Linga），通常被雕刻成男性生殖器的形象，而承托林伽是正方形的基座，则是象征女性生殖器的约尼（Yoni）。雕刻林伽的材料多以砂岩石为主，也有青铜、水晶等材质。无论林伽施以何种材质，若没有通过水赋予其象征与内涵的话，它只是蕴涵着原始生殖崇拜意味的雕刻而已。唯有仪式及其水的介入，通过水从林伽顶部流下，滋润沐浴林伽的基座（Yoni）之后而流出，林伽才可被赋予了神性的象征及内涵。若从宗教信仰的意义进行分析，吴哥都城及其重要寺庙的选址经营，皆是通过崇信湿婆神及林伽而与水关联到一起。通过用水沐浴林伽，水流过林伽及约尼，进而完成神圣的宗教仪式"提婆罗阇"（Devarâja）。湿婆神是印度教三大主神中最为重要的神，他象征着毁灭，但是毁灭之中又蕴涵着生生不息的轮回或重生。晚于印度教而兴起的佛教，汲取并吸收印度教的影响；水在佛教中依然继续传承着其特殊的内涵。今之柬埔寨，印度教虽早已烟消云散，奉上座部佛教（小乘佛教）为国教，但是蕴含在水中极其丰厚的象征与文化意义依然传承不息。

今之吴哥东北约40余公里的古伦山（Kulen Mountain，亦译作荔枝山），是多条流经吴哥地区主要河流的发源之地。正如前文所述，公元802年，开创吴哥时代的国王阇耶跋摩二世在古伦山顶通过举行提婆罗阇仪式而正式确立其王权统治。今之古伦山的海拔约为300米，山顶地势平坦，在河水源头处的岩石河床之上，耶输陀罗补罗城的缔造者耶输跋摩一世在此处雕刻了一千余座林伽，古伦山顶的河水之源由此而得名为"千林伽河"（River of the Thousand Lingas）。另外，位于古伦山北麓的高布思滨（Kbal Spean），不仅依然呈现出千林伽河的壮丽景观，而且当时的国王还将梵天神、毗湿奴神的雕像与这一千座林伽一并雕刻在河床之上，印度教三大主神"三位一体"（Trimurti）的宏大主题得以汇聚和展现。吴哥都城的选址及其规划意匠，也许正是利用源头之水流过千座林伽而变成圣水（sacred water），因而水源及其水系的环绕或连通则成为吴哥都城规划意匠与经营建设的肇端。

及至公元十二世纪末，古代高棉历史上最伟大的国王阇耶跋摩七世凭借全面战胜占婆入侵并且光复吴哥为契机而登上王位。在此后不到二十年的时间内，阇耶跋摩七世曾两度击溃占婆并将其征服和兼并，从而将高棉帝国的疆域扩张到空前辽阔的规模。

阇耶跋摩七世信奉大乘佛教，尊崇"佛陀、观音、般若波罗蜜多"三位一体（Trimurti）的信仰方式，他仿照其祖先亦在高布思滨（Kbal Spean）将大乘佛教奉祀的佛祖与观世音雕刻在石头河床之上；当源头之水流淌过佛陀及观世音菩萨的雕像之时，并将这些信仰之水悉心汇聚或调蓄至国王苦心经营的大吴哥城，从而产生出精神信仰与物质生产交融并峙的巨大功用，包括罗莱池、东池、西池、北池在内令人不可思议的吴哥时代伟大的水利工程应运而生。因此可以这么说，水滋养了信仰的源流，水汇聚了都城的理想；意匠或因信仰而发生改变，但渊深水阔却是亘古未变。总之，吴哥都城的规划意匠与建设格局就是汇聚或调蓄信仰之水，通过规划利用天然水源、河道及其水系，经由大小人工水池、运河系统进而最终皆与寺庙建设与都城经营紧密的关联在一起。

吴哥时代的城市与建筑，作为集中体现古代高棉复杂的宗教信仰体系的重要载体。在其经营过程中，一方面，水被赋予了极其核心的象征意义，城市与建筑，信仰与仪式，生存与生活，源自水的浸润和滋养无处不在；另一方面，水还是建设伟大城市与宏伟建筑的先决基础条件，稻米收获、建材运输、聚落迁徙，无不接受善水之利的恩泽。吴哥的城市与建筑，无论其选址布局，还是规划建造，抑或象征意义，流布于古代高棉社会生活的方方面面，皆将水之功用及其丰富的文化主题得以全方位的呈现。（图1-30，图1-31）

吴哥窟（Angkor Wat），又称小吴哥或吴哥寺。Angkor源自梵文（Nagara），意指城市或都城，Wat则是现代高棉语寺庙之意。吴哥窟即是"都城之寺"。在古代高棉语中，吴哥窟亦被称作"Vrah Vishnulok"，其意为"毗湿奴之神殿"。吴哥窟是现存吴哥古迹中保存状况较为完整的庙山建筑遗迹，以其建筑宏伟与雕刻精丽闻名于世，堪称古代世界规模最大的寺庙建筑组群之一。（图1-32，图1-33）

作为公元十二世纪高棉帝国的都城与国家寺庙，吴哥窟的兴建始于苏利耶跋摩二世（Suryavarman II，公元1112–1152年在位）统治时期。寺之平面格局以贯穿东西的中轴线左右对称，寺之入口位于四周环壕之外正西的十字平台上，以石雕蛇神（Naga）为栏杆的高架神道横跨200米宽的环壕，始与寺之第一重围廊塔门相接；进入第一重围廊，是极为弘阔开敞的内院，内院中央以长达800米的高架神道与第二重围廊的塔门相接，神道南北两侧各设藏经阁一座、水池一座（藏经阁之东）。通过第二道寺门进入第二重围廊，环绕围廊内壁是长达800米的精美石刻浮雕，雕刻题材主要源于古代印度史诗《罗摩衍那》、《摩诃婆罗多》中的神话故事，以及苏利耶跋摩二世征伐占婆及其宫廷生活的内容，其中以东回廊雕刻的印度神话"搅动乳海"最为壮阔。

在印度教神话中，作为宇宙中心的象征，须弥山是漂浮于大海之上的五座金色山峰。穿越吴哥寺第二重回廊进入庙山的内院，在其殿宇林立、廊庑交错的内院布局中央，作为须弥山崇高象征的五座石塔巨构矗立于高耸的砂岩

图1-30 吴哥都城选址与经营的源流之一：高布思滨的千林伽河

图1-31 吴哥都城选址与经营的源流之二：古伦山的千林伽河

图1-32 吴哥窟鸟瞰

台基之上，构筑起整座吴哥寺最为神圣庄严的建筑空间意象，古代高棉帝国崇拜的"神王合一"的终极象征，在此得以完满地呈现。根据婆罗门教森严的等级制度和宗教仪式，吴哥窟在某种意义上并不是作为公共祭祀场所而存在，可以推断其建筑空间与雕刻装饰是不对普通信众开放展示，只有在特定宗教仪式中，甚至是只有在某些具有巫术色彩的情境中，信众才可能有机会近距离地接触到寺庙内繁复精致的雕刻与装饰。因此，吴哥窟的建筑雕刻与装饰，其功能应该来源于建造者所要竭力灌输其中的叙事性概念，而决非纯粹用来加强建筑的美感的装饰意义。换言之，这些宏伟的庙山巨构，其建造目的不是为启发信众或游赏者，而是为极少数的王室成员或者高级神职人员服务，是信仰之神在世间所呈现出的具有象征意义的纪念碑。（图1-34）

图1-33　吴哥窟远眺

图1-34　吴哥窟中央五塔

吴哥通王城（Angkor Thom），亦称为大吴哥城，是由古代高棉历史上最伟大的国王阇耶跋摩七世（JayavarmanⅦ，公元1182－1219年在位）修建。在其统治的全盛时期，吴哥通王城以国寺巴戎寺（Bayon）为中心，四周围绕以高达8米的城墙，城墙之外环绕以宽达百米的城壕，都城的整体格局象征着漂浮在巨大海洋之上的须弥山，这是古代高棉帝国复杂的宗教信仰体系的重要载体。此城废弃之后，为热带丛林修藤巨木所掩盖，湮灭无闻。1564年（或1570年）葡萄牙传教士多尔塔（P. Antonia Dorta）始发现此遗迹，但未引人注意。1850年法国人布耶福（C. E. Bouillevoux）来此考察，重新发现，其后经过陆续考察与发掘，始大显于世。其中，1863年，法国博物学家亨利·穆奥（H. Mouhot）亲履其地，翌年发表著名游记，遂以重新发现吴哥闻名于世。（图1-35）

现存吴哥通王城遗址平面略呈方形，周长约计12公里，占地大约10平方公里。大吴哥城四面共设城门五座，北、西、南侧城墙各一座，唯东侧设两座城门（南侧之东门与北侧之胜利门）。大吴哥城中各个时期的都城和著名寺庙遗迹还包括：巴戎寺、巴方寺、皇宫（Royal Palace）、空中宫殿、大象平台（Elephant Terrace）、癞王台（Leppking Terrace）、普拉比图寺（Preah Pithu temple）、十二塔（Prasat Suor Prat）等。

巴戎寺位于大吴哥城的中心，是由阇耶跋摩七世创建的国家寺庙。在阇耶跋摩七世长达三十五年的统治期间，大乘佛教信仰盛极一时，并使之强化成为战胜一切痛苦的象征。对阇耶跋摩七世而言，吴哥时代之盛期的大乘佛教延续着"观音菩萨－释迦牟尼－般若波罗蜜多（Prajnaparamita）"之间"三位一体"的宗教观念，佛陀则取代毗湿奴和湿婆的角色，成为国王的化身和国家的象征。阇耶跋

图 1-35　吴哥通王城（大吴哥城）南门

摩七世大兴土木，恢复重建了吴哥通王城，不仅在都城中心位置苏利耶跋摩时代塔基之上重新创建国寺巴戎寺，并在吴哥通王城周围建设圣剑寺（Preah Khan）、达布隆寺（Ta Prohm）、蟠龙寺（Neak Pean）、班蒂克黛寺（Banteay Kedi）等著名大乘佛教寺院，将其视为高棉帝国王权统治与宗教信仰的根基。作为国寺的巴戎寺，以其五十四座脸形塔的独特形制开创了吴哥时代标新立异的庙山建筑形制。巴戎寺的每座脸形塔四面皆雕刻神态各异、面带微笑的佛陀、观音菩萨、祖先以及国王自己的面容，世称"吴哥微笑"即由此而得名。巴戎寺内回廊墙壁上的浮雕不仅展现阇耶跋摩七世与占婆激烈战斗的壮阔场面，也描绘出吴哥时期市井小民的日常生活场景。此外，由于在阇耶跋摩七世信奉大乘佛教之前，吴哥地区的宗教信仰以印度教为其主流，因此在巴戎寺内也保留了众多体现印度教和佛教和谐并存的特殊遗迹。（图1-36，图1-37，图1-38，图1-39，图1-40）

图 1-36　巴戎寺

图 1-37 阇耶跋摩七世（Jayavarman VII）雕像

图 1-38 巴戎寺脸形塔之局部

图 1-39 巴戎寺立面图

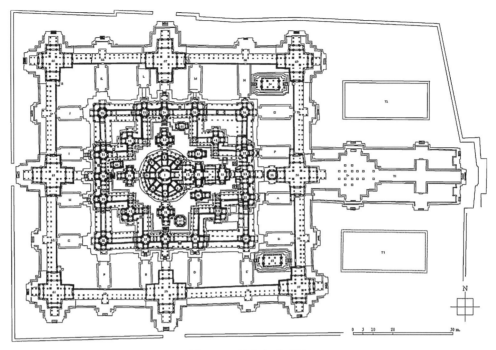

图 1-40　巴戎寺平面图

达布隆寺（Ta Prohm），位于大吴哥城东约 1 公里处，是吴哥地区现存古代寺庙遗迹中的规模较大的一座佛教寺院。该寺始建于公元 1186 年，是阇耶跋摩七世（公元 1182－1219 年在位）为纪念其母所建。寺内供奉的主尊为"般若波罗蜜多"（Prajnaparamita，意为"智慧之神"），据悉是以阇耶跋摩七世母亲的形象雕刻而成。达布隆寺内的石刻碑铭记载了兴建及供养这座寺庙的僧人与供养人的信息，是研究古代高棉历史文化的重要资料。（图 1-41）

若以现代的眼光来看，达布隆寺无疑是吴哥古迹中最具艺术特质的遗迹之一。与吴哥其

图 1-41　达布隆寺内景

他寺庙不同，达布隆寺依然保持着被热带丛林吞噬的原始状态与面貌，在时空交错、光影斑驳的密林深处，坍塌的石构殿宇被盘根错节的巨大树木根系所缠绕。如果说，以吴哥窟、巴戎寺、女王宫为代表的吴哥时代的寺庙建筑是古代高棉精神力量的象征，那么达布隆寺则让人们更多地感受到令人敬畏的自然之力。（图 1-42，图 1-43，图 1-44，图 1-45）

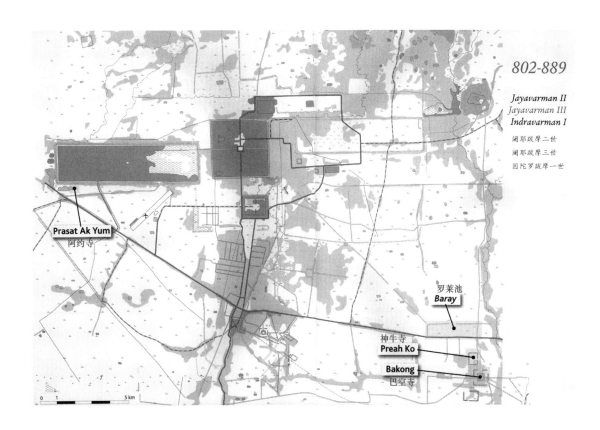

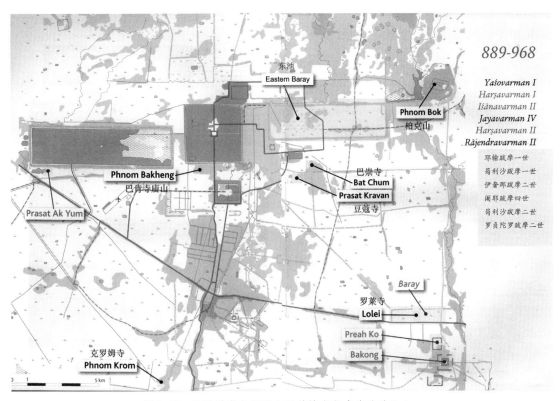

图1-42 吴哥时代主要君主及其城市与寺庙的建设之一

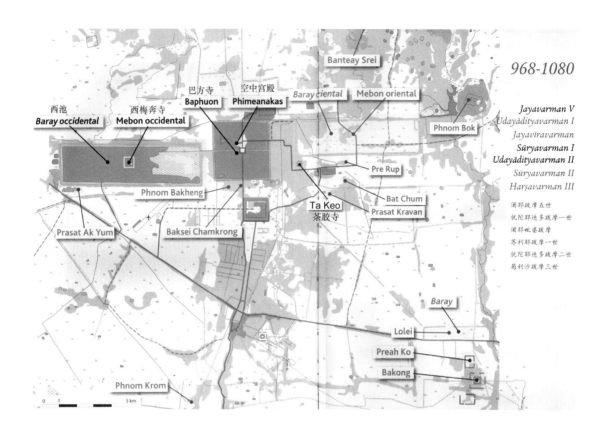

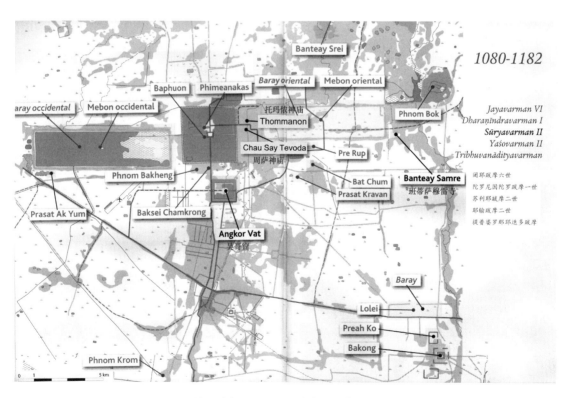

图1-43　吴哥时代主要君主及其城市与寺庙的建设之二

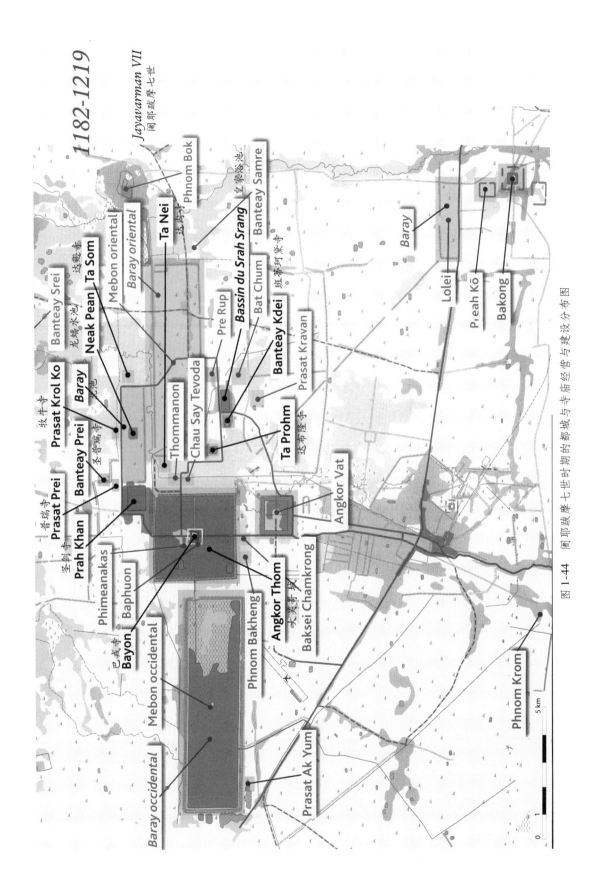

图 1-44 阇耶跋摩七世时期的都城与寺庙经营与建设分布图

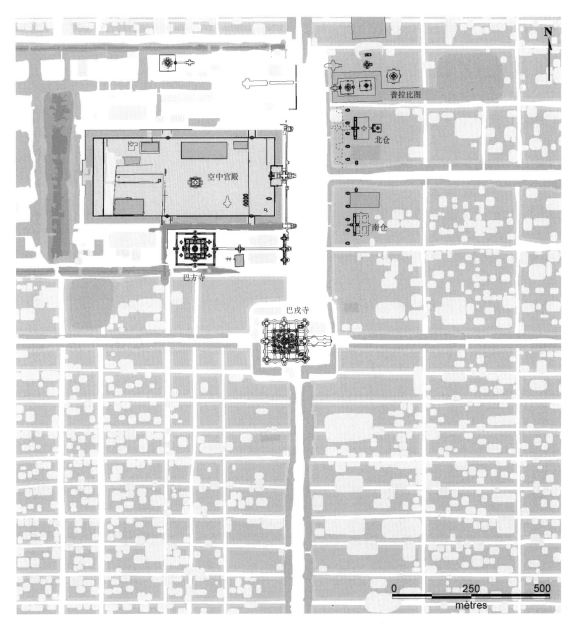

图1-45 大吴哥都城格局及考古遗址分布图

五、吴哥时代的历史终结（公元 1432 年）

纵观古代高棉史，阇耶跋摩七世是吴哥时代最后一位贡献卓著的国王。其后，吴哥时代的辉煌渐渐落下历史帷幕。

公元十三世纪初叶，高棉帝国的内外形势发生了显著的变化：一方面，暹罗人以湄南河流域为中心创立的素可泰王朝（Sukhothai），其势力逐渐扩张强大起来，动摇了高棉帝国对其西北部地区的统治，从此开始了柬泰两国之间长达四百余年的征战与战争。在某种程度上，暹罗人的入侵比盘踞在高棉帝国东部的宿敌占婆（Champa）更具威胁；另一方面，在阇耶跋摩七世死后，虽然印度教曾一度得以复兴，但此时上座部佛教亦开始从暹罗渗透传入，与印度教和大乘佛教盛行时期声势浩大的大兴土木有所不同，小乘佛教提倡简单俭朴与克己苦行，比较迎合民众解脱沉重负担的愿望，因而得到迅速广泛的传播；大规模的寺庙建设停滞下来，与印度教和大乘佛教紧密关联的梵文经典趋于消亡，逐渐被巴利文所取代。

阇耶跋摩七世以降，因陀罗跋摩二世（Indravaran II，公元 1210？－1243？年在位）、阇耶跋摩八世（Jayavarman VIII，公元 1243？－1296 年在位）、萨利·因陀罗跋摩（Çri Indravarman，或称因陀罗跋摩三世，公元 1296－1308 年在位）、萨利·因陀罗阇耶跋摩（Çri Indrajayavarman，公元 1308－1327 年在位）等国王相继继位，但皆未能摆脱高棉帝国内忧外患的统治格局。

公元十四世纪前半叶，随着素可泰王朝的衰落，高棉帝国的统治得以稍作喘息。而整个十四世纪，高棉帝国无论国力还是疆域皆呈现衰颓之势，迫于外族入侵而偏安的国王则被称为"小朝廷"（Minor Reign）。公元 1353 年，在湄南河流域肇兴的阿瑜陀耶王朝（Ayutthaya）的暹罗人以其更加强势的进攻数度入侵并最终攻陷吴哥，高棉国王阇耶跋摩波罗密首罗（Jayavarman Paramesvara，公元 1330－1353 年在位）及其太子皆被诛杀，此为吴哥首次遭到暹罗人的全面攻陷。

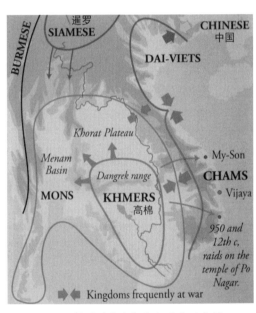

图 1-46　吴哥时代末期战争形势示意图

公元 1357 年，当时的高棉国王萨利·索里约旺一世（Çri Suriyavong I，约公元 1357－1363 年在位）率众出击，收复了暹罗人占领四年之久的吴哥。但此时高棉帝国已是元气大伤，国力及疆域日渐萎缩。公元 1393 年，阿瑜陀耶王朝的军队在国王腊梅萱（King Ramesuan）指挥之下，再度兴师围攻吴哥，高棉国王托玛·索卡（Thomma Soak，约公元 1373－1393 年在位）溃败而不知所终，吴哥再次陷落。暹罗人自此占据吴哥达八年之久，直到公元 1401 年方才撤退收兵[1]。面对迫在眉睫的频繁战事，仓促加冕登基的高棉国王萨利·索里约旺二世（Çri Suriyavong II，公元 1401－1417 年在位）在高

---

[1]（柬）梁合安：《柬埔寨历史简编》，载于中山大学东南亚历史研究所编：《东南亚历史译丛》第一辑，1976 年版，第 218 页。

棉帝国西北部继续与暹罗人征战，而其东部又遭受占婆的大规模入侵威胁直捣湄公河三角洲地区；由于腹背受敌，两线作战，高棉帝国的颓势终将不可挽回。

公元1431年，暹罗军队在围攻七个月之后第三次攻陷吴哥，都城遭受空前的劫掠和破坏。公元1432年，高棉帝国的末代国王蓬黑阿·亚特（Ponhea Yat，公元1393－1463年在位）撤离吴哥，并将都城迁移至湄公河东岸的巴桑（Tuol Basan），以凭据湄公河天堑阻挡暹罗进攻；由于河水泛滥，1434年都城又迁至扎多木（Chaktomuk，含有"四岔口"之意），即今之柬埔寨首都金边附近。

吴哥都城的最终陷落与放弃[1]，标志着辉煌灿烂的吴哥时代无可挽回的历史终结[2]，自此之后直至公元十九世纪末柬埔寨完全沦为法国保护国之前的历史，通常被称为"后吴哥时代"（Post Angkorian）。

吴哥的最终沦陷导致了高棉帝国近两个世纪衰落历程戛然而止的终结。观照吴哥文明覆亡衰落的原因，按其发生的时间顺序排列可分为以下几点。

其一，阇耶跋摩七世时代规模空前浩大的寺庙建设工程几乎耗尽了帝国及国民的财力和精力，从而可能招致整个民族内部的抵触情绪。当毫无休止的劳役、战斗或牺牲令其颇感困惑而又无力从内部变革之时，便引发了一种全新宗教信仰方式的适时传入。其二，外族的入侵特别是暹罗人的劫掠进而导致帝国财富与劳动力供应的锐减，财政状况及劳动力的匮乏使之已无法支持大型工程建设之需。其三，吴哥时代的建筑和艺术皆与印度教和大乘佛教诸神密切关联，而在其背后却是崇尚繁文缛节、糜费巨大的僧侣统治集团；然而小乘佛教的僧侣却安贫乐道，甘于布道行善，能够直接反映人民的诉求，因此小乘佛教在公元十三世纪中叶即已成为高棉帝国颇为盛行的一种宗教信仰；当时的某位高棉国王甚至还在其宫廷中举行过皈依小乘佛教的仪式，这也正式标志着上座部佛教取代大乘佛教或印度教而成为国家信仰之正统。

总之，毫无疑问的是"高棉帝国对上座部小乘佛教的皈依，比起暹罗入侵的蹂躏和洗劫更能直接地导致吴哥建筑与艺术的伟大时期的终结。其中的变革源自内部，这种温情脉脉的宗教的诱惑，经由阿瑜陀耶王朝战士的军事打击，更加具有征服作用。而来自外部的打击远远不足以说明在吴哥陷落以前数十年甚至上百年就已开始的那么多庙宇和雕塑处于半途而废的状态，以及吴哥随处可见的令人憎恨的神像被蓄意损毁的原因。或可仅凭一句简单的话即能将其归纳总结：绚丽的古代高棉文明的终结并不是由于高棉人遭到打击，而更有可能是由于他们找到了新的宗教。"[3]

---

[1] 事实上，吴哥本身并未被完全遗弃，寺庙里的僧侣们利用和维护诸如吴哥窟等庙宇直到公元十五至十六世纪。当时的皇室也曾在公元十六世纪及十七世纪短暂回迁至吴哥，但吴哥再也没有恢复和重现过往日的辉煌。（著者注）

[2] 对于吴哥衰亡的本质，有学者认为："我们必须分清吴哥的衰亡不等于柬埔寨的衰亡。在以后的几个世纪里，柬埔寨作为一个完整的政治实体，仍然是印度支那的主要国家之一不存在暹罗人统治时期，柬埔寨的领土似乎在这一时期也没有被肢解的迹象，暹罗人控制吴哥的时间好像没有超过一年。当失败的记忆逐渐淡忘以后，柬埔寨人可能并没有什么悲伤之感。毫无疑问，他们较过去辉煌岁月生活得更加自由、幸福和美满。仅从政治角度衡量，柬埔寨仍然是印度支那的一个重要国家。也许它比它的一些邻国缺少生机和创造力；但是，它仍然能够抵制强邻以各个方面的压力而保住自己的国土，如果后来它的领导人不执行错误的政治路线的话，也许它可以继续坚持下来，这尽管有些离奇。从另一方面看，吴哥的衰亡意味着具有古代柬埔寨文明特色的特定的文化形式的终结——这种文化主要表现为庄严的纪念碑和宏伟的雕塑，如诗似画的梵文碑铭。与中国文化不同，高棉人没有给这个世界留下行政、教育制度、伦理道德遗产；也不像印度人那样，创立了文学、宗教和哲学体系；但高棉人却把东方风格的建筑和装饰艺术成就推到了它的最高峰。这就是古代高棉人对人类文明的贡献。这也正是随着吴哥的衰亡而衰亡的东西。"引自Lawrence Palmer Briggs, *The Ancient Khmer Empire*, New Series Volume 41, Part 1, *The American Philosophical Society*, 1951, pp. 257－261.

[3] Lawrence Palmer Briggs, *The Ancient Khmer Empire*, New Series Volume 41, Part 1, *The American Philosophical Society*, 1951, pp. 257－261.

对于吴哥的陷落以及吴哥文明式微的原因，法国著名高棉史学者 L. 芬诺曾经有过如下透彻的评论："……（高棉帝国）危机的外因是暹罗人的入侵，用洪水泛滥这个词来形容暹罗人的进攻是很恰当的。卓越的暹罗民族行云流水一般，在其行进中不断掺入其他的元素，带着天空所有的色彩，饱览了五颜六色的河畔风光，保持了它既有多样性又有内在的共同语言和气质特征，又像一块巨大的花布挥洒在中国南部、东京（Tonkin）、老挝、暹罗，甚至缅甸和印度的阿萨姆等地。无论何处，只要有暹罗人，他们就会组成自己的小公国；他们只有在暹罗最后成功地建立起一个大国。在公元十二世纪吴哥寺的浮雕上人们可以看

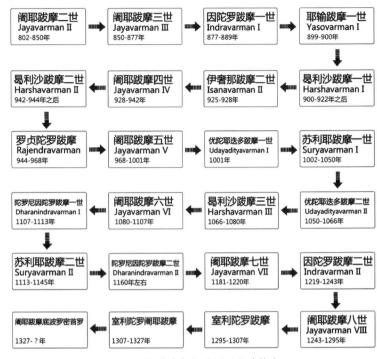

图 1-47　吴哥时代主要国王世系简表

见身穿野蛮人服装的暹罗军队，他们那时是为高棉帝国服役的战士，但不久他们便获得独立和解放，并成为了征服者。他们征服了老挝和马来半岛北部，他们最后向柬埔寨本土发起进攻，并且结束了它的灿烂文化。对于这一灾难的突然发生，起先总是用柬埔寨国家的异种组合来解释。来自国外的文化贵族，披着一层薄薄的宗教灵光，对一群粗鄙的高棉人进行熏陶和训导。现在确实如此，一些侵略行动并没有从精神上打击一个民族；但是，他们可以有效地摧毁一个精英集团，并最终断送由精英凝聚的文明，特别是如果借用在远东地区常用的办法，利用俘虏的大量倒戈则更能奏效。毫无疑问，随着社会中那一批既有思想内涵又刻苦勤勉的精英的消失，必然会导致建筑的突然终止、碑铭记载的中断和梵文的被遗忘。对于广大人民来说，他们对侵略似乎没有表现出强烈的反抗；或许他们会将其视为一种解救机会。他们不仅要为那些庞大的建设工程提供劳役，而且还要为遍布帝国大地如此众多的寺庙神殿提供服务和粮食（这犹如公元十一世纪的法国那样，整个国土如同穿上了一件布满寺庙的长袍），如果人们对以上事实有充分的了解的话，他就会毫不犹豫地认为，经过数个世纪这样的统治，广大的普通民众将会被抹杀和毁灭。所以人们对贪婪的帝王、奴隶主和征税人的事业漠不关心是毫不足怪的；对于寺庙有秩序的破坏活动很可能是那些被激怒了的普通民众干的。征服者为被征服者提供了这样一个珍贵的补偿机会，给他们带去了一个温和的宗教，而这一宗教的教义非常适合于这样一个十分疲惫、沮丧的民族；同时这一宗教又是十分经济的，它的教士安于贫穷，满足于陋居和勺饮。这还是一个遵守德行的宗教，它的戒律保证了人们的心灵平静和社会安宁。高棉民族毫不犹豫地皈依了它，从而高兴地卸下了光荣而沉重的历史包袱。"[1]

---

[1] Louis Finot, *Les etudes Indochinoises: lecon d'ouverture du cours d'histoire et de philologie Indochinoises faite au college de France*, le 16 Mai 1908, pp. 221 – 233.

Synopsis I Introduction to Ancient Khmer History and Culture

1. FU-NAN (ca. 1st century AD to ca. 550 AD)

The Indo-China Peninsula is the home of several distinct peoples who have or have had distinct, though sometimes related, cultures and civilizations and histories of sufficient importance to be worth recording in separately since the proto-historic times.

Among many ancient civilizations of the Indo-China Peninsula, Fu-Nan is probably the most ancient only from the reports of Chinese envoys who visited it early in the third century AD. Nevertheless, Paul Pelliot, an authoritative French sinologist has traced the foundation of the kingdom not later than the first century AD, and Chinese dynasties histories record that early in the third century a Chinese official was sent there to spread civilization and that an embassy from Fu-Nan subsequently appeared at the court of China. In generally, the name of Fu-Nan is apparently a Chinese transcription of a Khmer original, and the original was a Chinese transcription of ancient Khmer word Banam (in modern Khmer as Phnom), which means "mountain". Fu-Nan sovereign's name is Kurung Bnam as similar as Sanskrit words Parvatabhūpâla or Çailarâja, which also means the King of Mountain. For ancient Khmer people, the primitive worship or cult of mountain had been reflected owing to the original meaning of Fu-Nan. In the fourth century AD, the Gupta dynasty during the reign of the Samudragupta Gupta (reigned 335 – 380 AD) in India, expanded its conquest of the Ganges River and Southern India in addition to and the implementation of the policy of expansion to Malaysian Peninsula, Sumatra, and Java. At the same times, especially as the elite of intellectuals, the Brahman ecclesiastics of the Pallava dynasty in the southern India had been moving and expanding which started a prelude of the Indianized kingdom in the Southeast Asia.

Since then, Indianization had profound influence on the future of Fu-Nan in the various fields of politic, economic, and regional culture, or to be described as local foundation of Hindu superstructure. Even legends of the founding of Fu-Nan Kingdom in the Chinese accounts, Hun-Tian originated as King Kaundinya from a Brahmin family of Northern India. On the basis of comprehensive absorbing of Indian religious system, Fu-Nan had gradually established Shiva-Linga (Maheśvara-Linga) as the core of the national worship and royal sovereignty. As the ultimate symbol and the highest expression form with the supernatural power of Shiva, as well as the Brahmanism theocracy exerted their tremendous influence for national religion cults, Maheśvara-Linga had become either the protector of kingdom or the symbolism of monarch, additionally, Brahman ritual and state-cult were given the sacred dignity and the utmost status. It is therefore, as the spiritual power meaning, " Shiva-Linga" cults become a significant ritual owing to the king was being deified. In addition, owing to the influence of Brahmanism Trimurti which combined with Shiva, Vishnu and Brahma, Harihara (Vishnu and Shiva as one of the deities) worship and sacrifice had been not only very popular for a long time, but also it was one of the notable characteristics of the Fu-Nan religious belief.

Fu-Nan had been gradually dominating in the Indo-China Peninsula due to its vast territory and numerous vassal states from the fourth century AD to the middle of the sixth century AD. Referencing to the Chinese historical records, Liang-Shu (《梁书》), King Kaundinya Jayavaman, the king of Fu-Nan, had been ennobled as a general of Fu-Nan. Not only he was one of the most significant sovereigns but also his reign was the golden era of in the Fu-Nan history. The successor, King Rudravarman covenanted Sanskrit inscriptions in Ta Prohm of Bati temple which can provide some valuable information for the studies on Fu-Nan religion. Fu-Nan had always been the most formidable dominant state of the Indo-China Peninsula in the duration of five centuries, and it was still in the memory of later generations' reputation even after its fading a long period.

The ruins ofFu-Nan have faded away nevertheless the Chinese historical records can provide some demonstrations about overview of social lives and customs of Fu-Nan. The archaeological investigations and researches had already documented the location of early capital of Fu-Nan since the beginning of twentieth century, which roughly located in Baphnom or Banam, around today's Prey Veng province of Cambodia. And the sovereign centre of Fu-Nan, Navanagara, is roughly located in the Lower Mekong Delta region, and it is presumed at Angkor Borei and Phnom Da in present-day southern Ta keo province of Cambodia. Fu-Nan territory should also include today's southern Vietnam, Mekong River, most of the Chao Phraya River and the Malay Peninsula.

Forarchitecture and art of Fu-Nan, very little, it can be known with certainty. A French architect of EFEO, Henri Parmentier, has described the architectural outline as follows: The wooden building of Fu-Nan, according to the picture which has come down to us, had a single elevated storey, ornamented at the corners and surmounted by a vaulted roof with a gable at each end. The plan was like an elongated cross, the side wings being similar and of the same length. The doorway, which was simple and had no lintel, was placed at the shorter end of the cross under a light porch covered by an arch supported by slender columns. Often a decorative motif crowned this arch. A terrace, of symmetrical section, formed the base of the building. The walls were ornamented by plain pilasters, and were pierced at top and bottom by continuous balustered openings, which often provided the only means of lighting and ventilation. Sometimes, however, a whole wall panel was cut into a trellis. The eaves of the various gables terminated in makaras, with a decoration of human figures on the tympanum.

2. ZHEN-LA (ca. 550 AD to 802AD)

WheneverFu-Nan was toward extinction on the second half of sixth century AD, it seems clear that up to the sixth century AD, Zhen-La occupied the areas of the middle and lower of the Mekong River, where is located in present-day Stung Treng province of Cambodia. Accompanied with centralization and conquest, during the reign of King Bhavavarman I and King Mahendravarman, Zhen-La had gradually been strong and prosperous. Since then, due to the expedition of King Iśanavarman I and Jayavarman I, Zhen-La had expanded its territory and sovereign with wealth fame. In particular, the capital of King Iśanavarman I's sovereign, King Iśanapura had ever been glorious in addition to present-day Sambor Prei Kuk still remains of many important historical sites in Kampong Thom province of Cambodia.

Until the end of seventh century, Fu-Nan was eventually replaced and finally fallen down by Zhen-La's conquering. Chinese historical record Sui-Shu (《隋书》) in similar records, which is most precious documentary retrospective to Zhen-La replaced Fu-Nan historical facts, traditionally academia reliance. In addition, according to Chinese historical records also a general understanding of the basic look of Zhen-La era of religious. It shows that Brahmanism was most prevalent, especially enshrined Shiva-Linga, or Harihara, the adjacent of Vishnu and Shiva. In addition, it also was including Mahayana Buddhism, even many in the worship of ancestors and natural spirits.

As the successor of King Iśanavarman I, King Bhavavarman II can be dated at AD 639, which depended on the inscriptions. After King Bhavavarman II, King Jayavarman I inherited the expansion policy of King Iśanavarman I, known as " the kings of glory lion, victory of Jayavarman", then to conquer reaching as far as the border of southern China. However, the Zhen-La had been not peaceful owing to the rapid expansion of the territory, and planted the seeds of splitting. Finally, Zhen-La had plunged into chaos of the separate situation while King Jayavarman I's death.

At the beginning of the eighth century AD, Zhen-La had divided into two separate kingdoms, the northern part and the southern part. It is known as the Land Zhen-La where located in the mountainous jungle of northern part. And it is called Water Zhen-La where located in the Lower Mekong Delta region. Water Zhen-La possessed the old territory of Fu-Nan and its capital called Vyadhapura, its position is broadly in Angkor Borei, present-day southern of Ta Keo province in Cambodia. Land Zhen-La is located its ancestral homeland, including midstream of Mekong River and Kangrek Mountains of present-day northern Laos and its territory even extends north to the Chinese border.

WaterZhen-La had been trapped into the chaos since the second half of the eighth century AD, Java had raised forces in the Malaysian Peninsula and then invaded into Champa and occupied southern coast of Water Zhen-La in 774 and 787 AD. Ultimately, Java occupied present-day Sambor Prei Kuk, the capital of Water Zhen-La. With the invasion of both the Srivijaya dynasty in Sumatran and theśailendra dynasty in Java Since the eighth century AD as well as practicing Mahayana Buddhism the rise and expansion, the powerful and prosperous of former Zhen-La had gradually been on the decline until magnificent commencing of the Angkorian era.

In term of the history of Pre-Angkorian, the ancient documents have vanished away, but a few inscriptions have survived. The amount of Ancient Khmer inscriptions has been found approximately twelve hundreds, the vast majority in Sanskrit or ancient Khmer language, which have mostly been interpreted by the contribution of the Khmer historian predecessors such as Louis Finot's masterpiece of publication *Inscriptions du Cambodge*. These ancient Khmer inscriptions describe religious ceremonies, in addition to various forms of prayer congratulatory, but more credible lineages of pre-Angkorian dynasty have gradually been distinct through the study and interpretation of many scholars, as well as the social, economic, political and other aspects. However, the Chinese historical records retained much information of Fu-Nan and Zhen-La including the aspects as ancient scenery, social customs, political and economic, exchanges with China, beyond all doubt, which is the most precious references to the historical studies on the ancient history of Cambodia.

3. The Angkorian Period (802 AD to 1432 AD)

The Angkorian period is the most brilliant golden age of the history of the ancient Khmer since the ninth century to the fifteenth century. Nowadays, Angkor is one of the most important archaeological sites in Southeast Asia, which had been the flourished capital of Khmer Empire from the ninth century to the fifteenth century. The extensive archaeological complex of Angkor and environing sites are located in present-day Cambodia, ca. 20 kilometers north of the Great Lake, roughly 300 kilometers to the north by road from Phnom Penh. Including world-known Angkor Thom and Angkor Wat, there are closely ninety architectural groups with hundreds of built surroundings have scattered in about one thousand square kilometers of tropical jungle. The ancient Angkor plays an important role in the history of civilization in Southeast Asia owing to its splendid culture, which had also a profound impact on many kingdoms and national culture in the wide area of the Mekong River and Chao Phraya River. The mainstream of Hinduism and Buddhism religion imprinted deeply etched into the cultural context of the Angkor era, and which also constitute the subject of ancient Khmer art and architecture. As the epitomizing summit of the art and technology of ancient Khmer architecture, Angkor monuments consist of the space layout with symbolism in addition to the magnificent temple-mountains become significant symbol of the architectural characteristics in the Angkorian era. Not only the subtle carvings and decorations represent the excellent level of the ancient Khmer classic art but also demonstrate the endless charm of the masterpiece of ancient Khmer architectural style.

On the first half of the twelfth century, Angkor architectural art and technology is becoming more mature, and then up to the reaching the peak not in many shapes but artistic style. Theme of the art and architecture of Angkorian period, although its prototype from India, but, Khmer craftsmen had mixed the prevailing ethnological customs and the folk feelings in the process of following as well as they had seek for some changes and developments that were different from the Indian model.

For specially to be mentioned, Angkor monuments are also including such reservoirs and irrigation facilities as Indratataka Yaçodharatataka (Eastern Baray), Western Baray, Jayatataka and Srah Srang, in addition to the network of canal systems. In particular, some ancient reservoirs and irrigation network system still plays an important function in modern Cambodia society.

In addition, quite worth mentioning, Zhou Daguan (1266 – 1346 AD) was a Chinese envoy under the Temür Khan, Emperor Chengzong of Yuan. He is most well known for his accounts of the customs of Cambodia and the Angkor temple complexes during his visit there. He arrived at Angkor in August 1296, and remained at the court of King Indravarman III until July 1297. He was neither the first nor the last Chinese representative to visit the Khmer Empire. However, his stay is notable because he later wrote a detailed report on life in Angkor, Zhen-La-Feng-Tu-Ji (《真腊风土记》, The Customs of Cambodia in English translation). His portrayal about 8,500 words is today one of the most important sources of understanding of historical Angkor and the Khmer Empire. Alongside descriptions of several great temples, such as the Bayon temple, the Baphuon temple, Angkor Wat, and others, the text also offers valuable information on the everyday life and the habits of the inhabitants of Angkor. So traditionally, scholars of Cambodian history are attached great importance to this text. It is the brief chronological backtracking of Angkorian period as follows:

Due to the decline of Śailendra dynasty in Java, King Jayavarman II, probably himself imprisoned for a long time in Java, freed his occupied empire in the eighth century, led to Khmers gradually getting rid of Java colonization. At the beginning of his reign, King Jayavarman II was becoming a renovator who brought new concepts of Shiva belief, and has built a temple-mountain as his state temple probably as Ak Yum temple. Actually, it seems that, in spite of the relative lack of extant historical monuments, Indian religions, notably Hinduism, were already followed in Angkor in the early sixth century. Only in 802 AD, the cult of Devarâja is established on the summit of Mount Mahendra (Phnom Kulen) and King Jayavarman II was consecrated as supreme sovereign of the Khmers, Charkravartin. Devarâja is the God-King or God-Linga worship to become the symbol of power and domination of the ancient Khmer kings. It was the end of the split situation of ancient Khmer that continued more than a century, and the territory and glory of Fu-Nan or Zhen-La era had been retrocession.

As the successor of King Jayavarman II, King Jayavarman III, reigned AD 850–877, was mediocre conservatism, and then his cousin brother who was called King Indravarman I, reigned AD 877–889, usurped the sovereign and throne. King Indravarman I had modernized Hariharâlaya, completely restoring Bakong temple (among other things facing the whole temple with sandstones), reactivating Preah Ko temple as a sanctuary dedicated to ancient kings and establishing to the north the first major reservoir, a baray, which is 3,800 meters in length and 800 meters in width. Such two temples marked the beginning of the Angkor architecture which architectural form and style had influenced hereafter monuments of Angkorian period. The massive irrigation project of King Indravarman I also laid the foundation for the prosperity of Angkor.

King Yaçovarman I, King Indravarman I's son, was the founder of the city of Angkor, Yaçodharapura, a name used for centuries. Nevertheless, this city or state-temple was relocated several times. The city seemed to gather around the Phnom Bakheng, on top of which the king erected a huge pyramidal temple. Simultaneously, he developed Yaçodharatatāka or Eastern Baray, the first of the large Angkor barays which is totalling 7.5 kilometres length and 1.8 kilometres width, to the south of which he constructed the large āçrama, hermitages devoted to each of the main religions of the empire. Increasingly based on plenty of sovereign power, King Yaçovarman I expanded the territory to China border in the north, to Champa border in the east, to the west by the Indian Ocean, and to the northern part of the Malaysian Peninsula, and the vast territory area was closely to the territory of Fu-Nan or Zhen-La period. During that time, the ancient Khmers had conquered the most territory in the Southeast Asia owing to its sovereign power, splendid culture and artistic brilliance, known as the Khmer Empire.

The king was then replaced in succession by two of his sons, King Harshavarman I (910AD-ca. 920AD) and Īçānavarman II (ca. 920AD-ca. 928AD). The first erected Baksei Chamkrong, a small pyramid, dedicated to the worship of his parents. During his reign Prasat Kravan was also built. The following King Jayavarman IV, was reigning in Koh Ker when he seized the supreme throne, probably by force. Consecrated in 928AD, he remained in his capital where, with his ministries, they constructed around forty temples including the Prang, the highest pyramid of the Khmer country. He died in 940 AD, although the heir designated by King Jayavarman IV was quickly replaced by one of his brothers, King Harshavarman II, who only reigned for a few years. In 944 AD, when his first-cousin, King Rājendravarman, ascended the supreme throne on returning to Angkor, however, he settled his capital to the south of the Eastern Baray, around his consecrated (in 962 AD) state temple, Pre Rup. He had previously built the Eastern Mebon at the centre of the baray. He succumbed in 968 AD, most likely from an assassination attempt. In the same year, the Banteay Srei statues were consecrated. This was a temple erected by the King's guru, Yajñavarāha. King Jayavarman V, despite this turmoil, succeeded his father and seemed to have reigned peacefully. He built Ta Keo temple-mountain and his palace to the west of the Great Baray and died in closely 1000 AD. The aftermaths of these activities remain somewhat confusing: The first successor of King Jayavarman V was short-lived, as he died in 1002 AD or just before. Two men were then consecrated as "King of the Khmers" in 1002 AD, the same year, both boasting an unknown lineage: King Jayavīravarman settled in Angkor and continued the construction of Ta Keo temple-mountain, whilst King Sūryavarman I commenced his reign in the region of Battambang province. War broke out and following ten years of conflict, King Jayavīravarman was defeated. A triumphant King Sūryavarman I settled in Angkor, but erected solid ramparts along his royal palace estate—which later became Angkor Thom. Moreover, his palace, a modest-sized State temple, the Phimeanakas, was located within its precinct. He reigned until ca. 1050 AD, and developed the immense western Baray, which is 8 kilometres in length and 2.2 kilometers in width and most likely started the construction of the Baphuon temple-mountain, although he did not have sufficient time to complete it. King Sūryavarman I was succeeded by his son, King Udayādityavarman I. The latter continued the erection of Baphuon and built the Western Mebon located at the centre of the baray. Nevertheless, he had to fight against violent upheavals. After his death in 1066 AD, his brother King Harshavarman III, succeeded him on a potentially weakened throne. He kept on fighting the upheavals, with no great success. King Harshavarman III died in 1080 AD, leaving the throne to King Jayavarman VI, most probably his conqueror. Only a few details have emerged on the lineage of this king, simply that he originated from Mahīdharapura, a site that has yet to be located. King Jayavarman VI reigned until 1107 AD, but he is not known for having built any major monuments. Phimai temple in Thailand was also built during his reign.

Succeeded by his older brother, King Dharanīndravarman I, the latter was soon evicted by one of his grandnephews, King Sūryavarman II in 1113. He has been remembered by history as the builder of the Khmer art masterpiece, Angkor Wat. Yet, he dedicated most of his time to waging war, in particular against Vietnam. He died ca. 1145 AD. A renewed period of uncertainty commenced. Little is known of King Yaçovarman II, the king who assumed the supreme throne and died in 1165, during an ambush driven by his successor King Tribhuvanādityavarman. The latter's achievements and reign also remains vague.

In 1177 AD, a Cham king supported by a Khmer army conquered Angkor: It is commonly argued that he probably came with the intent of placing one of his Khmer friends on the throne. This was at the time when the future King Jayavarman VII was waging war against Cham (and Khmer) enemies and attempting to take control of a Khmer country, potentially already in a disintegrated state. He only managed to access the throne in 1182 AD, without having established a complete peace. He was a great organizer, and has been attributed among others with the revival of hospitals. A devoted Buddhist, but with a tolerant approach, he was a tireless builder in Angkor and the provinces, although it is most likely that he could not have built all the buildings that have been associated with his name. In Angkor are, among others: The temple and city of Ta Prohm; the temple and city of Preah Khan along with its baray; the Neak Pean; the city of Angkor Thom and the Bayon temple, its Royal Palace. He died ca. 1220AD, and the little-known King Indravarman II succeeded him. It is most likely that as he was also a devoted Buddhist; he continued during his long reign the construction commenced by King Jayavarman VII, although both did not seem to be of the same lineage. King Jayavarman VII was succeeded in 1270AD by King Jayavarman VIII who seemed to have dedicated his time to erasing any traces of the religion practiced by his immediate foregoers, by systematically destroying any of the Buddha figures that were abundantly disseminated. Simultaneously, he saw to the restoration of ancient Hindu monuments, in particular Angkor Wat and Baphuon. As mentioned above, it was towards the end of Jayavarman VIII reign that Zhou Dagan, a Chinese envoy, drafted a laudatory report on Cambodia and its wealth, Jayavarman VIII abdicated in 1298 AD to one of his sons-in-law, King Çrīndravarman, and a more open-minded spirit, who allowed both main religions to flourish.

4. Angkor: Hydraulic Urbanism

Since1950s, Bernard Philippe Groslier from EFEO and other scholars had already started to study the relationship between the hydraulic system and Angkor civilization. According to the investigation and survey activities for Angkor historical site in hydraulic facilities, they had especially analyzed the interaction between the Angkor changes and its special geographical location, natural climate impact, closely related to the proposed destruction of the Angkor civilization. No doubt to say, the rising, development, evolution, and declination of Angkor civilization, water had always played an unusually prominent and very crucial role in this process.

As the most important international river in Southeast Asia, the Mekong River is across from Laos into Cambodia, and throughout the whole territory of Cambodia, finally pours into the South China Sea in the Mekong Delta of southern Vietnam. Due to the influence of the tropical monsoon climate, Mekong River has abundant water. Especially in the annual rainy season, the heavy precipitation can cause the water level soared Mekong floods and then intrusion into the Tonle Sap Lake, the largest freshwater lake of Southeast Asia, also known as the Great Lakes. Mekong water intrusion is resulting in the water capacity of the Tonle Sap Lake can be extended more several times than that in the dry season. Consequently, it lead to the most rich and fertile alluvial plain of the Southeast Asia during the circulating of years by years.

As the history of the golden age of ancient Khmer Angkor historical process vicissitudes, seem closely related to the waters of the Great Lakes region, the cycle of retreat up. Since King Jayavarman II's reign, throughout, most of the power centers of the Khmer Empire or the capital planning are not beyond the regional range of the alluvial plain from Phnom Kulen to the Tonle Sap Lake. As the same way as the rise and fall of many civilizations in the history of mankind, which depends on the interaction between people and water, the course of the rise and fall of Angkorian civilization has been true.

The conflict between Hinduism and Mahayana Buddhism throughout is a shift in the long historical process of Angkorian era. However, during a consistent based on water as the context of the culture and heritage of the art. Both Shiva and Vishnu constituted the mainstream of ancient Khmer society, and water is given extremely rich and deep symbolism, religious and cultural connotations. In Hindu mythology system, as the ultimate symbol of the center of the universe, Mountain Meru that is floating in the endless ocean of the universe into the five golden peaks (Hemasringagri) is not only a symbol of the contest of good and evil for life elites, but also the dialectical reincarnation of " creation- Protection - destroy" is based on Brahma, Vishnu, Shiva as the background and highlight of the endless myth ocean. In the Hindu religion, Linga plays the highest incarnation of Lord Shiva, who is often carved as phallic image, and Yoni is a square base as a symbol of female genitalia. The materials of carved Linga are mainly sandstone, bronze, and even crystal. No matter what kind of Linga material, if there is no water given its symbolic connotation, it just implies a primordial worship of carving. Only the water flow down from the summit of Linga and then outflow from Yoni, Linga can only be given a symbol of divinity and connotations in the ceremony, that is to say, the water flow over Linga and Yoni, and it is completed for the core of sacred religious cult, Devarâja. Consequently, the religious significance of Angkor should be dedicated to God with water related. Shiva is the most important God in Hinduism, and he symbolizes the destruction, but destruction implies endless reincarnation or rebirth. The later rise of Buddhist had learned and absorbed the impact of Hinduism and water still continued to play a significant role with its special meaning in Buddhism. Nowadays, notwithstanding Hinduism is vanished, in addition to Theravada Buddhism is prevailing as the state religion in Cambodia, but water still contains extremely abundant meaning as a symbol of heritage and cultural significance remains endless.

Phnom Kulen is located in thenortheast of Angkor about 40 kilometers, where several major rivers are originated that flow through the Angkor region. As above-mentioned, Jayavarman II held Devarâja cults ceremony to formalize his monarchy ruled in 802 AD. On the top of Phnom Kulen, the founder of Angkor capital, King Yasovarman I carved thousands Lingas on the riverbed of water source that derived the River of the Thousand Lingas, as well as located in the northern foothill, Kbal Spean is not only still showing the magnificent landscape of thousand Lingas river, but also the King carved Brahma and Vishnu statues together on the riverbed, therefore, the grand themes of the three major gods of Hinduism Trimurti to be brought together. Artistic conception of Angkor and its planning, perhaps it is the source of the flowing water through Lingas and the normal water is turned into sacred water. Surrounded by such sacred water, either the complicated water system or the connectivity between artistic conception and urban planning, it should be the original of Angkor construction.

King Jayavarman VII, the greatest king in the ancient Khmer history, had recaptured Angkor with a comprehensive victory over Champa's invasion as an opportunity to take his throne until the twelfth century AD. After less than two decades, King Jayavarman VII had continuously defeated and conquered Champa, which expand the Khmer Empire's territory to an unprecedented scale. Due to King Jayavarman VII's Mahayana Buddhism belief, he consecrated Buddha, Avalokitesvara, Prajnaparamita as Trimurti and carved Buddha and Avalokitesvara in sandstone riverbed dedicating to his ancestors and kingdom in Kbal Spean. When water of the source is streaming down the Buddha statues and Bodhisattva statues, the belief-water is careful aggregation, regulation and storage to the construction of Angkor Thom, painstakingly resulting in a blend of spiritual beliefs and material production. On the other hand, the incredible hydraulic facilities, including a series of barays, canals, ponds and other great irrigation project came into being. That is to say, water nourishes the origins of faith, and water brings the ideal of capital of Angkorian period. Probably the artistic conception can be changed because of their beliefs, but the belief prototype of water is never varied. In short, for the planning of Angkorian capital, both artistic conception and construction pattern is for the aggregation, regulation and storage of the faith water, and it is planning to take advantage of the natural sources of water via the various size of baray and canal system, in addition to ultimately all works are closely associated with the urban planning and the temple construction together.

5. Declining of Angkor

In the ancient Khmer history, King Jayavarman VIIshould be the last King of the Angkorian period with outstanding contribution. Subsequently, the splendor of the Angkor has gradually been falling curtain of history. In the early thirteenth century, a series of significant changes of internal and external situation of the Khmer Empire had occurred, on the one hand, Siamese founded Sukhothai dynasty, its forces gradually expanded strong to the Chao Phraya River, has shaken the governing foundation of Khmer Empire. To a certain extent, as the enemies of the Khmer Empire, comparable to the invasion of the Siamese had more entrenched threatening than Champa. On the other hand, King Jayavarman VII's death, although once revived Hinduism, Theravada Buddhism also started to spread from Siam penetration. Comparing with Hinduism and Mahayana Buddhism prevailed during the massive construction projects, Theravada Buddhism advocated self-denial ascetic and cater to people's deliverance desire of the heavy burden, and rapidly and widely disseminated. Thus massive temple construction had been slow down, and classic Sanskrit associated with Hinduism and Mahayana Buddhist ended to wither and gradually replaced by Pali language.

After King Jayavarman VII dynasty, a series of kings such as King Indravaran II, reigned AD 1210 – 1243, King Jayavarman VIII, reigned AD 1243 – 1295, King Çrindravarman, or King Indravarman III, reigned AD 1295 – 1307, King Çrindrajayavarman, reigned AD 1307 – 1327, were unable to get rid of the internal and external deadlocks of the Khmer Empire. In the first half of the fourteenth century, with the decline of Sukhothai dynasty, the rule of the Khmer Empire had opportunities to take a breather. Regardless of national strength or territory, the Khmer Empire also called the Minor Reign had presented recession trend that forced by the alien invasion. The Khmer had suffered continuous strength and invasion of Ayutthaya dynasty in 1353 AD, King Çri-Lamphongsaraja and his Prince were killed due to Angkor was ultimately comprehensive captured by Siamese for the first time. King Çri Suriyavong I, approximately reigned 1357 – 1363 AD, was attacked and recovered Angkor that Siamese occupied for four years in 1357 AD. The Khmer Empire was badly weakened with shrinking national strength and territory. And then, the Ayutthaya armies under the command of King Ramesuan attacked and siege Angkor again, the Khmer King Thomma Saok had routed and disappeared and Angkor was fallen down again in 1393 AD and Siam army had just retreated until 1401AD. With hastily crown, reign 1401 – 1417 AD, King Çrisuriyavong II continued to fight with the Siamese in the northwest of the Khmer Empire. In addition to the frequent war, Khmer Empire has been suffering from the large-scale of invasion of Champa in the Mekong Delta region. Owing to two-front war Khmer Empire had eventually been irreparable decline. The third fall of Angkor was occurred after the siege of Siamese army in 1431AD, where suffered unprecedented looting and destruction. In 1432, the last king of the Khmer Empire, Ponhea Yat had to be evacuation from Angkor and moved the capital migrated to Tuol Basan where located in the east bank the Mekong River for blocking the Siamese. In 1434AD, the temporary capital moved toward Chaktomuk where near present-day Phnom Penh due to the heavy floods.

Eventually fall and abandon of Angkor, it is marking the irreversible end of glorious era of Angkor. It is often referred toas the post-Angkorian era from henceforth to the end of the nineteenth century AD, that is to say, the history before Cambodia became French protectorate.

第二章　吴哥古迹研究保护史概略

一、法国远东学院溯往

公元十五至十七世纪地理大发现时代的柬埔寨几乎引不起西方世界的兴趣。它既不强大又不富裕，还偏离"黄金之路"和"香料之路"，所以吸引不了雄心勃勃的商人和征服者。公元十六世纪中叶开始，唯有数量不多的探险家、传教士在此逗留[1]。1858 年，法国博物学家亨利·穆奥（Henri Mouhot）通过实地考察并发表《暹罗柬埔寨老挝安南游记》（*Voyage dans les Royaumes de Siam, de Cambodge, de Laos*），再度激发起西方世界对于隐秘于东南亚热带丛林之中的"吴哥壮丽废墟"的热情与向往。（图 2-1，图 2-2）

图 2-1　以"重现发现吴哥"而闻名的法国博物学者亨利·穆奥（Henri Mouhot，1826—1861）

图 2-2　1863 年法国政府驻柬埔寨代表杜达特德·拉格雷（Ernest Doudart de Lagrée）组建湄公河考察团

在近百年来柬埔寨吴哥古迹研究与保护的历史进程中，法国远东学院（École Française d'Extrême-Orient，亦简称为 EFEO）扮演着极为重要的角色。伴随着法国殖民体系及其影响在印度支那半岛的迅速扩张，以 1901 年法国远东学院的正式创立为肇端，西方世界对于东南亚国家历史文化以及柬埔寨吴哥古迹的关注和态度开始转入新的趋向，此前西方社会所热衷的以探险旅行、地理发现、丛林猎奇、

---

[1] 当时关于吴哥古迹时代最早且最为详尽的文字记录，很有可能出自于葡萄牙作家迭戈·多卡托（Diego do Couto）之手，在其作品中描述了公元十六世纪中叶的一位高棉国王在猎象时候偶然发现吴哥的经历。而事实上，迭戈·多卡托并未实地造访过吴哥古迹，他的著作主要参考了曾经于 1585 年访问吴哥的方济各教会（Capuchin Friar）修士安东尼·马格达勒那（Antonio de Magdalena）所提供的信息。参见 Bruno Dagens, *Angkor: Heart of an Asian Empire*, London, Thames&Hudson Ltd, pp. 26–29.

异域风情为潮流的趣味，逐渐被严谨务实的学术研究取代。[1] 对柬埔寨古代历史与吴哥古迹的研究与保护也开始进入了较为系统的科学研究阶段，一批具有建筑学、考古学、艺术史等专业背景的研究人员崭露头角，成为引领柬埔寨吴哥建筑研究和保护实践的开拓者和先驱者。[2] 百余年来，正是基于以法国远东学院为代表的一大批学者对于吴哥古迹庚续不辍的学术研究活动与古迹保护修复实践，开创并奠定了吴哥学术研究与古迹保护的根基，其深厚的学术积淀与成就一直影响至今。

1887年10月，法属印度支那联邦成立，其领土包括南圻（亦称交趾支那Cochin Chine）、中圻（亦称安南保护国）、北圻（亦称东京）、柬埔寨，随后在1893年爆发的法暹战争中，湄公河东岸的老挝也被收入法属印度支那联邦之中。法属印度支那联邦总督是联邦中央行政首脑，印度支那联邦行政管理权基本操纵在法国国内殖民部手中，宗主国国民会议掌握殖民地立法大权，联邦的总预算、税收定额等重大经济政策与重要基本建设项目皆由法国国内政府决策实施。

与之相应的是，法国远东学院的历史可以追溯至由时任法属印度支那联邦总督保罗·杜梅尔（Paul Doumer）于1898年倡导在越南西贡创立的法属印度支那联邦研究所。作为法国殖民体系的重要组成部分，这所学术机构由以下两个部门组成：其一，考古与史迹研究保护部；其二，自然科学研究部。与英国殖民政府于1861年在印度创立印度考古调查局（Archaeological Survey of India，亦简称ASI）致力于推动有关印度考古学研究与古迹保护的策略相类似，法国学术研究机构也与法国殖民体系的扩张有着千丝万缕的联系，法属印度支那考古与史迹研究保护部即是与法国碑铭学文学研究院（Académie des Inscriptions et Belles-Lettres）联合创办的。创立之初，来自法国国立图书馆从事梵文研究的路易斯·芬诺（Louis Finot 1864—1935）被任命为印度支那考古与史迹研究保护部的首任负责人（图2-3）。在1899年至1900年期间，路易斯·芬诺的足迹遍及印度支那半岛及爪哇群岛，主要从事历史古迹的考古调查工作。在此调查工作的基础上，印度支那考古调查队（Mission Archéologique de l'Indochine）得以正式成立。按照路易斯·芬诺的研究计划，印度支那考古调查队的建置还应包括建立一座图书馆与一座博物馆，其研究区域也不应仅限于印度支那半岛，而是包括中国、印度以及涵盖整

---

[1] 自19世纪60年代起，随着法国在印度支那地区殖民体系的建立与扩张，柬埔寨及其吴哥古迹更全面、更深入地为西方社会所知，在此期间著名的探险家和旅行者包括：E. 杜达特德·拉格雷（Ernest Doudart de Lagrée），鲁内特·德·拉云魁尔（Étienne Lunet de Lajonquière），以及埃廷内·艾莫涅尔（Étienne Aymonier）等。其中以1863年法国政府驻柬埔寨代表E. 杜达特德.拉格雷组建的湄公河考察团的影响较大。通过对吴哥地区历经多年的考察之后，考察团绘制完成了主要遗迹的分布图，并撰写题为《印度支那探索之旅》（Voyage d'exploration en Indochine）的考察报告。然而，杜达特德·拉格雷于1868年死于在中国云南的考察途中，未能见到其为之付出生命的考察报告的正式发表。该考察团中的另外两名成员弗朗西斯·加内尔（Francis Garnier）和路易斯·德拉蓬特（Louis Delaporte）最终完成了全部考察任务，湄公河考察团至今仍被视为法国从事吴哥研究的肇端与先驱。加内尔于1885年发表题为《印度支那探索之旅》（Voyage d'Exploration en Indo-China）考察报告；而德拉蓬特则于1880年发表了其考察随笔《柬埔寨之旅：高棉的建筑》（Voyage au Cambodge：L'Architecture Khmer），书中由德拉蓬特绘制的吴哥古迹系列铜版画至今仍然享有很高的声誉和知名度，因为它真实再现了19世纪末叶吴哥的风貌。德拉蓬特的著作曾广泛流传，成为诸多描绘吴哥文学作品的灵感来源。另外，当湄公河考察团调查吴哥古迹时，适逢足迹遍至亚洲各地的苏格兰著名摄影师约翰·汤姆逊（John Thomson）也正在造访吴哥，在其著作《十年海外探险之旅：马六甲海峡、暹罗与印度支那》（The Straits of Malacca, Siam and Indo China or Ten Years Travels, Adventures and Residence Abroad）中发表的吴哥窟、巴戎寺等吴哥寺庙的照片，应是迄今为止最早的关于吴哥古迹的影像资料。引自Pierre-Yves Manguin, The EFEO in Cambodia: a century-long partnership, Archaeologists in Angkor, EFEO & Musée Cernuschi, Paris, 2010. p. 25.

[2] 1879年，来自荷兰的学者亨德瑞克·科恩（Hendrik Kern）开始研究古代高棉碑铭，而法国学者奥古斯特·巴斯（Auguste Barthe）和伯尔盖涅（A. Bergaigne）则在科恩所提出的研究方法基础上，更加系统地整理和翻译古代高棉碑铭。1890年，E. 艾莫涅尔（Étienne Aymonier）主持编纂《柬埔寨古迹系列》（Le Cambodge et ses monuments）。参见Coedès Georges. Etienne-François Aymonier (1844—1929). In: Bulletin de l'Ecole française d'Extrême-Orient-BEFEO. Tome 29, 1929. pp. 542—548.

个东亚、南亚及东南亚在内的广大地区。另外，印度支那考古调查队的研究目标也不仅局限在建筑与考古专业，还应更多地涉及历史学、语言学、宗教学、民族学等诸多研究领域。

1900年1月10日，印度支那考古调查队正式更名为 École Française d'Extrême – Orient，即法国远东学院，1902年法国远东学院总部由西贡移至河内。（图2-4）作为法兰西学院体系（法国在希腊、意大利、西班牙、埃及也建有研究机构）所属的一所专门从事东方学研究的学术机构，其研究领域主要包括：考古调查与发掘、历史文献收集、古迹保护修复、民族学与人类学调查、语言学研究等。早期的法国远东学院在路易斯·芬诺之后的历任院长，诸如阿尔弗雷德·菲澈（Alfred Foucher）、鲁内特·德·拉云魁尔（Lunet de Lajonquière）[1]、亨利·帕蒙蒂埃（Henri Parmentier）等（图2-5，图2-6，图2-7，图2-8），皆是在柬埔寨历史文化及吴哥古迹研究方面皆颇有建树的著名学者，这不仅奠定了柬埔寨吴哥保护修复事业的基础，而且在组织运作方面，他们也大力推动与当时法属印度支那政府的配合，以创造充满活力的学术生活来吸引一流学者。

图2-3　法国远东学院首任院长路易斯·芬诺（Louis Finot, 1864–1935）

图2-4　1900年成立于越南河内的法国远东学院旧影

图2-5　法国远东学院拉云魁尔（Lunet de Lajonquière）及其考察队

在长达一个多世纪的历史进程中，伯希和（Paul Pelliot，1878–1945）、马伯乐（马司帛洛 Henri Maspero，1883–1945）、戴密微（Paul Demiéville，1894–1979）的汉学研究，路易斯·芬诺（Louis Finot，1864–1935）、乔治·赛代斯（George Cœdès，1886–1969）的印度支那碑铭学研究、亨利·帕蒙蒂埃（Henri Parmentier，1871–1949）的考古学研究、保罗·穆斯（Paul Mus，1902–1969）的宗教史研究等，法国远东学院在诸多研究领域培养并涌现出一批享誉世界的著名学者。正如法国远东学

---

[1]　拉云魁尔关于吴哥古迹的调查报告及其编号（共计910处）一直沿用至今。（著者注）

图 2-6　法国远东学院高棉历史学家乔治·赛代斯（George Cœdès，1886–1969）

院的首任院长路易斯·芬诺在《法国远东学院创建章程》（*The École Française d'Extrême–Orient Founding Charter*，1901）中的阐述："……法国远东学院在印度支那开展古迹调查与学术研究的主要目标，就是要绘制出完整细致的吴哥和安南的考古学图录、搜集并整理那里的古代碑铭以及各类文物；与此同时，法国远东学院还须创建一座名副其实的博物馆，因为与我们追求好奇或声望的兴趣相比，学术研究的系统性与专业性则是法国远东学院更重要的主旨和方向；法国远东学院在对印度支那半岛重要历史遗存进行的考古调查和研究过程中，必须摒弃那些掠夺和攫取资源的方式，而应付出我们最大的努力去保存与保护那些伟大的史迹。"[1]

图 2-7　法国远东学院考古学家亨利·帕尔芒捷（Henri Parmentier，1871–1949）

图 2-8　法国远东学院早期学者合影

近百余年来，正是基于对上述目标的不懈坚持，自创立之初，法国远东学院通过对东南亚历史文化研究不遗余力的推动和开拓，以其在 1907 年创建吴哥古迹保护处（Conservation des monuments d'Angkor / the Angkor Conservation Office）[2]、1912 年创建巴利文学校（the School of Pali）、1917 年创建国立博物馆（the National Museum）、1925 年创建皇家图书馆（the Royal Library）、1930 年创建佛教

---

[1] Pierre–Yves Manguin, *The EFEO in Cambodia: a century–long partnership*, Archaeologists in Angkor, EFEO & Musée Cernuschi, Paris, 2010, p. 25.
[2] 1907 年，通过法国殖民政府的政治斡旋，吴哥地区被归还给柬埔寨，同时归还的还有在过去一个多世纪中被暹罗兼并的所有省份。吴哥古迹回归柬埔寨之日，即是法国远东学院负责对其实施大规模保护之时，作为"永久性干预"的新任务取代了此前的局部使命。此后六十余年间，法国远东学院几乎垄断了吴哥古迹的研究与保护领域，无论人员与技术如何变化，法国远东学院始终以其无可取代的工作的延续性使得吴哥古迹得以重现旧日荣光。参见 Bruno Dagens, *Angkor: Heart of An Asian Empire*, London: Thames &Hudson, 1995, p. 84.

研究所（the Buddhist Institute）和波威寺博物馆（Vat Po Veal Museum）[1]等学术机构作为重要标志，并通过1901年对巴戎寺的科学调查，以及1920年创立吴哥考古遗址公园（Angkor Archaeological Park）等一系列大规模具有开创性的学术研究和保护实践活动，法国远东学院以其建立和积累的学术理论与研究方法，奠定了东南亚历史研究，尤其是古代高棉历史、吴哥建筑历史与艺术、吴哥古迹保护修复等研究领域的重要地位并保持至今。

尤为值得关注的是，百余年来，法国远东学院保存积累了数量丰厚的历史档案资料，特别是涉及吴哥古迹保护与研究方面的档案资料包括：各类考古发掘报告、建筑测绘调查报告、建筑保护设计方案以及数目庞大的历史照片。这些现存于巴黎法国远东学院图书馆弥足珍贵的档案文献，对于任何从事吴哥古迹研究与保护的学者而言，无一敢置其于不顾。

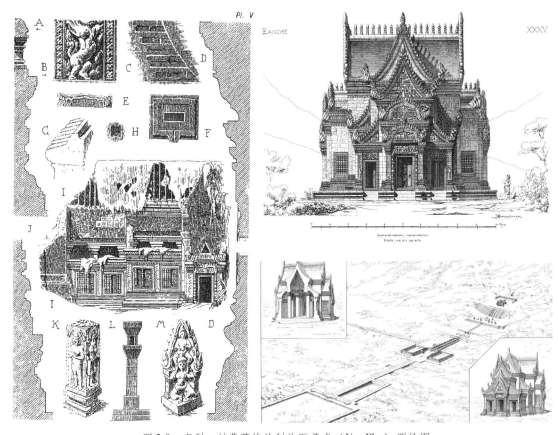

图 2-9　亨利·帕蒙蒂埃绘制的瓦普寺（Vat Phu）测绘图

---

[1] Olivier de Bernon, *The role of the École Française d'Extrême-Orient in establishing scholarly institution in Cambodia*, Archaeologists in Angkor, EFEO & Musée Cernuschi, Paris, 2010, pp. 29–34.

二、吴哥古迹研究史概述

如前文述及,正是基于百余年来以法国远东学院为代表的一大批法国学者对吴哥古迹庚续不辍的学术研究活动与古迹保护修复实践,开创并奠定了吴哥学术研究与古迹保护的根基,其深厚的学术积淀与成就一直影响至今。作为吴哥古迹研究与保护事业的开拓者和先驱者,他们通过大规模的吴哥古迹考古清理与发掘工作,以及对大量建筑实物的测绘与调查,佐之以考古学、碑铭学、语言学、艺术史、建筑史等诸多学术领域的研究方法并将其融会贯通,大致可以确定古代高棉建筑形制与艺术风格的主要类型及其基本特征。

概括言之,一个多世纪以来,以法国学者为主流的吴哥古迹学术研究,尤其在建筑与艺术方面,或可大致分为四个阶段。

其一,由于以艾莫涅尔(Étienne–François Aymonier)为代表的早期研究者对著名的斯达卡通碑铭(Sdok Kak Thom)的释读存在较大偏差,故将此碑铭中关于公元九世纪末耶输跋摩一世(Yasovarman I,公元889－900＋年在位)营建耶输陀罗补罗(Yasodhrapura)及其中央寺庙(Yasodharâsrâma 或 Vnam Kantal)的记载,与位于大吴哥城和巴戎寺内碑铭所提及的耶输陀罗补罗及其中央寺庙互相参证,遂将巴戎寺以及与巴戎寺建筑风格相似的其他寺庙(诸如班蒂克黛寺、达布隆寺、圣剑寺、斑蒂奇玛寺(Banteay Chmar)等创建于公元十二世纪末期的大乘佛教寺院)推断为吴哥时代早期的建筑形式。[1] 艾莫涅尔还相信巴戎寺是一座印度教寺院,脸形塔系梵天的雕像,另外,他也并未认识到巴戎寺浅浮雕中的战争场景的双方是高棉和占婆。

大约在1900－1904年,艾莫涅尔通过研究神牛寺(Preah Ko)和罗莱士寺(Rolous)内的古代高棉文碑铭,虽然已经将位于今之暹粒市东南大约15公里处包括巴空寺(Bakong)在内的罗莱士遗址组群的始建年代推断为公元九世纪因陀罗跋摩一世(Indravarman I)和耶输跋摩一世时期,但其并未意识到罗莱士遗址组群即是斯达卡通碑铭中所提及的诃里诃拉罗耶(Hariharalaya)。诃里诃拉罗耶乃耶输跋摩一世创建迁都吴哥之前的重要都城之一。艾莫涅尔却认为诃里诃拉罗耶是位于离"耶输陀罗补罗"(吴哥)不远处的圣剑寺(Preah Khan)。另外,艾莫涅尔还通过在吴哥窟内发现的碑铭推断其历史年代概为公元十二世纪末期。他还认为无论从古代碑铭的记载,还是从建筑艺术风格来看,吴哥窟与巴戎寺都是迥然相异:"吴哥窟表现出沉稳特质且不带有梵天的面容,暗示其建造时代明显晚于大吴哥城、巴戎寺等密林深处雕有四张面容的建筑。"[2]

由于艾莫涅尔的学术影响所及,在1928年之前,当时的学术界始终确信吴哥窟应是吴哥时代所建造的最后一座大型庙山建筑。虽然关于吴哥时期重要遗址始建年代的早期研究出现了重大混淆和偏差,但是在艾莫涅尔的研究及其后的近二十年里,针对阇耶跋摩七世时代建筑遗址的研究一直都是吴哥建筑与艺术研究的主流。(图2-10,图2-11)

---

[1] 1884年由艾莫涅尔最早发现的斯达卡通碑铭(Sdok Kak Thom,编号K.235)位于今之泰国南部与柬埔寨接壤的阿然亚蓬埃特(Aranyaprathet)境内,该碑铭刻于公元1053年,其内容记录了公元十世纪至十一世纪前后共八代吴哥国王世系情况,其间历经包括从阇耶跋摩二世到优陀耶迭多跋摩二世(Udayadityavarman II,公元1049－1065年在位)在内的共十二位吴哥国王。斯达卡通碑铭是已调查发现的古代高棉碑铭中最为著名的一方,具有极高的学术价值。

[2] *Le Cambodge. Le groupe d'Angkor et l'histoire*, E. Leroux, Paris, 1900－1904.

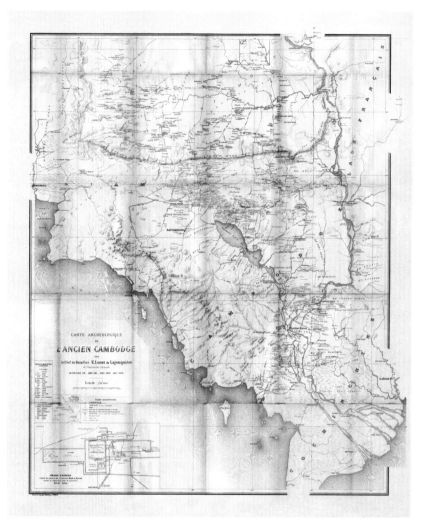

图 2-10　1909 年拉云魁尔主持调查并绘制的柬埔寨古迹分布图

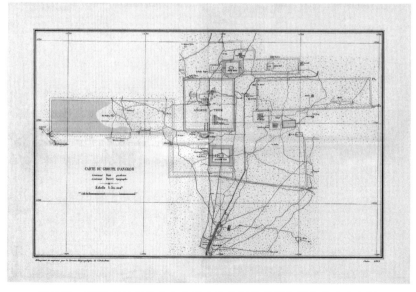

图 2-11　1909 年拉云魁尔主持调查并绘制的吴哥地区的古迹分布图

1906年，赛代斯通过研究达布隆寺内的古代高棉碑铭（编号K.273），将此碑铭中所记录的准确时间（公元1186年）推定为阇耶跋摩七世统治时期，以此将寺院的建造年代也确定为公元十二世纪末叶阇耶跋摩七世在位时期。即便如此，由于赛代斯还是受到斯达卡通碑铭与巴戎寺碑铭所共同提及耶输跋摩补罗及其中央寺庙困惑的影响，当时他仍然坚信巴戎寺的始建年代应在公元九世纪，并将其视为一座以供奉林伽及其提婆罗阇（Devarâja）为核心的印度教寺院，[1]并推测巴戎寺的脸型塔系梵天或湿婆的雕像。[2]

法国远东学院的简·康迈勒（Jean Commaille）在其1912年出版的《吴哥古迹指南》（*Guide aux ruines d'Angkor：ouvrage illustré de cent cinquante - quatre gravures et de trois plans*）中对吴哥地区主要寺庙的始建年代进行了综述。在继承艾莫涅尔观点的基础上，简·康迈利认为公元802年阇耶跋摩二世举行提婆罗阇仪式则是滥觞于圣剑寺，而位于罗莱士地区的神牛寺（Preah Ko）、巴空寺（Ba Kong）与巴戎寺同属因陀罗跋摩一世统治时期创建，而由其继任者耶输跋摩一世最终完成了巴戎寺及吴哥通王城的兴建；而位于马德望北部的斑蒂奇玛也是由阇耶跋摩二世时期所创建的。

其二，以1925年路易斯·芬诺出版专著《印度支那的观世音信仰》（*Louis Finot, Lokeçvara en Indochine, Etudes Asiatiques*）为标志，吴哥古迹研究进入了一个新的历史阶段[3]。1927年，正是基于路易斯·芬诺的重要影响和启发，菲利普·斯特恩发表论文《吴哥巴戎寺及高棉艺术之演进》（*Philippe Stern：Le Bayon d'Angkor et l'évolution de l'art khmer*）[4]。作为法国巴黎印度支那博物馆（Indochinese Museum at the Palais du Trocadero in Paris）的研究人员，斯特恩并未实地造访吴哥，仅凭亨利·马绍尔提供的现场照片以及馆藏文物的研究，他针对吴哥时代建筑装饰风格提出了自己的观点："建筑风格"限定在材料、平面、回廊、柱头、山墙、抱框、檐部等，而"装饰风格"则涉及雕像、浮雕、狮像、大象等。斯特恩从装饰样式演变的观点出发，并将吴哥时代的艺术风格分为"第Ⅰ期"与"第Ⅱ期"两类，并以此作为建筑历史编年的重要参照系。此前的研究认为巴戎寺、圣剑寺、达布隆寺、斑蒂克黛寺和斑蒂奇玛寺皆是建于公元九世纪的印度教寺庙，而斯特恩通过斑蒂克黛寺大乘佛教造像艺术风格和装饰细节的深入研究，最终推断巴戎寺、大吴哥城、圣剑寺、达布隆和斑蒂克黛寺等吴哥地区的著名寺庙皆为供奉佛像或观音的大乘佛教寺院。然而，由于当时的条件所限，斯特恩仅凭建筑装饰细节的规律，遂将巴戎寺的建造年代确定在优陀耶迭多跋摩二世统治时期（Udayadityvarman II，公元1050－1066年在位）。在这一阶段，吴哥窟建于巴戎寺之后的观点依然没有改变，当时学术界普遍认为吴哥窟标志着古代高棉建筑历史空前绝后的巅峰与绝响。[5]

斯特恩提出阇耶跋摩七世时期的巴戎寺风格可以分为三个阶段：达布隆寺、圣剑寺和斑蒂克黛寺等皆是第一时期的主要寺庙，佛陀构成大乘佛教寺庙尊崇的中心，建筑入口的山花或门楣上都有以本

---

[1] Stern Philippe. *Le temple - montagne khmèr, le culte du linga et le Devarâja*. In：Bulletin de l'Ecole française d'Extrême - Orient. Tome 34, 1934. pp.611 - 616.

[2] Henri Dufour. Charles Carpeaux. Jean Commaille：*Le Bayon d'Angkor Thom：bas - relief*, France. Ministér de l'éducation nationale. Commission Archéologique de l'Indo - Chine. 1913.

[3] Goloubew Victor. *Louis Finot*. In：Bulletin de l'Ecole française d'Extrême - Orient. Tome 35, 1935. pp.515 - 550.

[4] Philppe Stern：*Le Bayon d'Angkor et l'évolution de l'art khmer；Étude et discussion de la chronologie des monuments Khmers*. Annales du Musée Guimet, Biblio. de vulgarization, 47. Paris：Paul Geuthner, 1927.

[5] Marchal Henri. Ph. Stern：*Le Bayon d'Angkor et l'évolution de l'art khmer*. In：Bulletin de l'Ecole française d'Extrême - Orient. Tome 28, 1928. pp.293 - 305.

生故事为题材的雕刻；第二阶段则以"观音化"（Lokesvarisation）作为重要的宗教革新的标志，观世音造像大量地出现在各主要寺庙入口山花的雕刻之中，这一时期的重要变化还包括在此前建造的都城城门和主要寺庙的外围墙塔门之上设置了脸形塔；巴戎寺风格的第三阶段则以大规模兴建巴戎寺为代表。

斯特恩始终认为吴哥艺术史的年代学框架是通过解读分析雕刻与建筑装饰的细节得以确立的。然而，巴戎寺风格不应与巴戎寺建造年代混淆，而这些也正是斯特恩没有仔细研究的。受此影响，他的学生柯瑞尔·雷慕沙将斯特恩的研究方法逐渐深入扩展至建筑构件与装饰细部（门楣、壁柱、山花、假门、浮雕），并分门别类地将公元七世纪至十三世纪的吴哥艺术史的年代学演进序列进行了细化。[1]

其三，1928年，赛代斯在其论文《巴戎寺的年代学研究》（La date du Bàyon, Etudes cambodgiennes）中提出巴戎寺始建于阇耶跋摩七世时期，与巴戎寺建筑艺术风格多有相似之处的寺庙，诸如达布隆寺、圣剑寺、斑蒂克黛寺和斑蒂奇玛寺等，亦应属于阇耶跋摩七世时期的作品。正如赛代斯所言："最紧迫的任务是能够建立完整的年代学框架，并以此确认古代高棉国王的世系和与之对应的寺庙名称以及在其统治下发生的事件。"[2]赛代斯关于巴戎寺年代学研究是迄今为止最具权威性的结论，由此也终结了关于吴哥窟是否作为吴哥建筑史上最后巅峰的争论。尽管赛代斯的高棉碑铭学研究无人出其左右，但他却未能从碑铭记录的内容中获取与巴戎寺相关的直接证据，也只是在建筑形制、构造工艺、装饰艺术等方面与巴戎寺极为相似的达布隆寺、圣剑寺的碑铭中，他间接地发现并分析了这些寺庙与阇耶跋摩七世密切相关的线索或佐证。

其四，此阶段大约始于二十世纪六十年代，主要包括菲利普·斯特恩关于阇耶跋摩七世时期的艺术史研究[3]、雅克·杜马西（Jacques Dumarcay）关于巴戎寺、茶胶寺等寺庙建筑学研究[4]、伯纳德·菲利普·格罗斯利埃（Bernard Philippe Groslier）关于阇耶跋摩七世历史与巴戎寺碑铭学研究、达庚斯（Bruno Dagens）的巴戎寺图像学的研究以及雅克·克劳德（Claude Jacques）巴戎寺历史研究等。上述研究主要是围绕阇耶跋摩七世统治时代与巴戎寺之间的互动关系而展开，皆从各自不同的领域和视角讨论了巴戎寺的历史年代、建筑形制、艺术风格、造像题材、宗教信仰等内容，并且还论证了诸多阇耶跋摩七世时期重要寺庙与巴戎寺的关联性。

另外，法国高棉艺术史家简·布瑟利耶（Jean Boisselier）则系统地论述并归纳了吴哥时期建筑艺术各时期"风格"的特征，至今仍是研究吴哥装饰艺术的权威著作[5]。在其著作中，布瑟利耶将装

---

[1] Gilbettee de Coral – Remusat: *L'art Khmer; Les grandes étapes de son evolution. Étud. d'Art et d'Ethnologie Asiatique I.* Paris: Van Oest, 1951.

[2] G. Cœdès: *Some problems in the ancient history of the Hinduized states of Southeast Asia.* Journal of Southeast Asia History 5/ii. pp. 1–14.

[3] 关于吴哥古迹的艺术史研究主要集中体现在以菲利普·斯特恩（philips Stern）、柯瑞尔·雷慕沙（Gilbettee de Coral – Remusat）、皮埃尔·杜邦（Pierre Dupont）、简·布瑟利耶（Jean Boisselier）等学者的研究工作中。他们通过对建筑形制演变过程的细致研究，并依据建筑材料、建造技术、平面布局、装饰艺术、造像风格等方面的特点，按照时代顺序将古代高棉建筑划分为十个艺术风格时期，并以每个时期的代表性建筑命名其风格。（著者注）

[4] 二十世纪六十年代末至七十年代初，杜马西主持开展了多项重要吴哥古迹的大规模建筑测绘与研究工作，主要包括巴戎寺、茶胶寺、巴肯寺等，先后出版了十余本全套图集及研究报告：如法国远东学院1967年出版的《巴戎寺：寺庙建筑历史研究》（*Le Bayon, histoire architecturale du temple*）、1970年出版的《茶胶寺：寺庙的建筑研究》（*Ta Kèv: étude Architecturale du Temple*）、1971年出版的《巴肯寺建筑研究》（*Phnom Bakheng: étude architecturale du temple*）等。（著者注）

[5] J. Boisselier: *Ben Mala et la chronology des monuments du style d'Angkor vat*, BEFEO, XIVI, 1952, pp. 187–226, Pl. I–XI.

饰性门楣、花柱、山花、假门、壁柱、台基、檐口以及其他装饰元素的装饰风格与艺术特征进行归纳，主要通过类型学的方法，逐步构建起吴哥时代艺术"风格"的概念，并据此将吴哥各个艺术时期进行分期，归纳了每个阶段"风格"的特点[1]，较为全面地论述了吴哥地区的建筑与艺术，并将其进行分类讨论[2]，其中包括：（基座）线脚、线脚元素、门、门框、门楣、花柱、山花、壁柱、假门、墙面、檐壁、壁衣、浮雕、窗户、假层、滴水瓦、塔顶、塔楼、拱背、脊梁纹样、侧垛、界石、栏杆、女儿墙等方面，基本涵括了吴哥时期建筑形制与艺术风格所可能涉及的方面[3]。

雅克·克劳德则坚持认为巴戎寺的建造过程一直持续到阇耶跋摩七世的继任者因陀罗跋摩二世时期（Indravarman，公元1218？－1270年在位），后者主持修建了脸形塔，在他之后的阇耶跋摩八世（Jayavarman Ⅷ，公元1270－1295年在位）以湿婆教的复兴来反对阇耶跋摩七世时期的大乘佛教。阇耶跋摩八世在巴戎寺内部回廊雕刻浅浮雕并毁坏了十六条连接内外层围廊空间的通道，并最终形成巴戎寺的第三阶段。

观照上述历经近一个世纪可大致划分为四个阶段的吴哥学术史，高棉碑铭学家、历史学家、建筑史家、艺术史家通过百家争鸣，各领域推陈出新的研究成果相互佐证，迄今已经基本奠定了吴哥研究全面准确且丰富厚实的学术基础，包括罗莱士建筑遗址群、巴戎寺、圣剑寺、达布隆寺、吴哥窟等主要吴哥寺庙在内，诸多涉及其始建年代、建筑形制、艺术风格、碑铭研究等方面的主要问题得以基本廓清。（图2-12，图2-13，图2-14）

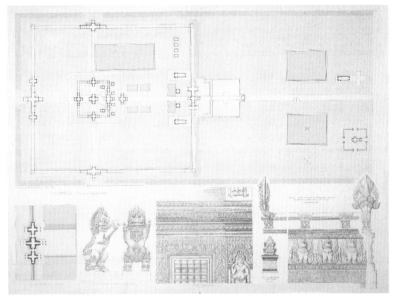

图2-12 德拉蓬特绘制的今之柏威夏省圣剑寺测绘图

---

[1] 对于"风格"之内涵和外延，布瑟利耶并未给出明确的界定。但根据其总结各个时期风格特点的论述，可以推断，对于建筑装饰而言，主要包括装饰性门楣、花柱、山花、假门、壁柱、台基、檐口等主要的建筑装饰部位的特点和其他重要或独特的装饰元素。（著者注）

[2] 此前，布瑟利耶于1952年撰写的《奔密烈以及小吴哥风格的年表》（Ben Mala et la chronology des monument du style d'Angkor Vat），从建筑、结构和装饰三个方面对奔密烈寺进行了年代学研究。其中，装饰部分的风格包括如下：1. 装饰性门楣 2. 花柱 3. 山花 4. 假门 5. 壁柱 6. 台基 7. 檐口 8. 其他装饰元素。随后总结了这一阶段的风格特点，发表《小吴哥风格的定义》（Précisions sur la statuaire du style d'Angkor Vat）。

[3] J. Boisselier: Le décor architectural in Asie du Sud-est, Tome I Le Cambodge, A. et J. Picard, Ed., 1966, XVI-480, p.64.

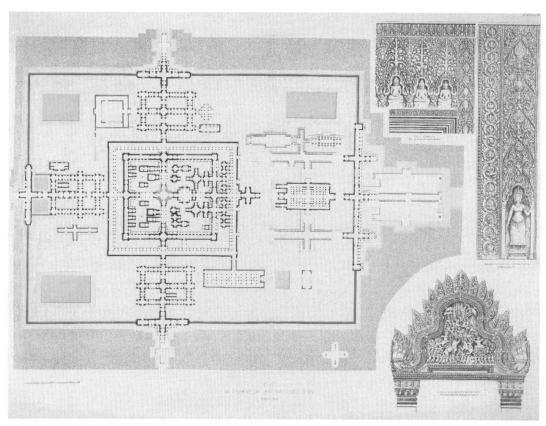

图 2-13　德拉蓬特绘制的吴哥圣剑寺测绘图

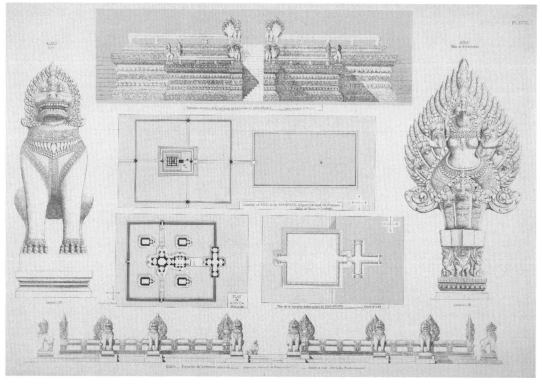

图 2-14　德拉蓬特绘制的皇家浴池及其雕刻测绘图

三、吴哥古迹保护修复史概略

1907年3月,根据法国与暹罗王国签订的条约,柬埔寨的暹粒省正式纳入法属印度支那联邦,而对于其境内吴哥古迹的研究保护开始进入了以法国远东学院为全面主导的时期。当时,法国远东学院考古研究部负责人亨利·帕蒙蒂埃借此创建吴哥古迹保护处(Conservation des monuments d'Angkor),进而全面推动吴哥古迹的学术研究与保护修复实践,吴哥古迹的保护修复工作也逐渐成为法国远东学院的最为重要的工作领域之一。由于吴哥古迹保护的成绩卓著,帕蒙蒂埃曾经先后两度出任法国远东学院的院长(1909–1910任期、1918–1920任期)。此后直至1970年3月柬埔寨朗诺–施里玛达集团发动政变而导致工作中断为止,吴哥古迹保护处始终都处于柬埔寨吴哥古迹保护与研究的最前沿。

作为柬埔寨吴哥古迹研究与保护领域的先驱者,吴哥古迹保护处的历任负责人,简·康迈勒(Jean Commaille)、(图2-15)亨利·马绍尔(Henri Marchal)、乔治·特鲁维(Georges Trouve)、莫瑞斯·格莱兹(Maurice Glaize)、简·布瑟利耶(Jean Boisselier)、伯纳德·菲利普·格罗斯利埃(Bernard Philippe Groslier)等,由于他们的专业背景皆为建筑学、考古学或艺术史,因此贯穿于二十世纪上半叶吴哥古迹的学术史与保护修复史,亦是主要围绕考古研究与建筑修复两大主题而展开的。在当时殖民政权的庇护和支持之下,直到1954年法国远东学院在河内总部被迫关闭、1961年其在法国巴黎重新开放之前,法国远东学院在柬埔寨的考古调查与保护修复活动,一共涉及大约1500多处柬埔寨的历史遗迹,其中规模较大的吴哥古迹保护修复项目包括吴哥窟、巴戎寺、巴方寺、女王宫等著名的吴哥寺庙。

纵观二十世纪柬埔寨的现代政治史,从争取民族独立到殖民体系瓦解,从红色高棉暴政到多年连绵内战,吴哥古迹保护与修复曾经深受影响。即便如此,法国远东学院还是通过薪火相传、庚续不辍的学术研究与保护实践,逐步形成了吴哥古迹保护的理念与方法,并将其大规模地付诸实践。

1920年,法国远东学院正式明确提出建设"吴哥考古遗址公园"的计划,主要针对吴哥古迹遗址范围内多数寺庙倾圮、植被覆盖的状况,通过古迹保护区规划设置,分阶段实施吴哥古迹的保护与修复项目,今日环绕吴哥遗址主要寺庙的道路即是在这个时期规划建设的。其中,在吴哥古迹保护与修复的理念上,"保存现状"与"恢复原状"的争论与博弈仍然是其核心问题。十九世纪末至二十世纪初欧洲文物建筑保护的主流,主要是基于对历史古迹艺术价值的认同,特别是对一些标志性、纪念性的历史建筑更是将艺术价值和艺术风格作为其出发点,表现出强烈的艺术至上倾向。由于受到法国学派的影响,二十世纪初的早期吴哥古迹保护修复项目主要以清除覆盖遗址废墟的植被和腐殖沉积物,并以加固支护存在坍塌风险的建筑结构为主;另外,法国学派还强调修复工作必须是建立在深入研究的科学基础之上,古迹

图2-15 吴哥古迹保护处首任负责人简·康迈勒(Jean Commaille, 1868–1916)

修复不仅是形式风格的保护，更重要的是结构的修复。因此，当时法国远东学院吴哥古迹保护修复理念的主旨是加固与支持，并反对古迹的复原或重建。例如，当时法国远东学院吴哥古迹保护处采用的原则是绝对禁止以材质大小相同的新石块代替原物，以免混淆古迹的真实性；在加固维修古迹时，必须严格保持其在清理工作完成后所展示的废墟状态，尽量不要实施带有明显修复痕迹的工程项目。所以，当时实施的吴哥古迹保护修复项目，多是在清理工作完成之后，采用钢筋混凝土结构进行不加任何修饰的临时性支护，属于抢险加固的性质。

伴随着二十世纪二十至三十年代现代建筑思潮的兴起[1]，欧洲古迹保护修复理念也逐渐发生变化，那种追求艺术至上、保存古迹遗址废墟状态的观念和倾向，逐渐遭到质疑和摒弃。一种被称为"Anastylosis"的古迹保护修复方法在欧洲古迹保护修复领域，特别是在希腊和罗马的古迹保护中得到推广和应用。[2] 所谓"Anastylosis"，因其具有建筑解体、测绘记录、构件修、复原位归安等涵义，或可较为简洁地翻译为"原物归位法"，是指在古迹落架解体之后，通过精确的测绘记录，利用建筑原有构件，根据建筑结构的形制，进行构件修补并原位归安；修复过程之中，允许非常慎重地添配新构件替换缺失部分。

1930年以降，以亨利·马绍尔主持女王宫保护修复工程为代表，在法国远东学院柬埔寨吴哥古迹保护修复实践中，"原物归位法"大行其道[3]，一度被认为是保护吴哥古迹最好的方法之一，此法也逐渐取代了此前以清理加固为主的保护策略，成为当时吴哥古迹保护修复理念与技术的主流，其产生的深刻影响一直流传至今。以法国远东学院为主导的吴哥古迹保护修复工作，使之成为当时整个亚洲乃至全世界规模最大的历史古迹保护修复项目之一。

第二次世界大战之后，东南亚地区殖民体系的瓦解，政治格局发生重大变革，在某种程度上也影响到了法国远东学院的吴哥古迹保护修复工作。但无论政局如何变化多端，法国远东学院的吴哥保护修复工作始终没有停止。而且时任吴哥古迹保护处负责人伯纳德·菲利普·格罗斯利埃还坚持提出，吴哥古迹保护修复也需要革新，应该使之成为融合学术研究与工程技术的综合性项目，应以科学的调查研究为主旨，学术研究贯穿保护工程之始终；另外，吴哥古迹保护修复除了建筑师、工程师、考古学家、历史学家的密切配合之外，还需要生物学、岩石学、气象学等多学科的共同参与，以期通过科学保护创造出具有最高艺术和科学境界的古代遗址作为目标。

二十世纪三十年代初，时任吴哥古迹保护处负责人的亨利·马绍尔（图2-16），首次成功地采用"原物归位法"完成女王宫保护修复工程。作为建筑师，亨利·马绍尔是二十世纪上半叶吴哥古迹保护修复工程领域的最重要的先驱者之一，他用"原物归位法"对女王宫的成功修复，成为吴哥古迹保护修复史中影响最大且具有里程碑式意义的重要实例。（图2-17，图2-18，图2-19，图2-20）

---

[1] 1933年，国际现代建筑协会（CIAM）《雅典宪章》表明，当时的建筑师、城市规划师已经认识到历史古迹中所蕴涵的人类价值的统一性，并将史迹视为人类共同的文化遗产，并一致呼吁古迹保护修复指导原则应在国际上得到公认并做出相应的规范，引自温玉清、吴葱：《文物建筑"真实性"再认识》，载于《天津大学学报》（社科版），2001年第1期。

[2] 希腊古迹修复专家巴兰诺斯在修复雅典卫城古迹时的阐述："原物重建法意指古迹解体之后，以建筑物本身的材料，依据其原有的建筑形制与构造方式予以修复并复原重建。修复过程中，允许以非常谨慎准确方式使用新材料制作添加新构件以替换缺失部分，否则原有构件也无法完全重归原位。"转引自 Glaize M, Les monuments du groupe d'Angkor, 3rd edition, Paris, 1963.

[3] 1936-1947年，吴哥古迹保护处负责人莫瑞斯·格莱兹利用"原位归安重建法"主持维修了东梅奔寺（Eastern Mebon 1937-1939）、克罗姆寺（Phnom Krom 1938）、柏克寺（Phnom Bok 1939）、巴肯寺（Bakheng 1936-1944）、龙蟠水池（Neak Pean 1938-1939）、巴戎寺（1939-1946）、吴哥窟（1939-1946）等著名的吴哥史迹。（著者注）

图 2-16　吴哥古迹保护修复的先驱者亨利·马绍尔（Henri Marchal，1876–1970）© EFEO Archives

图2-17　吴哥古迹保护修复"原物归位法"（Anastylosis）的最早工程实例：女王宫（1934年）© EFEO Archives

图2-18　吴哥古迹保护修复工程实例：大吴哥城南门神道（1961年）© EFEO Archives

图2-19　吴哥古迹保护修复工程实例：龙蟠水池（1938年）© EFEO Archives

图2-20　吴哥古迹保护修复工程实例：圣剑寺（1930年代）© EFEO Archives

1916年，亨利·马绍尔进入法国远东学院，成为吴哥古迹保护处负责人简·康迈利的助手，自此开始了其长达五十余年的吴哥古迹保护生涯，亨利·马绍尔也由此而获得法国远东学院的终身职位。在1933－1937年期间，亨利·马绍尔出任吴哥古迹保护处建筑保护部的负责人，1946－1953年马绍尔接替莫瑞斯·格莱兹成为吴哥古迹保护处的负责人。在此期间，亨利·马绍尔及其助手负责的吴哥古迹保护修复项目，其中包括女王宫、巴肯寺、巴方寺、圣剑寺、巴戎寺、班蒂萨穆雷寺、巴空寺以及吴哥通王城的三座城门（北门、南门、胜利门）等，都以不同程度的"原物归位法"实施过保护修复，如此大规模应用"原物归位法"并取得成功，极大地提升了法国远东学院在吴哥保护修复领域的声望和影响。

"原物归位法"与吴哥古迹保护修复的渊源，或可追溯至1929年荷兰考古学家凡·斯坦恩·卡隆菲尔斯（Pieter Vincent van Stein Callenfels）对柬埔寨及吴哥古迹的访问。在1907－1911年，一直致力于东南亚历史与考古研究的卡隆菲尔斯博士，利用"原物归位法"较为成功地完成了爪哇婆罗浮屠的修复保护项目。其修复过程包括古迹解体，拆落构件现场保存、编号记录、测量绘图、拼对修补、原位归安等主要步骤；另外，在不影响古迹外观的前提下，在其构造内部添加了钢筋混凝土构件的加固与支撑。卡隆菲尔斯对吴哥古迹的访问，直接促使马绍尔于1930年前往印度尼西亚实地学习"原物归位法"，当他返回吴哥以后，首次在吴哥古迹保护修复实践中，利用"原物归位法"实施女王宫保护修复工程。

女王宫是一座兴建于公元十世纪中叶的小型印度教寺庙，坐落于吴哥东北方向大约40公里的荔枝山脚下的暹粒河畔。由于女王宫距离吴哥中心区域较远，开展田野调查较为困难，所以早期由艾莫涅尔（Etienne Aymonier）与拉云魁尔（Etienne Lunet de Lajonquière）编撰的吴哥古迹目录中未提及该寺。1924年，为消除"马尔罗事件"（马尔罗被指控参与盗掘走私吴哥文物）的不良影响，安德鲁·马尔罗[1]出版了关于女王宫的一部以在吴哥的丛林中探险为主题的著作；1926年，法国远东学院出版了的《伊塞婆罗补罗的神庙》（Le Temple d'Içvarapura），这是关于女王宫的第一部系统的考古学著作，在这部著作中，路易斯·芬诺、维克多·格罗布维、亨利·帕蒙蒂埃等从女王宫的碑铭、雕刻以及建筑形制等方面进行了较为全面的研究[2]。此前的1924年3月，亨利·帕蒙蒂埃和维克多·格罗布维曾经主持清理加固了女王宫的东寺门和第二层围墙。1931年，由于女王宫的规模较小，装饰精美，且砂岩构件的保存状况较好，亨利·马绍尔因而选定女王宫作为法国远东学院首次采用"原物归位法"进行吴哥古迹保护修复的对象。正如亨利·马绍尔所言："女王宫是柬埔寨现存所有古代建筑遗存中最适合于'原物归位法'的一座，因为其石块保存状态较好，石构件上精美的雕刻与建筑装饰使得保护修复工作堪称真正的拼图游戏。"[3]亨利·马绍尔在其日记中记录了女王宫的修复过程，例如："……本月我们开始雕刻精美的中央圣殿的工作，（建筑全部解体完成后）我们提升了建筑的基座，事实上，最困难的工作是找到其原始地坪以及结构水准，由于地基下沉，我们不知道其原始地坪，这是确定其他构件位置的基础数据。在砂浆垫层之上，我们更换了部分角砾岩，以替换铺设中央圣殿基台的砂岩石块。另外，我们还利用铁件和混凝土来修补碎裂的石块。"[4]遗憾的是，女王宫保护修复工程结束之后，亨利·马绍尔却未能将其修复工程报告及时出版或公之于众。但是，瑕不掩瑜，亨利·马绍尔依据"原物归位法"的原则将散落石构件归安至其原位，通过较为完整地修复整体格局与复原建筑形制，使女王宫这座曾经被丛林吞噬的伟大高棉史迹得以重现天日。

在成功修复女王宫之后，法国远东学院将"原物归位法"在巴方寺修复工程中表现得更为全面而彻底。自1908年起，法国远东学院就一直反复对巴方寺进行清理和加固，但这座庙山却又接连不断的遭遇崩塌与破坏，1948年，亨利·马绍尔不无幽默地在日记里写道："眼前的这座寺庙（巴方寺），与其说是一座建筑，倒不如说是一片茂盛的山林。"[5]1960年，时任法国远东学院院长的伯纳德·菲利普·格罗利埃以"原物归位法"在钢筋混凝土的框架上，重建巴方寺。将巴方寺残损的石块拆落，并详细记录拆解过程，在基台内部增设钢筋混凝土挡土墙作为围护支撑结构，然后再将拆解或散落的石构件重新归安原位，将"原物归位法"的原则和做法发挥到极致。

由于战争爆发，自1972年法国远东学院全部撤离吴哥，直至1995年学院重返吴哥，二十余年的

---

[1] 安德烈·马尔罗（Andre Malraux，1901－1976年），法国小说家、艺术史家和政府官员。在其写作和政治生涯中，马尔罗的小说大多涉及革命及其哲学内容，马尔罗还曾在巴黎研习考古学和亚洲语言学，在法属印度支那，他与一个民族主义团体一起工作，支持中国的反殖民主义运动，并因为走私吴哥文物而受到指控。二十世纪三十年代，他积极投身反法西斯斗争，支持国际共产主义运动。在西班牙内战中，马尔罗组织了反独立的空军力量。第二次世界大战时期，他在法国军队服役，法国沦陷后成为一名抵抗战士。法国解放后，他支持查尔斯·戴高乐的政党。1959－1969年，马尔罗担任文化事务部部长。他的其他小说还有《王家大道》（1930年）和《阿尔腾堡的胡桃树》（1943年），其他著作有《砍倒的橡树》（1971年）、《毕加索的面具》（1974年），以及自传体的《反回忆录》（1967年）和《拉扎尔》（1974年）等。（著者注）

[2] L. Finot, H. Parmentier, V. Goloubew：*Mémoires archéologique I*：*Le Temple d'Içvarapura*, EFEO, Paris, 1926.

[3] Eric Bourdonneau, *Henri Marchal and the preservation of the splendour of Banteay Srei*, Archaeologists in Angkor, EFEO & Musée Cernuschi, Paris, 2010, p.100.

[4] Henri Marchal, *RCA Documents of Activities*, 1931. Archaeologists in Angkor, EFEO & Musée Cernuschi, Paris, 2010, p.104.

[5] Pascal Royère, *The Baphuon：last attempts to save a monument*, Archaeologists in Angkor, EFEO & Musée Cernuschi, Paris, 2010, p.125。

战乱不仅使巴方寺保护修复的原始记录被毁，而且曾经参加施工的当地技术人员和工人也多在战乱中丧生，约计 30 万块拆落的石构件散布丛林之中。法国远东学院再次面临难以想象的挑战和困难，其修复任务是按照既定的"原物归位法"，继续完成二十世纪六十年代利用钢筋混凝土建造的基台内部结构加固部分，甚至还包括修复重建一尊十六世纪利用巴方寺的石构件堆砌的巨大卧佛。此后将近二十年的保护修复工作，法国远东学院通过将巴方寺大部分散落构件确定原位、修复基座、回廊和卧佛，基本排除了寺庙存在的险情和隐患，并尽最大可能地恢复了寺庙原状。2011 年末，巴方寺保护维修工程正式竣工。

毋庸讳言，在吴哥古迹保护修复领域，法国远东学院以其"原物归位法"为标志，如此大规模地采用钢筋混凝土改变古迹修复材料与原有形制的做法，也经常遭到各种质疑。然而，正如曾经主持巴方寺修复工程的小格罗利埃所言："这些重建工程的决定与设计，其最终目的只有一个，这就是挽救古迹，这种外科手术式的修复方式，只能用在最紧要的时刻与部位，而且也一定会留下疤痕。"[1]

---

[1] 同前注。

Synopsis II Introduction to Researches and Conservation Activities on Angkor Monuments

1. The Role of the École Française d'Extrême-Orient in Research and Conservation Activities on Angkor Monuments

Tracing back to the subtle and various modern history of Cambodia, especially regarding as the history of researches and conservation activities on Angkor monuments, obviously, the École Française d'Extrême-Orient, also referred to as EFEO, has always played the most important role in the study, promotion and conservation of the archaeological heritage of the ancient kingdom of Cambodia.

The attentions and attitudes of the western world, towards the culture and history of the Southeast Asia, has been transferred to a new attitude or direction, along with the expansions of the French Indochina system and its impact on the Indo-China Peninsula in 1880s, it became evident that the new colony required an institution capable of administering the extraordinary heritage already revealed by early explorers and functionaries including Ernest Doudart de Lagrée, Étienne Lunet de Lajonquière and Étienne Aymonier. Especially, since the beginning of the establishment of the EFEO in 1901, which means the interests in the field of adventured tours, discovery of the new land, excited searching in the jungle, and the exotic, had been replaced by academic researches, which are much more rigorous and pragmatic. The studies on the ancient Khmer history and the conservation activities for Angkor monuments had commenced a more scientific stage. Meanwhile, many scholars with the background of architecture, archaeology, and art history had prominently emerged and become the pioneers and precursors who were leading the research and conservation activities for Angkor monuments. Based on the outstanding role of EFEO, especially continuing scientific studies and conservation activities on Angkor monuments, the foundation of academic history of Angkor site had been established, as well as nowadays the profound accumulations and achievements has influenced to the researching fields of Angkor monuments.

French authority of Indochina was established in 1887, which was a part of the French colonial mechanism in the Southeast Asia, a federation of the three Vietnamese regions, Tonkin (North), Annam (Central), and Cochin-China (South), in addition to Cambodia. Laos was added in 1893 during the Franco-Siamese War. The proconsuls of the French Indochina were the heads of central administration, the administrative rights keeping in the hands of the domestic French colonial section. The federation had been lasted until 1954. In the four protectorates, the French formally left the local rulers in power, who were the Emperors of Vietnam, Kings of Cambodia, and Kings of Luang Prabang, but in fact gathered all powers in their hands, the local rulers acting only as figureheads.

Correspondingly, the history of EFEO could be dated back to 1898, when the proconsul of the French Indochina, Paul Doumer, initiated to establish the French Indochina Institute in Saigon, Vietnam. As a significant organization of the French colonial system, the institute contained the two departments: the first one was archaeological mission; the other was natural science exploration. Quite similar as the Archaeological Survey of India (abbreviation: ASI) founded in 1861 under British colonial administration by Sir Alexander Cunningham with the help of the then Viceroy Charles John Canning, the French academic institute also had quite close connection with the expansion, the archaeological mission of the French Indochina was actually set up by the Académie des Inscriptions et Belles-Lettres. In the first years, Louis Finot (1864 – 1935), an assistant librarian with the French National Library, was designated as the first director of the archaeological mission in Indochina. During 1899 to 1900, Louis Finot explored all around French Indochina, result in all his contribution to the study of Khmer history, architecture and epigraphy. And thanks to all his works, Mission Archéologique de l'Indochine has been formally established. Up on his research plan, there should be a library and a museum in the, the range of study not limited in French Indochina, but also China, India, and the whole East Asia, South Asia, and Southeast Asia. Besides, the target of the research should be further than in the field of the architecture and the archaeology, but limited more to the history, philology, religion study and ethnology.

On the 10thJanuary 1900, the Mission Archéologique de l'Indochine, was renamed as École Française d'Extrême-Orient (EFEO), the headquarters moving from Saigon to Hanoi. As a academic institute special on Orientalism in the system of L'institut de Française, EFEO missions were mainly on the fields of archaeology investigation and excavation, and collection of the historic resources, protective conversation of the monuments and sites, the investigation of ethnology and anthropology, researches on philology, etc. During the one-century history of EFEO, a quantity of the experts on archaeology, Inscriptions et Belles-Lettres, and religion histories, played quite important roles. For example, Paul Pelliot, Henri Maspero, Paul Demiéville and George Maspero in the field of ancient Southeast Asia history; Louis Finot and George Cœdès in the Inscriptions; and Henri Parmentier in Archaeology; Paul Mus in the history of religions, etc. And quite a lot of well-known scholars came out from EFEO.

In the recent hundreds of years, based on the unremitting courting of the aims before, ever since the beginning of the establishment, EFEO never stopped the promoting and spreading works of the Asian history and culture studies, which settled the quite solid foundation of the study of the Southeast Asian history, especially the Khmer's, the history of Angkor architecture and art, and the conservative restoration of Angkor's monuments and sites, through the founding of Conservation des monuments d'Angkor / the Angkor Conservation Office, the School of Pali in 1912, the National Museum in 1917, the Royal Library in 1925, the Buddhist Institute and Vat Po Veal Museum in 1930, and the scientific investigating in the Bayon temple in 1901, and the originating of Angkor Archaeological Park in 1920. With all these above, EFEO organized a serious of great creative academic researches and protective practices. After one hundred years of academic studies and conservation activities, EFEO accumulated a huge number of historical archives, especially in the aspect of conservation and research on Angkor monuments, seems like the reports of the archaeology excavation, the survey and drawing of the architectures, the design scheme of the conservation activities, and amounts of the historic photographs. Up to whoever involved in the research and conservationactivities of Angkor monuments, all these above, preserved in the archive of École Française d'Extrême-Orient Library, where is the top level and really fascinating institution for the Southeast Asia historical studies all over the world.

2. HistoricalOutline of the Researches on Angkor Monuments

As above-mentioned, based on the continued academic and conservation activities of EFEO as well asmany French scholars during more than a century, the academic foundation of Angkor monuments has been established, which academic achievements have been affected till today. Regard as the pioneers of research and conservation for Angkor monuments, they had carried out a large number of surveying, clearance, and archaeological excavation missions which were accordance with the studies on inscriptions, linguistics, art history, architectural history, and many other academic field research methods and mastery, and it can roughly be determined that the basic characteristics of evolution of ancient Khmer architectural form and the artistic style.

In generally, as the mainstream of academic researches on Angkor monuments, especially in terms of the French scholars' studies on the ancient Khmer architecture and art, it can be broadly divided into four stages in the past more than a century as follows:

Firstly, as the representative of the early researchers, due to the deviations of his interpretation on the famous Sdok Kak Thom inscriptions, Étienne-François Aymonier suggested that Yasodharâsrâma or Vnam Kantal in the Sdok Kak Thom inscriptions should be consistency with the inscriptions of Angkor Thom and Bayon temple, and then he deduced that Bayon temple should be the early architectural style of Angkorian era, in addition to the other temples (such as Banteay Kdei, Ta Prohm, Preah Khan) with the similar styles to Bayon temple. Aymonier also believed that the Bayon temple with its four-face towers should pertain to Hinduism Brahma statue. In addition, he had ever not recognized both Khmer and Champa of the war scenes in the bas-relief of Bayon temple.

Aymonier studied on the ancient Khmer inscriptions of Preah Ko temple of the Rolous group in 1900 – 1904, and he had already deduced that the Rolous group including Bakong temple was built during the reign period ofKing Indravarman I and King Yasovarman I which is inferred to be the ninth century AD, but he did not realize that Rolous group is exactly Hariharalaya which is mentioned in Sdok Kak Thom inscriptions. Hariharalaya had always been one of the significant capitals before King Yasovarman I moved his capital to Yasodharâpura (Angkor), nevertheless Aymonier deduced that Hariharalaya should be related to Preah Khan temple. Due to Aymonier's academic influence, the scholars had always known that Angkor Wat should be the last temple-mountain of the Angkorian era before 1928. For the dating of the significant Angkor monuments, although it had occurred a series of confusions and deviations, the studies on the ruins of King Jayavarman VII era had always been the mainstream for nearly two decades since Aymonier's publications.

In 1906, G. Cœdès deduced that the exact time (1186 AD) of Ta Prohm temple inscription (No. K273) was defined during the reign of King Jayavarman VII, in addition to it was also determined that the chorological dating of the temple's construction should be at the end of the twelfth century AD, which was the reign period of King Jayavarman VII. Even so, G. Cœdès had still convinced that the dating of the Bayon temple's construction should be in the ninth century AD because he was influenced by the confusion of Yasodharâsrâma or Vnam Kantal, which was all mentioned in the inscriptions of both the Sdok Kak Thom and the Bayon temple. Meanwhile, he suggested that the Bayon temple should be as a Hinduism temple to worship the Linga or the cult of Devarâja, and he deduced that the four-face tower of the Bayon temple should be the statue of Brahma or Shiva.

In 1912, Jean Commaille summarized the chorological results of the main Angkor monuments in his publication of *Guide aux ruines d'Angkor*: *the ouvrage illustré de the cent cinquante-quatre gravures et de trois plans*. In inheriting Aymonier's viewpoints, Jean Commaille mainly inferred that Devarâja, the national cult of King Jayavarman II, was originated in the Preah Khan, perhaps it was held in 802 AD. He also inferred the Preah Ko temple and the BaKong temple in Rolous region were similar to the Bayon temple, which were all belonged to the reign of King Indravarman I, in addition to King Yasovarman I, his successor, carried out the final completion of the construction of the Bayon temple and Angkor Thom, whereas Jean Commaille suggested the Banteay Chhmar temple in Battambang province was also created by King Jayavarman II.

Secondly, as a significant symbol, Louis Finot published the monograph of Lokeçvara en Indochinae in 1925. Subsequently the studies on Angkor monuments had promoted into a new historical stage. In 1927, based on an important influence and inspiration of Louis Finot, Philippe Stern published his outstanding publication namely *Le Bayon is a d'Angkor et l'évolution de l'art khme* and he suggested the point of view for the architectural style of Angkorian era as following aspects as the architectural style and the decoration style. On the one hand, Philippe Stern suggested that the architectural style was assigned to the aspects as material, plan, corridors, pilaster, lintel, pediments, cornice, roof etc., and on the other hand, the decoration styles were involving statues, bas-reliefs, lion statues, elephant statues etc. Regarding as the evolution of decoration styles, for the architectural style of Angkorian era, it can be divided into the Phase I and Phase II categories, which were referenced to the chorological foundation of the Angkorian architectural history. Although in the previous studies such temples as Bayon, Preah Khan, Ta Prohm, Banteay Kdei were belonged to Hinduism temples which should be built in the ninth century AD, Stern eventually inferred that the famous temples of Bayon, Preah Khan, Ta Prohm, Banteay Kdei as well as Angkor Thom in the Angkor region were belonged to Mahayana Buddhist temple for the worship of Buddha or Lokeçvara owing to his in-depth studies on the details of architectural decoration and the characteristics of Mahayana Buddhist statues. However, due to the prevailing conditions, Stern proposed that the construction dating of Bayon temple could be during the reign of King Udayadityvarman II (1050-1066AD) depending on the decorative details. In this phase, it was still no changes that Angkor Wat was built after Bayon temple, and then academia was generally believed that Angkor Wat marked the unprecedented summit of the architectural history in the ancient Khmer.

Philippe Stern proposedthat the Bayon style should be divided into three stages: firstly, Ta Prohm temple, Preah Khan temple, and Banteay Kedi temple are defined to the main temples of the first period, as well as the Mahayana Buddha is enshrined as the center of the temples and the Jataka themes have carved on the lintel of the Gopuras. Secondly, as a sign of Lokesvarisation or religious innovation, Lokeçvara statues were widely appeared in the main temples and the emergence of the face-tower was prior to the construction of both the capital gates and the gopuras of temples. The third stage of Bayon style was the large-scale construction of the Bayon temple represented. Philippe Stern always believed that the chronological framework of Khmer art history has been established through the interpretation of the details of the sculpture and architectural decoration. The Bayon style, however, should not be confusion with the Bayon temple's construction era, which is exactly what Philippe Stern did not carefully studied. But Gilberte de Coral-Rémusat, one of Stern's students, was affected by his research methods and gradually deepening extended to architectural components and decorative detail (lintels, pilasters, pediment, false-doors, relief) and categorized the chronological evolution sequence from the seventh century to the thirteenth century in addition to the Angkorian art history were refined.

Thirdly, in 1928, G. Cædès suggested that the Bayon temple was built during the reign period of King Jayavarman VII in his paper *La date du Bàyon*, *Etudes cambodgiennes*, and such as Ta Prohm temple, Preah Khan temple, Banteay Kedi temple and Banteay Chhmar temple, which the architectural styles had many similarities to Bayon temples, also should belong to the period King Jayavarman VII. For the Bayon temple's chronological studies, G. Cædès proposed the most authoritative conclusions, and thus also the end of a debate about the Angkor Wat whether the final summit in the Angkorian architectural history.

Fourthly, it is mainly including Philippe Stern's studies on the art history of King Jayavarman VII period, Jacques Dumarcay's architectural studies on Bayon temple and Ta Keo temple, Bernard Philippe Groslier's studies on King Jayavarman VII history and Bayon temple's inscriptions, Bruno Dagens' studies on Bayon temple's images, as well as Jacques Claude's Bayon historical research since 1960s. The studies above-mentioned is commenced around the interaction between the King Jayavarman VII reign and Bayon temple, meanwhile, the historical dating, architectural form, artistic style and statues themes of Bayon temple are focused on different perspectives, religion and other content, and also demonstrated the significant relevance of the important temples and Bayon temple during King Jayavarman VII's reign.

3. Introduction to Conservation and Restoration Activities at Angkor Site

In March 1907, according to the treaty between French and Siam kingdom, Siem Reap province of Cambodia had formally entered into French Indochina federal, and the research and conservation works of Angkor Sites had began to enter a period leading by EFEO.

At that time, Henri Parmentier, the director Archaeology Mission of EFEO created Conservation des monuments d'Angkor, and then promoted the academic research and conservation of repair practice in Angkor Sites, which also gradually became the most important work for EFEO. Due to the distinguished conservation works in Angkor Sites, Henri Parmentier worked successively as the director of EFEO twice (1909 – 1910, 1918 – 1920).

As the pioneers in the field of the study andconservation of Angkor Sites in Cambodia, Jean Commaille, Henri Marchal, Georges Trouve, Maurice Glaize, Jean Boisselier, Bernard Philippe Groslier, etc. all their professional backgrounds are architecture, archaeology or history of art, so throughout the 20th century, the history of academic researches and conservation activities in Angkor, is also mainly developed around two themes both archaeological research and conservation activities.

From that time, under theconservation and support of colonial regime, until EFEO's headquarter in Hanoi were forced to shut down in 1954, and before which reopened in Paris, France in 1961, the archaeological investigation and restoration activities of EFEO in Cambodia referred to a total of about 1500 historical sites in Cambodia, including Angkor Wat, Bayon, Baphuon, Banteay Srei and other famous Angkor temples.

EFEO formally proposed to build the Angkor Archaeological Site Park in 1920, and to take forward the conservation and restoration missions of Angkor site on different stages, which was depending on the situation of Angkor monuments, with most temples collapsed and covered by vegetation. Today's main roads around Angkor temples were just planning and construction in this period. Among all these works, in the concept of conservation and restoration on Angkor sites, the debates and the mediation between " preserved status" and " restoration" was always the core problem. At the end of the nineteenth century and the early twentieth century, the mainstream of European conservation conceptions and methods is mainly aimed to the artistic values or the aesthetic identity of the historical sites, especially for some historical landmarks, monumental buildings, art value and artistic style as its starting point, and showing a strong preserved artistic tendency. Due to French influence, at the beginning of the twentieth century, the early restoration activities on Angkor sites mainly paid close attention on the clearances of the vegetation covering on the ruined sites in addition to the humus soils, and mainly confirmed on the reinforcement of the collapsing existence. In addition, the French school also stressed that the restoration works should be based on in-depth studies on scientific foundation, the restoration mission was the conservation process not only for the artistic style but also for the significance of structural consolidation. For example, at that time, Conservation des monuments d'Angkor, also referred to as CA had adopted the conservation principle which was absolutely forbidden to use the new stones, even of the same material and size to take place of the original one, so as to not confusion of the authenticity. Consequently, in process of the reinforcement of maintenance, the ruins should be strictly preserved as the decay states. It is therefore that the restoration missions for Angkor monuments were mainly undertaken after the completion of the clearance works, reinforced with concrete structures, and without any modification of the temporary supports, which could be recognized as rescuing reinforcement at that time.

In the twentieth century, according to the rising of modern trendy of architecture, the conception of European conservation and restoration had also gradually changed, the pursuit of art, and the ideas and tendency to preserve the historical relics sites as the ruined state, were gradually questioned and abandoned. One method of conservation and restoration called " Anastylosis" has been promoted and applied in European sites, especially in Greek and Roman Sites. The so-called " Anastylosis", because it focuses on the surveying and documentations and its process is including dismantling, repairing, matching and assembling of the components to its original position according to the accurate documentations and mapping, and it is allowed to add new elements to replace missing parts very cautiously in this process.

As the commencing of Anastylosis method, it had leaded by Henri Marchal in EFEO's conservation and restoration works for Angkor monuments on Banteay Srei temple since the 1930s. It was approbatory once even considered as the best method for safeguarding Angkor monuments, and this method had also gradually replaced the previous strategies, which was including the treatments of clearance and reinforcement, and then became the mainstream of the conservation and restoration activities for Angkor monuments, in addition to its influence has profoundly been impacted on the conservation activities at today's Angkor site.

During the 1930s, Henri Marchal, who was in charge of the conservation activities for Angkor monuments, had firstly utilized anastylosis method in successfully to complete the conservation and restoration works of Banteay Srei temple. As an architect, Henri Marchal was one of the most significant pioneers in the field of conservation and restoration works for safeguarding Angkor in the twentieth century. Due to the anastylosis utilization of Banteay Srei temple in successfully, it had became one of the most important landmark in the history of conservation and restoration activities at Angkor site. Banteay Srei temple is dedicated to the Hinduism God Shiva, which was built in the tenth century and its location is nearby Phnom Kulen around 25 km in the northeast of the main group of Angkor temples that belonged to the medieval capitals of Yasodharapura and Angkor Thom. The incident stimulated interest in the site, which was cleared the following year, and in the 1930s Banteay Srei temple was restored in the first important utilization of anastylosis at Angkor site. In March 1924, Henri Parmentier and Victor Goloubew ever carried out the charge of clearance and consolidation at the east gopura and the second enclosure of Banteay Srei temple. In 1931, due to its appropriate scale size, exquisite decorations, as well as the relatively good condition, Henri Marchal decided to select Banteay Srei temple as the primary project to use anastylosis at Angkor. Unfortunately, after the conservation and restoration of Banteay Srei temple, Henri Marchal failed to publish the final report timely.

Due to the influence ofWorld War II, the colonial system in Southeast Asia had disintegrated, and the political situation had changed in significantly. To a certain extent, the EFEO's conservation and restoration for Angkor historic sites was also affected. But no matter how differently the politics changed, the works of EFEO never stopped. In addition, the director of CA, B. P. Groslier had still adhered to propose that the conservation and restoration activities for Angkor monuments also should be innovated and integrated both the academic researches and the engineering technology, which was based on a scientific investigation with academic research going through all the conservation works. The conservation and restoration works for safeguarding Angkor sites should be the professional knowledge of biology, petrology, meteorology and multidisciplinary, in addition to architects, engineers, archaeologists and historian with close cooperation, as to create the highest realm of art and science of the ancient site through scientific conservation activities.

第三章 茶胶寺庙山源流考

一、概述

茶胶寺（Ta Keo temple – mountain），亦称茶胶寺庙山，是吴哥古迹中最为雄伟且具有鲜明特色的庙山建筑之一。(图3-1)

图3-1 茶胶寺庙山西南侧外观

茶胶寺庙山坐落于今之柬埔寨西北部暹粒省（Siem Reap Province）首府暹粒市北郊的吴哥考古遗址公园（Angkor Archaeological Park）之内，位于大吴哥城（Angkor Thom）胜利门（Victory Gate）外东约1公里处。茶胶寺庙山西侧的环壕距暹粒河（Stung SiemReap）约280米，庙山东侧环壕之外以长度约为500米的神道与东池（Eastern Baray）相接。在茶胶寺庙山的四周，分布着诸如塔布隆寺（Ta Prohm）、班蒂克黛寺（Banteay Kdei）、塔内寺（Ta Nei）、龙蟠水池（Neak Pean）、周萨神庙（Chausay Tovada）、托马侬神庙（Thommanon）等多座吴哥时代的重要庙宇。(图3-2)

在现代高棉语的发音中，不加区别地将茶胶寺庙山称为"Preah Kèo"、"Prasat Kèo"、"Prasat Preah Kèo"、"Ta Kèo"、"Prasat Ta Kèo"、"Banteay Ta Kèo"等，而在上述称谓的发音之中皆含有音节"Kèo"，这个音节的意义原型与发音可能皆源自古代暹罗语，其中含有诸如"琉璃"、"水晶"、

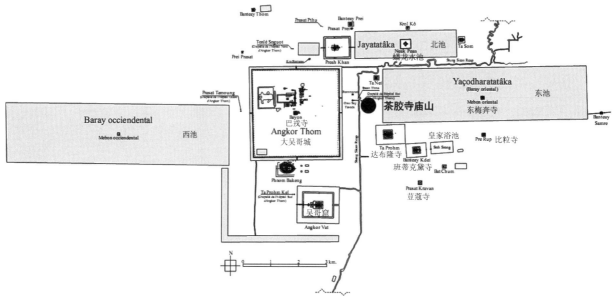

图 3-2　茶胶寺庙山的区位

"珍宝"、"宝石"、"宝物"等涵义;[1] 或因茶胶寺庙山中央五塔的塔基以上部分全部由硬质砂岩砌筑而成,质地密实且有光泽,"Ta Keo"由此而得名。需要指出的是,在茶胶寺庙山内发现的多块古代高棉碑铭(编号K275、K276、K277、K278)中曾经出现的"Hemaçringagri"或"Hemaçrṅga"等梵文音节,可译为"金角山"或"金色的山峰"之意,推测应与茶胶寺的寺名及形制的源流有关,亦可以佐证茶胶寺庙山建筑按照印度教"须弥山"意象进行选址、设计和建造的意匠。[2]

另外,由于现代高棉语中"Ta Kèo"的拼写和发音都与今之柬埔寨茶胶省(Ta Keo Province)的地名发音甚为接近,故在汉译中也通常将这座庙山之名译作茶胶寺。[3]

作为早期吴哥庙山建筑的重要遗构,茶胶寺庙山依然保留着千年前创建之初尚未完工的状态,并以其完全以砂岩砌石构筑庙山中央五塔的做法,或因其巨形砌石的使用、十字形平面且四面开敞皆出抱厦的塔殿以及环绕须弥台回廊平面格局的出现,皆开创了吴哥时代庙山建筑风气之先的形制;加之古代高棉宗教与文化赋予茶胶寺庙山独特的建筑风格和艺术魅力,使之成为见证古代高棉历史变迁的最为重要的吴哥古迹之一。

然而,关于茶胶寺庙山的历史源流及其建造背景,在相当长的时间里却始终知之甚少。自二十世纪二十年代以来,围绕茶胶寺庙山建筑源流及其变迁的探讨,诸多法国学者尝试从平面布局、建筑装饰以及碑铭内容等诸方面入手,针对茶胶寺庙山的建筑形制及其始建年代开展了甚为详缜的梳理与考证,以期追溯茶胶寺庙山的历史源流与变迁历程。其中,尤以菲利普·斯特恩(Philippe Stern)、柯瑞尔·雷慕沙(Coral – Rémusat)、维克多·格罗布维(Victor Goloubew)、乔治·赛代斯(Georges Cœdès)等人的研究考证成果最为显著。[4]

---

[1] Jacques Dumarçay. *TA KEV*: *ETUDE ARCHITECTURALE DU TEMPLE*, l'Ecole française d'Extrême – Orient EFEO: 1971, pp. 10 – 15.
[2] Lawrence Palmer Briggs, *The Ancient Khmer Empire*, New Series Volume 41, Part 1, *The American Philosophical Society*, 1951, p. 31.
[3] 另有"塔高寺"、"达高寺"等汉译名,本书则以"茶胶寺"译名为准。(著者注)
[4] Coral – Rémusat, V. Goloubew, G. Cœdès, LA DATE DU TÀKÈV, *Bulletin de l'Ecole française d'Extrême – Orient – BEFEO*, Tome XXXIV, 1934, pp. 401 – 427.

二、茶胶寺庙山建造年代的考证

1927年，菲利普·斯特恩出版专著《巴戎寺与吴哥高棉艺术之演进：关于吴哥古迹的年代学的讨论》(*LE BAYON D'ANGKOR ET L'EVOLUTION DE L'ART KHMER：Etude Et Discussion De La Chronologie Des Monuments Khmers*)[1]，在吴哥艺术史研究领域，这是一部具有里程碑意义的学术专著。斯特恩通过系统梳理古代高棉建筑装饰题材与表现技法，特别是依据山花（pediment）、门楣（lintel）和花柱（colonnette）等建筑构件及其装饰题材的演变规律与时代特征进行分类比较。

在斯特恩创建的吴哥建筑装饰类型及其年代序列中，茶胶寺庙山被归为第二期类型（The Second Style），并推测其始建年代大致介于比粒寺（Pre Rup）和巴方寺（Baphuon）之间[2]。斯特恩还通过分析比较茶胶寺庙山内花柱的形制所出现的一些极其细微的变化，进而推测其始建年代或介于比粒寺和女王宫（Banteay Srei）之间[3]。在此基础上，斯特恩还提出：在茶胶寺的塔门（Gopura）形制与皇宫（Royal Palace）的塔门之间存在诸多相似之处，以此相类似的山花形制推断茶胶寺始建年代还应大致介于女王宫和皇宫塔门之间。总之，斯特恩考证分析茶胶寺庙山的始建年代应为公元十世纪末或十一世纪初，并且晚于有明确历史纪年的始建于公元952年的东梅奔寺（Eastern Mebon，位于东池中央的小岛之上，供奉毗湿奴神）[4]。（图3-3，图3-4，图3-5，图3-6，图3-7，图3-8，图3-9）

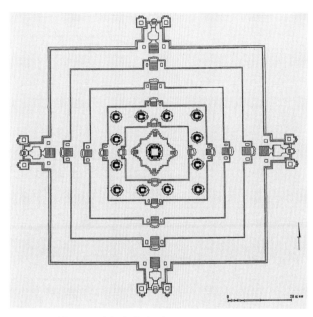

图3-3 巴空寺庙山（BaKong）平面图

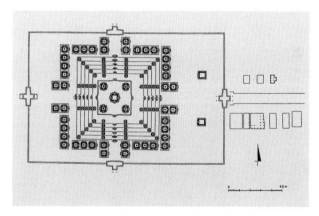

图3-4 巴肯山（Bakheng）平面图

---

[1] Philippe Stern, *Le Bayon d'Angkor et l'evolution de l'art khmer：Etude Et Discussion De La Chronologie Des Monuments Khmers*, Paris, Librarie Orientaliste Paul Geuthner, 1927.

[2] Henri Marchal, Philipe Stern：*Le Bayon d'Angkor et l'évolution de l'art khmer* In：Bulletin de l'Ecole française d'Extrême - Orient. Tome 28 N°1, 1928. pp. 293 - 305.

[3] P. Stern, *L'évolution de l'architecture khmère*, JA., avril - juin 1933.

[4] Coral - Rémusat, V. Goloubew, G. Cœdès, *LA DATE DU TÀKÈV*, Bulletin de l'Ecole française d'Extrême - Orient - BEFEO, Tome XXXIV, 1934, pp. 401 - 413.

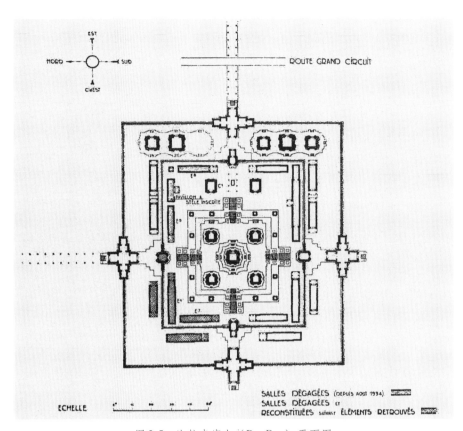

图 3-5 比粒寺庙山（Pre Rup）平面图

图 3-6 比粒寺庙山（Pre Rup）西北侧外观

第三章 茶胶寺庙山源流考

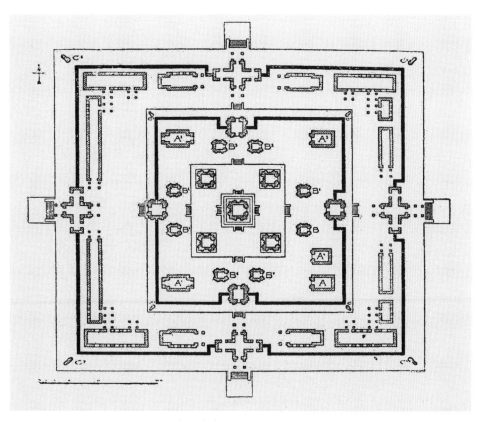

图 3-7 东梅奔寺（Eastern Mebon）平面图

图 3-8 东梅奔寺（Eastern Mebon）中央五塔外观

1934年，柯瑞尔·雷慕沙则从吴哥时代庙山建筑形制的源流及其装饰细节入手，较为详细地考证了茶胶寺庙山始建年代。吴哥时代庙山建筑形制的主要特征是在人工砌筑的巨大多层石构基台上（a stepped pyramid in a stupa），由数量不等的塔殿（Prasat）、藏经阁（Library）、长厅（Long Hall）和回廊（Gallery）等不同类型的单体建筑组合构成具有复杂宗教象征意义的建筑空间组合，其中又以坐落于须弥台之上的中央主塔构成统领整个建筑组群的核心。

例如，因陀罗跋摩一世（Indravarman I，公元877－889年在位）时期经营诃里诃拉罗耶（Hariharalaya）并创建巴空寺（Bakong）作为其国寺（Vnam Kantal），即以14

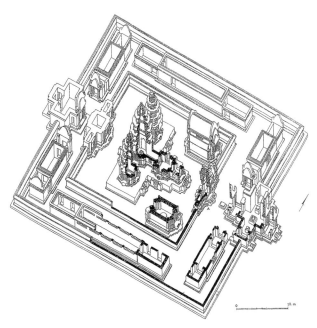

图3-9 女王宫（Banteay Srei）轴测图

座小塔（包括2座砖塔和12座砂岩石砌筑的石塔）围绕在60米×60米见方的石砌基台（15米高）四周，唯其核心却是坐落于须弥台顶面之上由砂岩砌筑而成的中央主塔。

如本书第一章所述及，吴哥时代大规模的都城建设或可追溯至因陀罗跋摩一世的继任者耶输跋摩一世（Yasovarman I，公元889－900年在位）统治期间将其都城迁移至耶输陀诃罗补罗（Yasodharapura），并在位于该城中心的巴肯山（Phnom Bakheng）上选址并经营建设巴肯寺庙山（Bakheng）作为其国寺。巴肯寺庙山矗立于自然山体之巅，以砂岩砌筑的108座小型石塔环绕在石砌须弥台及其中央五塔的四周且呈星形对称布局，吴哥时代独树一帜的庙山建筑形制至此皆已成奠基之功。

公元十世纪上半叶，阇耶跋摩四世（Jayavarman IV，公元928？－941年在位）由于不为人知的原因再次将其都城从耶输陀诃罗补罗迁移至贡开（Koh Ker），并在贡开延续着建造庙山的传统并以形制简洁的中央大塔（Prasat Thom）作为当时都城的中心，而诸多规模及形制各异的寺庙建筑组群则环绕中央大塔分布。公元十世纪中叶，罗贞陀罗跋摩二世（Rajendravarman II，公元941－944年在位）重返吴哥之际所建庙山建筑中，坐落于石构基台之上的单体建筑通常以红砖砌筑而成，东梅奔寺或比粒寺即是其中的最为典型者。

值得关注的是，自此之后，作为高棉建筑艺术新的变革元素，长厅开始成为庙山建筑组群中重要的单体建筑类型。通过东梅奔寺、比粒寺和空中宫殿（Phimanakas）[1]的平面布局的比较可以看出：环绕东梅奔寺的第一层基台对称设置了14座长厅，而比粒寺第一层及第二层基台四周则对称设置了14

---

[1] 有研究者认为，空中宫殿（phimanakas）的基台部分的起源可以追溯至公元八世纪下半叶，基台顶部坐落着木结构建筑；第二阶段是阇耶跋摩五世（Jayavarman V，公元968－1001年在位）及苏利耶跋摩一世（Suryavarman I，公元1002－1049年）在位时代所增建的回廊与塔门，由于苏利耶跋摩一世及其继任者优陀耶迪耶跋摩（Udayatityavarman II，公元1049－1066年在位）规划建设巴方寺作为国庙，因此当时的空中宫殿有可能已经成为皇宫（Royal Palace）内的一座寺庙而不再是都城的中心，而今之空中宫殿顶部的中央圣殿应是阇耶跋摩七世扩建大吴哥城时所改建，时间约在公元1190年。参见 Jacques Dumarçay. ARCHITECTURE AND ITS MODELS IN SOUTHEAST ASIA, Bangkok: Orchid Press, 2002, p. 74.

座长厅（第一层台南、北、西侧各有 2 座，第二层台的每侧各有 2 座），这些长厅多以砂岩砌筑，成为构成当时庙山建筑组群的最为重要的建筑类型之一。及至始建年代稍晚于东梅奔寺和比粒寺的空中宫殿，至迟在公元十世纪末或十一世纪初，吴哥时代庙山建筑以须弥台为核心的平面布局逐渐走向成熟，形制及其装饰也得以基本确定：即以砂岩砌筑的围合拱廊环绕作为庙山主体部分的须弥台与中央圣殿，即中央主塔。因而柯瑞尔·雷慕沙亦将这个时期的建筑艺术风格归纳为南北仓风格（Style of Kleang）且主要表现在以下方面：人工砌筑的须弥台承托塔殿、环绕四周的石构回廊、回廊覆以砌石叠涩拱以及等臂十字形且四面向外开敞的塔殿平面布局。[1]

较之茶胶寺庙山建筑组群，位于第一层基台东侧南北对称布局两座石构长厅、第二层台东侧南北对称布局的两座石构长厅、第二层基台上的围合石构回廊以及砂岩砌筑的石塔完全取代砖塔等特征，皆可以推断茶胶寺的始建年代亦应晚于东梅奔寺和比粒寺。而始建于公元十一世纪中叶的巴方寺（Baphuon），其砌石构造的回廊不仅环绕第一层和第二层基台，而且以十字交叉形的回廊布局成为连接中央五塔的重要标志，至此以吴哥窟（Angkor Wat）为典范的吴哥时代庙山建筑形制业已初现雏形。

在上述年代序列之中，茶胶寺庙山的建造年代应大致介于比粒寺和巴方寺之间。而在上述庙山建筑形制序列之中，作为一种过渡形式，砖砌屋顶、石砌叠涩拱回廊的出现以及长厅的逐渐消亡都是非常重要的标志[2]。因此，茶胶寺庙山内塔门的砖砌屋顶与回廊等遗迹愈显弥足珍贵，或是佐证茶胶寺庙山形制处于承上启下的特殊阶段的重要实物遗存。[3]

总而言之，吴哥窟所体现出庙山建筑的完美形制并非一日之功，"中央塔殿供奉国王的林伽，其他的塔殿则供奉祖先，这样的配置已经成为惯例。由此也可以看出庙山源流与祖先崇拜之间的相互影响、相互融合"。[4] 古代高棉哲匠的灵感和创造在历经近三个世纪（公元九世纪至公元十二世纪）的传承和变迁之后，吴哥时代庙山建筑的完美形制才得以全方位的呈现。

另外，柯瑞尔·雷慕沙的研究还注意到一种被她誉为"公元九世纪至十世纪上半叶罕有宝石"的装饰纹样，这为更为精准地推断茶胶寺的始建年代提供了非常有价值的线索。这种被称为"花萼"（stem with leaves）的装饰纹样，所描绘的是一种由植物花瓣和缠枝卷叶（foliage）组合而成花萼状的装饰性图案，线条流畅，造型饱满，多见于花柱和山花的雕饰。这种花萼式样的装饰纹样"也许在公元九世纪初的高棉雕塑艺术中即已初露端倪（罗莱士寺的花柱雕饰），但其形式尚较为粗糙模糊以至于难以确认。此后，类似的装饰纹样却一度沉寂消失，不仅在耶输跋摩一世时期的巴肯山、克罗姆山（Phnom Krom）、波克山（Phnom Bok）诸寺内皆未曾出现，而且在贡开时期的寺院中也颇为罕见（在巴肯山、巴克山和克罗姆山诸寺庙的山花边缘多以装饰一系列连续小拱形的三叶草作为装饰），及至罗贞陀罗跋摩时期东梅奔寺和比粒寺遗存的装饰纹样中更是难觅其踪（较之东梅奔寺和比粒寺的

---

[1] Lawrence Palmer Briggs, *The Ancient Khmer Empire*, New Series Volume 41, Part 1, *The American Philosophical Society*, 1951, p. 152.
[2] 柏威夏寺（Preah Vihear）和空中宫殿是石砌叠涩拱顶回廊的最早实例。（著者注）
[3] 维克多·格罗布维（V. Goloubew）根据巴塞特寺（Vat Baset）、埃克寺（Wat Ek）、吉索山（Phnom Gisór）以及大塔卡姆寺（Prasat Ta Kam Thom）等公元十一世纪的建筑遗构，推测茶胶寺的回廊及两座藏经阁应始建于苏利耶跋摩一世时期（Suryavarman I，公元 1002 – 1049 年在位）。
[4] Philipe Stern, *Le Temple – Montagne Khmer Le Culte Du Linga Et Le Devaraja*, Bulletin de l'Ecole française d'Extrême – Orient – BEFEO, Tome XXXIV 1934, pp. 612 – 616.

山花装饰,虽然残损严重,但是仍能够看出类似连续小拱形的三叶草装饰的端倪);然而在大约沉寂了近半个世纪之后,这种以花萼为主题的装饰纹样却突然回归于女王宫异彩纷呈的雕刻之中且随处可见"。[1]

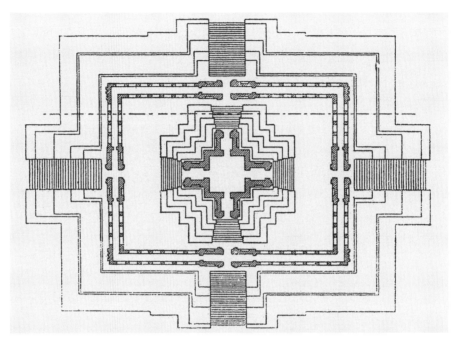

图 3-10　空中宫殿(Phimanakas)平面图

图 3-11　空中宫殿(Phimanakas)外观

---

[1] Coral‐Rémusat, V. Goloubew, G. Cœdès, LA DATE DU TÀ KÈV, *Bulletin de l'Ecole française d'Extrême‐Orient‐BEFEO*, Tome XXXIV, 1934, pp. 401–413.

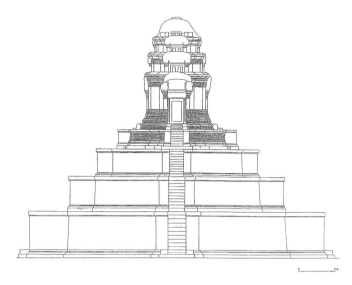

图 3-12 巴塞增空寺（Baksei Chamkrong）立面图

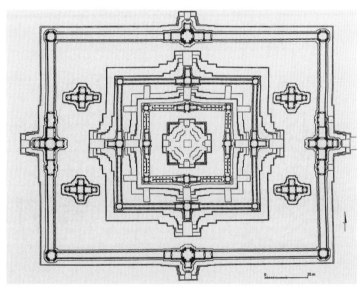

图 3-13 巴方寺（Baphuon）平面图

公元十一世纪末所建的巴方寺，尤其在寺庙塔门山花雕饰的顶端镶边处，类似的花萼与卷草纹样的组合形式依然非常流行，但在巴方寺之后却随即消逝无踪。在茶胶寺保存数量不多的雕饰之中，类似的花萼纹样亦是频频出现。纵然，由于茶胶寺庙山建造工程没有最终完成，其建筑装饰也未能最终呈现出"在最后一刻，这座建筑像是为了节日而盛装，卷叶构成的花环从檐口垂下"[1]的国寺盛况，但是柯瑞尔·雷慕沙依然认为，茶胶寺庙山的建造年代，应是吴哥建筑装饰艺术重要转型期的先导。这主要表现在以下两个方面：一方面，茶胶寺庙山极为繁复地运用花萼等植物图案作为装饰的主题，造型流畅饱满的枝叶和花萼充斥着诸如山花、假层、须弥台等建筑装饰的重要部位；而另一方面，源

---

[1] "The garlands were ephemeral, it was desired to make them last longer by rooting ways of the shrines probably also solidify an ephemeral decoration, that of temporary structures erected in front of the temples in India." Jacques Dumarçay and Pascal Royère, *Cambodian Architecture*: *8th to 13th centuries*, Brill Press, 2001, p. 27.

自宗教或神话故事的雕刻题材在茶胶寺庙山内却极少出现，只有摩羯（Makara）和蛇神（Naga）的造型偶尔闪现于曲线繁复的植物枝蔓之间。

图3-14　推测茶胶寺庙山年代的"花萼"纹样

因此，正是由于这种花萼式样的装饰题材带有明显的时代演变特征，或可将其作为把上述几座建筑关联起来的重要依据。在这其中，女王宫、东梅奔寺及比粒寺可以归为第Ⅰ期风格，皇宫塔门则可归为第Ⅱ期风格。[1]颇为值得注意的是，皇宫塔门的花柱皆由一系列琐细且等尺寸的缠枝卷叶纹样构成装饰母题，无论尺度抑或组合方式皆与女王宫不同，应该是此类花萼纹样演变发展的一个重要标志。据此，在上述几座寺庙建筑年代序列的最上端，首先可以确定比粒寺与具有明确纪年的东梅奔寺（公元952年）之间的关系，接下来的应该是大约公元969年创建的女王宫；而纵观皇宫塔门门楣和花柱纹样产生的细微变化，应该将其归于装饰风格的转折时期。而较之茶胶寺的情况，其装饰纹样基本延续女王宫的风格特征，尚未出现诸如皇宫塔门的风格转折或改变的痕迹，或可推测茶胶寺庙山的始建年代应大致介于女王宫（公元969年）和皇宫塔门（公元1011年）之间。

---

[1]　同上注。

三、茶胶寺庙山碑铭研究述略

茶胶寺庙山的碑铭遗存共计六块，编号分别为 K275（Ⅰ）、K276（Ⅱ）、K277（Ⅲ）、K278（Ⅳ）、K535（Ⅵ）、K536（Ⅴ），主要镌刻于庙山东侧的东内塔门、东外塔门的门道抱框石之内侧。关于这些碑铭的时代及内容，学界历来争讼纷纭。例如，来自法国远东学院的著名高棉碑铭学家芬诺（Louis Finot）考证认为茶胶寺庙山碑铭应为公元九世纪末叶耶输跋摩一世时期所刻，但是艾莫涅尔（Etienne-François Aymonier）则以茶胶寺碑铭中所出现的"Çivaçakti（Śivācārya）"为线索[1]，推断其雕刻年代应为公元1047年，并推测"Çivaçakti（Śivācārya）"可能是当时一位地位非常显赫的婆罗门，甚至也有可能就是茶胶寺庙山碑铭的作者。

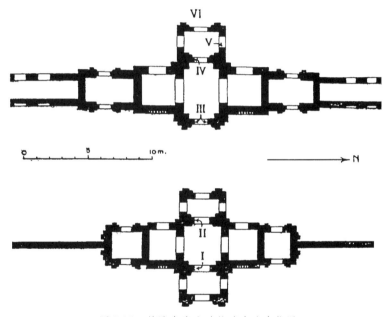

图3-15　茶胶寺庙山碑铭遗存分布位置

然而，对于茶胶寺庙山碑铭内容的释读和考证，则以赛代斯的研究最为显著[2]。赛代斯首先提出，茶胶寺庙山碑铭的雕刻文字系古代高棉文并非梵文，且似与苏利耶跋摩一世（Suryavarman Ⅰ）时期的柏威夏寺（Preah Vihear）碑铭（编号K382）多有关联，因而其雕刻时代应远远晚于耶输跋摩一世时期[3]；另外，他还对芬诺考证的罗贞陀罗跋摩二世（Rajendravarman Ⅱ）至优陀耶迪耶跋摩二世

---

[1] G. Cœdès. Etienne-François Aymonier (1844-1929). In: Bulletin de l'Ecole française d'Extrême-Orient. Tome 29, 1929. pp. 542-548.

[2] B. Dagens, *Étude iconographique de quelques fondations de l'époque de Suryavarman Ier, Arts asiatiques XVII*, Paris 1968, p. 175.

[3] G. Cœdès: *Inscriptions du Cambodge IV*, Paris, 1952, pp. 152-160. 赛代斯编著的《柬埔寨古代碑铭全集》（Inscriptions du Cambodge, Hanoi/Paris, EFEO-Textes et documents sur l'Indochine, 3）全书共分为8卷，在1937年至1966年的近三十年时间里编纂完成并陆续出版。

图3-16 茶胶寺庙山碑铭（编号K.275）局部

图3-17 茶胶寺庙山碑铭（编号K.278）局部

（Udayadityavarman）时期六块碑铭的结论提出了质疑。[1]

---

[1] L. 芬诺考证茶胶寺庙山碑铭K275的时代大约是公元九世纪末，而其余五块碑铭时代皆为公元十世纪中叶至十一世纪末期。参见 Coral – Rémusat, V. Goloubew, G. Cœdès, LA DATE DU TÀ KÈV, *Bulletin de l'Ecole française d'Extrême – Orient – BEFEO*, Tome XXXIV, 1934, pp. 401–413.

第三章　茶胶寺庙山源流考

在对茶胶寺编号为K275[1]、K276[2]、K277、K278[3]等四块重要碑铭进行详细考证之后，赛代斯再次首先提出，上述四块碑铭皆刻于苏利耶跋摩一世时代，而其中在编号K275和编号K278的碑铭中皆提及的"ici"一词，应与茶胶寺庙山建造过程中曾经遭受的一次由于雷击而引起的火灾相关，而在碑铭中还叙述了当时为消除这个不详征兆而举行的一次赎罪仪式，并且在仪式之后购买些新的石材和大象继续工程建设。因而赛代斯借此推断茶胶寺庙山的始建年代不应晚于公元1007年。[4]

正如赛代斯所言："（茶胶寺庙山K275碑铭）最后六行残损较少且意味深长，这座寺庙让我们不应该忽略对湿婆卡亚（Çivàcârya）和瑜伽阇罗若盘迪陀（Yogiçvarapanditâ）这两个婆罗门的关注，他们虽然来自不同的婆罗门家族，但是仅就茶胶寺庙山的建造工程而言，他们可能具有某些重要的共同特征，他们似乎皆与金山或金角山（Hemagiri或Hemaçrngagiri）相关。首先，他们都参与建立或完成了在金角山上称之为'盘卡素拉'（pancaśula，五座塔殿之意）的工程；其次，他们都曾在茶胶寺庙山建造过程中出任过级别很高的督查工程的官员。另外，金角山（Hemaçrngagiri）可能意指须弥山（Meru mountain），其建筑表现形式即是茶胶寺庙山建筑的核心——须弥台上的中央五塔。茶胶寺庙山的碑铭告诉我们，'金角山'（Hemaçrngagiri）的建造肇始于阇耶跋摩五世（Jayavarman V，公元968－1001年在位）时期，而完成于苏利耶跋摩一世（Suryavarman I，公元1002－1049年在位）时期。因此，从茶胶寺庙山碑铭内容亦可以确认的始建年代与柯瑞尔·雷慕沙的研究是一致的。"[5]

---

[1] 茶胶寺庙山内编号为K275的梵文碑铭系根据赛代斯法文译本转译，译文如下："瑜伽阇罗若盘迪陀（Yogiçvarapandita）或许是作为苏利耶跋摩（Suryavarman）的婆罗门或上师（guru），也有可能是一位协助国王建设金角山（hemagiri）得力的执行者。他不仅创建金角山的五座石塔（pancaçula hemagiri）并将罗贞陀罗亚娜（râjendrayâna）敬献给湿婆神，而且还创建了一座石头赛拉卡拉纳（ciracarana）、两座神牛（Nandi）雕像、两座饕餮（Kala）雕像、两座狮子雕像并此敬献给毗湿奴神。公元1002年，适逢苏利耶跋摩的统治时期，无尚地赞美婆罗门阇那帕达（Janapadâ）嫁给了婆罗门瑜伽阇罗若盘迪陀（Yogiçvarapandita）的一位信徒，并在耶输陀诃罗补罗城（Yogiçvarapura）的东部为其养育后代。咒语……"参见Coral-Rémusat, V. Goloubew, G. Cœdès, LA DATE DU TÀ KÈV, Bulletin de l'Ecole française d'Extrême-Orient-BEFEO, Tome XXXIV, 1934, pp. 401-413.
[2] 茶胶寺庙山内编号为K276的古代高棉文碑铭系根据赛代斯法文译本转译，译文如下："无比珍贵的礼物敬献给林伽罗补、安纳·达尼利（Anne Danle），瓦伽·艾克·占佩赛瓦拉（Vak Ek Câmpeçvara），卡姆达瓦特·迪科（Kamdvat Dik），斯瑞·阇耶喀塞塔（Cri Jayaksetra），威南·普瓦（Vnam Pûrvva），占婆让司（Chpâr Ransi），以及金角山（Hemaçrnga）。咒语……"参见Coral-Rémusat, V. Goloubew, G. Cœdès, LA DATE DU TÀ KÈV, Bulletin de l'Ecole française d'Extrême-Orient-BEFEO, Tome XXXIV, 1934, pp. 401-413.
[3] 茶胶寺庙山内编号为K278的梵文碑铭系根据赛代斯法文译本转译，译文如下："阇耶跋摩五世（Jayavarman V）任命湿婆卡亚（Çivâcârya）作为建设金角山（Hemaçrngagiri）的工程督造官；公元1007年，苏利耶跋摩一世统治期间，在诃里诃罗耶补罗（Hariharalayapura）有四座雕像是为Patit（神）的儿子湿婆闻度（Çivavindu）所立，他的名字被称为克斯廷德罗帕卡拉帕（Ksitindropakalpa）。"参见Coral-Rémusat, V. Goloubew, G. Cœdès, LA DATE DU TÀ KÈV, Bulletin de l'Ecole française d'Extrême-Orient-BEFEO, Tome XXXIV, 1934, pp. 401-413.
[4] G. Cœdès, Inscriptions du Cambodge vol. IV, Paris, l'Ecole française d'Extrême-Orient, 1952, pp. 152-160.
[5] 参见Coral-Rémusat, V. Goloubew, G. Cœdès, LA DATE DU TÀ KÈV, Bulletin de l'Ecole française d'Extrême-Orient-BEFEO, Tome XXXIV, 1934, pp. 401-413.

四、茶胶寺庙山历史沿革

如前文所述，通过斯特恩、柯瑞尔·雷慕沙、格罗布维、赛代斯等法国学者分别从茶胶寺庙山的建筑装饰、平面布局、碑铭等几方面探讨其建造年代，从而基本确定了茶胶寺庙山创建于公元十世纪末叶阇耶跋摩五世（Jayavarman V，公元968－1001年）在位期间，庙山的建造工程直至公元十一世纪初苏利耶跋摩一世（Suryavarman I，公元1002－1050年）统治时期可能仍处于兴建之中，但却由于某种尚未确定的原因导致庙山工程未能最终完工。

公元802年，阇耶跋摩二世在摩诃因陀罗山（Mount Mahendraparvata，今之吴哥东北约40公里的荔枝山地区）通过举行提婆罗阇仪式[1]（Devarâja），通过对象征国王神性的林伽的崇拜仪式，使之成为古代高棉国王权力与统治的最高象征，借此开创了古代高棉黄金时代的历史滥觞。

在数度迁移都城之后，公元944年，高棉帝国的国王罗贞陀罗跋摩二世再次将其国都从贡开迁回旧都耶输陀河罗补罗，"重建荒芜已久的圣城耶输陀河罗补罗，恢复其壮丽与魅力，装饰着黄金和宝石的殿宇，犹如因陀罗建在大地上的宫殿"，[2]并将位于耶输陀河罗补罗城中心的国寺巴肯寺废弃，并在巴肯寺南北方向轴线与东池东西方向轴线的交汇之处创建空中宫殿作为新的国寺以奉祀象征"神王合一"的提婆罗阇（Devarâja）。[3]

公元968年，阇耶跋摩五世承继大统，登基之初年幼的国王或许暂由其亲属或高级官员辅佐摄政[4]，因此最初可能仍以空中宫殿作为其国寺[5]。大约在公元975年左右，阇耶跋摩五世开始规划建设茶胶寺庙山作为新的国寺。在其继任者优陀耶迪耶跋摩一世（Udayadityavarman I，约公元1001－1002年在位）时期的短暂停滞之后，茶胶寺庙山的建设迎来了新的国王苏利耶跋摩一世。在历经近十年与军阀政权阇耶毗罗跋摩（Jayaviravarman，公元1002－1011年）之间的征伐较量之后，来自高棉帝国西北部的世族首领苏利耶跋摩一世全面攻占吴哥地区并自立为王。通过对茶胶寺庙山形制及其碑铭内容的分析和解读，推测或许由于改朝换代过程中所涉及的诸多变革，皆成为促进苏利耶跋摩一世统治时期建筑及艺术革故鼎新的重要因素，而处于变革时代的茶胶寺庙山或为其中的

---

[1] "与其说国王是作为行政管理者出现，不如说他更像一个活在世上的神。他那筑有城墙掘有壕沟的都城是宇宙的缩影。都城的中央以一座象征须弥山的庙山为标志，山顶是借助于婆罗门从湿婆神那里领受的国王之林伽——提婆罗阇（代表王权的林伽）。我们不清楚是否这座包含着王权本质的林伽，经相继的这几位国王统治的时期，始终都是独一无二的？反过来说，国王在自己登基时举行仪式使之神圣化了的、刻有他们尊号的林伽，与提婆罗阇是否即为同一座林伽，抑或不是。每一位有足够时间和财力的国王，都在他们都城的中央修建自己的庙山。"参见参见G. 赛代斯著，蔡华等译：《东南亚的印度化国家》（G. Cœdès, LES ÉTATS HINDOUISÉS D'INDOCHINE ET D'INDONÉSIE），北京：商务印书馆2008年版，第206－207页。

[2] G. Cœdès, *Inscription of Bat Chum*, Journal of Asia, Volume 10, Issue VIII, 1908, pp. 213－234.

[3] 在古代高棉社会中，国寺建设、新王加冕、提婆罗阇仪式，三者相互关联，成为王权统治最为重要的象征。

[4] 关于阇耶跋摩五世统治时期的碑铭记载，可以参见Lawrence Palmer Briggs, *The Ancient Khmer Empire*, New Series Volume 41, Part 1, The American Philosophical Society, 1951, pp. 141－144.

[5] 根据贡波（Kok Po）的碑铭记载，公元978年，阇耶跋摩五世曾经委派一位称为湿婆卡耶（Sivacarya）的婆罗门负责空中宫殿的修建工程以监督工程进度及质量，更重要的是由他来筹集工程建设所需的资金。另外，根据在圣恩克赛寺（Preah Enkosei）内发现有明确纪年（公元968年，阇耶跋摩五世统治时期）碑铭"在月圆之夜，国王下令寺院将其所有的奴隶、物产以及财产统一列入清单"的记载或可了解当时的庙山建设的经济背景。转引自Jacques Dumarçay and Pascal Royere: *Cambodian Architecture: Eighth to Thirteenth Centuries*. Brill Academic Pub, 2001, p. 7.

最典型者。

　　根据茶胶寺庙山内编号为K275的梵文碑铭，可以大致了解苏利耶跋摩一世时期茶胶寺庙山历史背景的诸多细节：一位被称为瑜伽阇罗若盘迪陀（Yogiçvarapandita）的婆罗门，或许即是茶胶寺庙山工程的实际负责人；茶胶寺庙山供奉的主尊是湿婆神以及毗湿奴神；来自阇那帕达（Janapadâ）婆罗门家族的一位女性成员成为瑜伽阇罗若盘迪陀的信徒，而她奉命嫁给瑜伽阇罗若盘迪陀的另外一位信徒卡萨婆（Kesava），也有可能是苏利耶跋摩一世的外戚或祭司。[1]而根据在塔拉蓬如寺（Prasat Trapang Run）内所发现的碑铭则记载，军阀阇耶毗罗跋摩（Jayaviravarman）曾经莅临空中宫殿建造工程现场，而当时空中宫殿建造工程的督造官湿婆卡亚（Çivàcârya）可能在苏利耶跋摩一世时期继续承担了在金角山上建造盘卡素拉（pancaśula，五座塔殿）工程，可能即为茶胶寺庙山的建造工程。[2]

　　另外，对于今之所见茶胶寺庙山未完工状态的解释大多源自赛代斯对于茶胶寺庙山内编号为K275、K278的两方碑铭的研究。苏利耶跋摩一世时期的茶胶寺庙山建设工程由于遭遇一次雷击引发火灾，国王可能为祛除不祥之兆而举行了一个重要的救赎仪式。尽管如此，好大喜功的国王最终还是丧失了继续茶胶寺庙山建造的兴趣[3]。

　　事实上，在苏利耶跋摩一世之后，建筑主体结构业已完成的茶胶寺庙山并未完全遭到废弃，可能在很长一段时间内仍然作为寺院被继续使用。在二十世纪二十年代所进行的考古清理中，就曾经出土过数目不菲的吴哥窟乃至巴戎寺风格的雕像，甚至还包括有阇耶跋摩七世时期的碑刻等。总之，至于茶胶寺庙山最终废弃的确切时间及其后世的使用状况，历来鲜有踪迹可循，诸多情形皆不得而知。

---

[1] B. Dagens: *Étude iconographique de quelques fondations de l'époque de Suryavarman Ier*; *Arts asiatiques XVII*, Paris 1968, p. 175；另可参见 Lawrence Palmer Briggs, *The Ancient Khmer Empire*, New Series Volume 41, Part 1, *The American Philosophical Society*, 1951, p. 149.

[2] Lawrence Palmer Briggs, *The Ancient Khmer Empire*, New Series Volume 41, Part 1, *The American Philosophical Society*, 1951, p. 152.

[3] G. Cœdès: *Inscriptions du Cambodge IV*, Paris, 1952, pp. 152–160.

五、茶胶寺庙山研究保护简史

自十九世纪六十年代起，随着法国在印度支那地区殖民体系的建立与扩张，柬埔寨及其吴哥古迹更全面、更深入地为西方社会所知，在此期间著名的探险家和旅行者包括：杜达特德·拉格雷（Ernest Doudart de Lagrée），鲁内特·德·拉云魁尔（Étienne Lunet de Lajonquière），以及埃廷内·艾莫涅尔（Étienne Aymonier）等。其中以1863年法国政府驻柬埔寨代表E. 杜达特德·拉格雷组建的湄公河考察团的影响较大。通过对吴哥地区历经多年的考察之后，考察团成员陆续绘制完成了主要遗迹的分布图，并编撰出版题为《印度支那探索之旅》（Voyage d'exploration en Indochine）的考察报告。（图3-18）

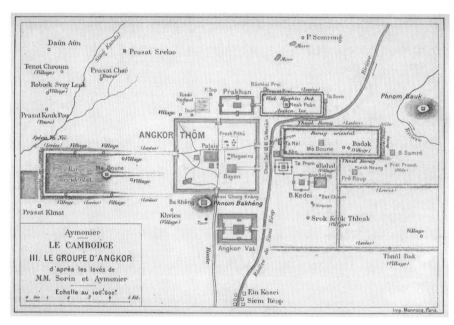

图3-18　艾莫涅尔（É. Aymonier）绘制的吴哥地区古迹分布图（1904年）

1863年，杜达特德·拉格雷（Ernest Doudart de Lagrée）在其绘制的地图上标出了茶胶寺庙山的位置，此为近代以来关于茶胶寺庙山最早的发现与记录。1873年，弗朗西斯·加内尔（Francis Garnier）《印度支那探索之旅》报告中则对茶胶寺进行了简略的描述[1]。湄公河考察团的另一位主要成员德拉蓬特（L. Delaporte）则于1880年发表了湄公河考察团系列考察报告之《柬埔寨之旅：高棉的建筑》（Voyage au Cambodge：L'Architecture Khmer），由于较为真实地表现了十九世纪末吴哥古迹的风貌，由德拉蓬特绘制的吴哥古迹系列铜版画至今仍然享有很高的声誉和知名度，在此书中首次发表了茶胶寺庙山的总平面图，系由湄公河考察团的测量员拉特（Ratte）于1873年所绘制完成。1883年，莫拉（J. Moura）编纂出版两卷本的《柬埔寨王国》（Le Royaume de Cambodge），书中通过实地测量茶胶寺庙山的部分尺寸，较之此前发表的资料则更为全面详细地对庙山的保存状况进行了描述记录，并对庙山

---

[1]　参见本书第二章相关内容。

的名称的来源进行了初步的梳理与考证；另外，莫拉还曾注意到茶胶寺庙山的东西方向轴线有所偏移的现象，只不过他所标注的偏移尺寸很显然误差过大，而且部分尺寸数据也多有谬误。1890年，在卢森·弗内日奥（Lucien Fournereau）与雅克·波日彻（Jacques Porcher）共同编著的《柬埔寨北部地区高棉古迹艺术与历史的研究》（*Les Ruines d'Angkor, étude artistique et historique sur les monuments kmers du Cambodge siamois*）中，简明扼要地引述了莫拉的研究和观点，但遗憾的是并未对其所引用的数据资料的错误进行更正或说明。

1904年，艾莫涅尔在对茶胶寺庙山的碑铭进行初步研究的过程中，重新绘制了一张较为简略的茶胶寺庙山平面示意图。1911年，拉云魁尔（Étienne Lunet de Lajonquière）在其编著的《柬埔寨古迹名录》（*Inventaire descriptif des monuments du Cambodge*）中亦将茶胶寺庙山正式收录在内，并将之编号为第533号，可能是由于当时茶胶寺庙山尚处于密林荫翳之中，加之调查及测量条件所限，所以《柬埔寨古迹名录》的记录与描述并不十分准确，错误或存疑之处颇多。即便如此，《柬埔寨古迹名录》中所绘制的茶胶寺庙山总平面图和剖面图亦是当时最为重要的实测记录。1920年，德拉蓬特（L. Delaporte）在《柬埔寨古迹名录》提供的茶胶寺总平面图基础上，再次结合拉特（Ratte）绘制的平面图，对茶胶寺庙山的平面图进行了较大的修改和更正；经过此番修订校正之后，除庙山的第一层基台建筑及其围墙的交接关系等细节尚须完善之外，这在当时已是质量及准确度最高的实测平面图；另外，虽然德拉蓬特曾经注明他仅凭拉特（Ratte）和劳埃德瑞奇（Loedrich）的研究报告所进行的复原研究是很不完整的，但这毕竟是首次对于茶胶寺庙山所开展的复原研究并且绘制完成了一幅较为完整的复原立面图。（图3-19，图3-20，图3-21）

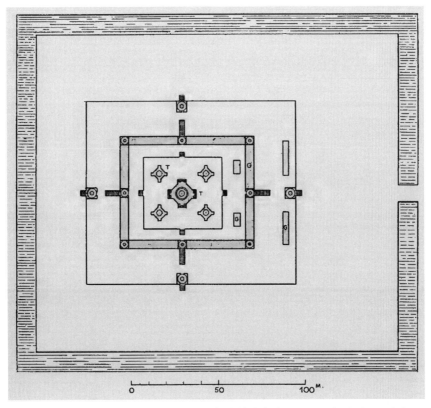

图3-19　艾莫涅尔（É. Aymonier）绘制的茶胶寺庙山平面示意图（1904年）

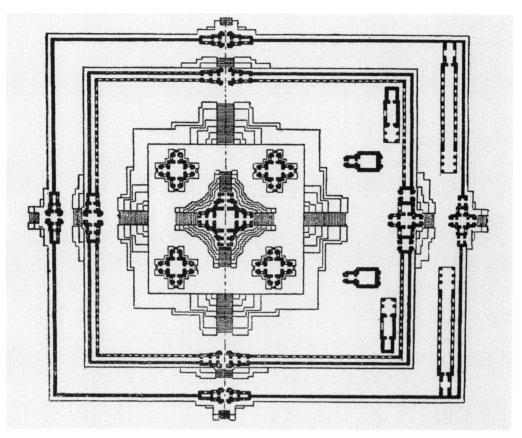

图 3-20 拉云魁尔（Lajonquière）绘制的茶胶寺庙山平面图（1911 年）

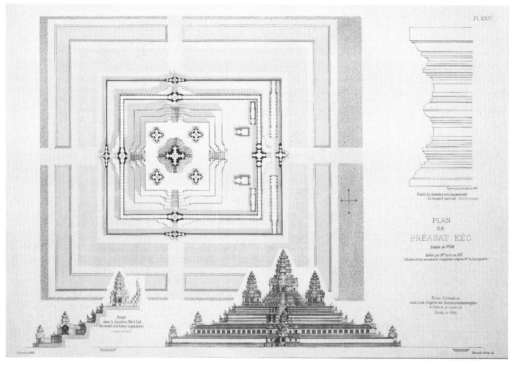

图 3-21 德拉蓬特（L. Delaporte）完成的茶胶寺庙山平面图及其复原图（1920 年）

第三章 茶胶寺庙山源流考

1920年，法国远东学院正式明确提出建设"吴哥考古遗址公园"的计划，主要针对吴哥古迹遗址范围内多数寺庙倾圮、植被覆盖的状况，通过古迹保护区规划设置，分阶段实施吴哥古迹的保护与修复项目。如本书第二章所述，当时的吴哥古迹保护处（Conservation des monuments d'Angkor）在亨利·马绍尔（H. Marchal）的具体指导下，与建筑师巴特尔（C. Batteur）合作开始着手对吴哥地区的寺庙遗址进行系统清理，该项工作一直持续至1923年末（清理工作在当时的工作人员的日志中进行了详细的记载）；其中，在对茶胶寺庙山进行清理的过程中，不仅将覆盖于庙山建筑上的大量积土和植被清理运出，并对散落石构件进行了简单的分类、整理及摆放工作。今之所见茶胶寺庙山四周及各层基台顶部地面被摆放整齐的石构件即是当年清理工作的结果。

与此同时，除清理工作之外，还针对茶胶寺庙山建筑的危险部位进行了加固，采取的方法有两种：一种是用钢筋混凝土制成的柱子支顶上部行将坍落的结构，或支撑倾斜的墙体；另外一种是用扁铁将断裂变形的石构件箍紧，防止其发生进一步的变形和损坏；与茶胶寺庙山的清理加固过程同步进行的考古调查工作也取得了不少收获；例如，首先在茶胶寺庙山的东南方向发现了一座与茶胶寺中央五塔结构特征类似的砂岩砌筑的石塔，该塔在马绍尔绘制的吴哥古迹平面图和特洛威（G. Trouvé）绘制的该地区的平面图上皆被编号为第54号。1927年，帕蒙蒂埃（H. Parmentier）亦对茶胶寺庙山进行过一次较小范围的测绘与记录，但是其最终的调查报告并未正式发表，唯余大量手稿及草图存世。1929年4月，马绍尔及其助手还在东池的西堤上调查发现了一座位于茶胶寺庙山东西方向轴线上的砌石平台遗址；而位于茶胶寺庙山东北方向的其他较小的遗迹则迟至1936年才被发现并标记出来。（图3-22，图3-23，图3-24）

如前所述，围绕茶胶寺庙山建筑的始建年代及其源流，自二十世纪二十年代以来，诸多法国学者尝试从平面布局、建筑装饰以及碑铭内容等诸方面入手，针对茶胶寺庙山建筑形制及其始建年代开展了甚为详缜的梳理与考证，以期追溯茶胶寺庙山的历史源流与变迁历程。其中，尤以菲利普·斯特恩（Philippe Stern）、柯瑞尔·雷慕沙（Coral-Rémusat）、维克多·格罗布维（V. Goloubew）、赛代斯（G. Cœdès）等人的研究考证及成果最为显著。[1]

1934年，柯瑞尔·雷慕沙、格罗布维、赛代斯三位学者合作在法国远东学院学报发表《茶胶寺年代考》（*LA DATE DU TÀKÈV*, *Bulletin de l'Ecole française d'Extrême-Orient-BEFEO*, Tome XXXIV, 1934）通过对茶胶寺庙山的形制、装饰、碑铭等诸方面的分析研究，茶胶寺庙山的始建年代得以基本明确。这篇论文的发表堪称茶胶寺庙山研究乃至吴哥古迹学术研究的典范之作，围绕茶胶寺庙山的始建年代的考证，来自不同研究领域的学者皆以其各自不同的学术视角和研究方法共同探讨，互为佐证，并最终取得了令人信服的研究结论，而这种研究的范式对于吴哥古迹的研究与保护影响至深。1952年，赛代斯继续研究茶胶寺庙山的碑铭，将其全文译为法文并收录于其编著的《柬埔寨古代碑铭全集》之中。

二十世纪四十至五十年代，莫瑞斯·格莱兹（Maurice Glaize）和马绍尔（Henri Marchal）在各自关于吴哥古迹的论著[2]中都简要介绍了茶胶寺庙山，皆是以拉特（Ratte）所绘制的平面图为底稿进行了

---

[1] Coral-Rémusat, V. Goloubew, G. Cœdès, LA DATE DU TÀKÈV, *Bulletin de l'Ecole française d'Extrême-Orient-BEFEO*, Tome XXXIV, 1934, pp. 401-427.

[2] Maurice Glaize: *Les monuments du groupe d'Angkor*, A. Portail, Paris, 1sted. 1944, 2nd ed. 1948, 3rded. 1963, 4thed. 1993; Henri Marchal: *L'architecture comparée dans l'Inde et dans l'Extrême-Orient*, Paris, 1948; *Le décor et la sculpture khmers*. Paris, 1951; *Le décor et la sculpture khmers*, Paris, 1951.

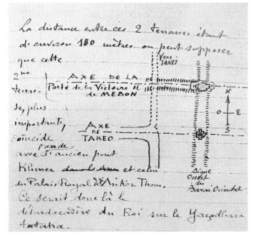
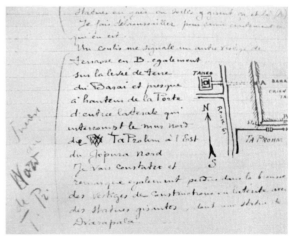

图 3-22　1920－1923 年法国远东学院茶胶寺庙山清理发掘历史照片
以及当时调查记录手稿　© EFEO Archives

图 3-23 十九世纪二十年初法国远东学院吴哥古迹保护处茶胶寺清理发掘现场历史照片之一　© EFEO Archives

图 3-24 十九世纪二十年初法国远东学院吴哥古迹保护处茶胶寺清理发掘现场历史照片之二　© EFEO Archives

重新绘制，保留了原图中庙山在东西方向轴线上的偏移，并改正了一层平台附属建筑的位置。与此同时，小格罗斯利埃（B. P. Groslier）还曾绘制完成一张茶胶寺庙山的轴测图，但在这张轴测图中他未将庙山第二层基台上的几座附属建筑进行注明。

时任法国远东学院院长的简·费里奥扎（Jean Filliozat）与吴哥古迹保护护处负责人小格罗斯利埃（B. P. Groslier）委托建筑师雅克·杜马西（Jacques Dumarçay）先后于1967年、1969年两次对茶胶寺庙山进行实地调查与测绘。1970年，法国远东学院出版了杜马西关于茶胶寺庙山的调查报告《茶胶寺建筑研究》（*Ta Kèv：étude Architecturale du Temple*），对茶胶寺庙山进行了准确而详细的描述和记录，并附有大量实测图，这是在中国政府援助吴哥古迹保护工作队（Chinese Government Team for Safeguarding Angkor – CSA）开始承担茶胶寺保护修复工程之前对茶胶寺庙山所进行的一次最为完善而翔实的实测记录工作。（图3-25）

图3-25　雅克·杜马西（Jacques Dumarçay）绘制的茶胶寺庙山立面图

自二十世纪九十年代，柬埔寨内战结束伊始，在联合国教科文组织的帮助下，柬埔寨政府有关部门也曾经组织力量对茶胶寺庙山进行过一些局部的场地清理和临时保护加固工作。例如，在第二层基台转角坍塌部位所形成的空隙处砌筑红砖与混凝土支架以支撑上部的角楼，局部还用木结构斜撑对倾斜的墙体和松散结构进行临时支撑。

援助柬埔寨保护吴哥古迹的国际行动始于二十世纪九十年代。包括中国在内的十几个国家或国际组织共同参与吴哥古迹保护研究工作。2006年4月，中、柬两国政府正式确认茶胶寺庙山作为中国政府援助柬埔寨吴哥古迹保护的二期项目[1]。自2006年末开始，在国家文物局、柬埔寨吴哥古迹保护与发展管理局（APSARA Authority）、联合国教科文组织吴哥古迹保护协调委员会（ICC – Angkor）的具体指导下，中国文化遗产研究院及中国政府援助吴哥古迹保护工作队作为茶胶寺保护修复工程的实施单位，确定了茶胶寺庙山保护修复工程"科学研究贯穿保护修复工程全过程，排险加固，局部维修与全面修复相结合"的总体指导思想，并先后组织各类专业技术人员开展实地调查、测绘与勘察，工作主要涉及建筑、考古、测绘、结构工程、岩土工程、保护科学等学科，先后完成各类研究报告与技术文件三十余项，这是迄今为止针对茶胶寺庙山所开展的最大规模的综合研究与保护修复工作。[2]

2010年11月，茶胶寺保护修复工程举行了开工典礼。根据两国政府换文规定，茶胶寺保护修复工程主要包括建筑结构加固、建筑材料修复及考古研究等三个部分。项目总经费为4000万元人民币，整个保护修复工程计划将于2018年结束。

---

[1] 在联合国教科文组织的积极协调下，中国国家文物局和柬埔寨吴哥古迹保护与发展管理局（APSARA Authority）密切合作，由中国文化遗产研究院（中国文物研究所）主持勘察、设计及施工。自1998年至2008年末，历时十年，完成了中国政府援助柬埔寨吴哥古迹保护（一期）周萨神庙保护修复工程，基本恢复了寺庙原有建筑格局与艺术风貌，赢得了柬埔寨政府、国际组织以及国际同行的一致赞誉。（著者注）

[2] 本书的编著即源自中国文化遗产研究院承担的《柬埔寨吴哥古迹茶胶寺保护修复工程综合研究》之研究子项《柬埔寨吴哥古迹茶胶寺建筑形制与复原研究》、《柬埔寨吴哥古迹茶胶寺保护修复史研究》、《柬埔寨吴哥古迹茶胶寺建筑调查、测绘与记录》等课题的初步研究成果。（著者注）

Synopsis Ⅲ Historical Context of Ta Keo Temple-mountain

1. Introduction

For the historical context of Ta Keo temple-mountainin addition to its construction background, is always poorly clearance in quite long time, but many French scholars had started to study the aspects of architectural layout, decoration and inscriptions of Ta keo temple-mountain since the first half of twentieth century. Especially, Philippe Stern, Gilberte de Coral-Rémusat, Victor Goloubew and Georges Cœdès with their dating results are very detailed and the most significant that concerning the origins and changes of Ta keo temple-mountain such as architectural forms, decorations and inscriptions.

In spite of the unknown reasons for Ta Keo temple-mountain remaining unfinished, for it was abandoned soon after the start of its ornamentation, based on the remaining fragments, the currently accepted view is that Ta Keo temple-mountain dates to the end of tenth century AD and the early years of the eleventh century AD. The inscriptions in Sanskrit engraved on the entrance jambs of the inner eastern gopura, relating to donations made to the temple (but not to its foundation) indicate a date of 1007 AD.

"Ta Keo" is used for spelling the name of the temple under discussion. In fact, the temple-mountain has different Cambodian names with indistinguishable meanings: Preah Kèo, Prasat Kèo, Prasat Preah Kèo, Takèo, Prasat Takèo, Banteay Takèo. The last letter "o" of Kèo could also be replaced by the semi-vowel "v", as in Jacque Dumarçay's treatise Ta Kèv (1971). According to the French scholars, Ta Keo means the ancestor of Keo, while the word "Keo" with its Siamese origin signifies glass, crystal, jewellery, gem, treasure or similar. Moreover, the temple might be named after the visual quality imbued by its greywacke construction, which gives the upper towers an impressive look as if they are "darker, denser, and more lustred than others" even offering a glazed appearance.

The inscription in Sanskrit with its serial number K277 is the most important of historical records on the naming of Ta Keo temple-mountain. According to one interpretative study of inscription number K277, the true meaning of the two words "Hemadringagiri" and "Hemagiri" is "Mountain of the Golden Horn" which is a reference to the symbolism of Phimeanakas and Ta Keo temple-mountain in the ancient Khmer spiritual world.

2. Dating of Ta Keo Temple-mountain

In 1927, Philippe Stern publishedLE BAYON D'ANGKOR ET L'EVOLUTION DE L'ART KHMER: Etude Et Discussion De La Chronologie Des Monuments Khmers, which is a landmark academic monographs on Angkor art history. Stern had systematically classification of decoration theme of pediment, lintel and colonnette as well as analysis of the evolution and characteristic of ancient Khmer architecture with its techniques. In the chronology sequence of architectural decoration that created by Stern, Ta Keo temple-mountain is roughly classified as The Second Style and speculated the dating of built between Pre Rup temple and Baphuon temple. Stern also speculated the dating of built between Pre Rup Temple and Banteay Srei temple according to the analysis of comparative subtle divergence of colonnette style, and then on this basis, Stern also proposed that there had many similarities of Gopura style between Ta Keo and Royal Palace.

In short, Stern generally speculated the dating of Ta Keo temple-mountain was the tenth century AD or the beginning of the eleventh century, and later than Eastern Mebon Temple dedicated to Vishnu which historical chronology was built in 952 AD.

In 1934, as a student of Stern, Coral-Remusat continued to study the architectural forms and decorative details of temple-mountain in Angkor era. It is mainly architectural features of temple-mountain that is an artificial stepped pyramid in a stupa, which embraced by varying amounts the Prasat, Library, Long Hall and Gallery as well as the combination of architectural space and complex religious symbolism.

For example, King Indravarman I, reigned AD 877 – 889, established Hariharalaya and Bakong Temple as Vnam Kantal. Totally 14 towers (including 2 brick-towers and 12 sandstone-towers) around the pyramid which is 60 m × 60 m (15 meters high), but the core is the central sanctuary which located in the top platform of pyramid.

As described in the first chapter of this book, the massive capitals of Angkor can be traced back to the successor of King Indravarman I, King Yasovarman I (reigned AD 889 – 900) established Bakheng Temple on the summit of the mountain in the natural, which located in the center of capital Yasodharapura. 108 small sandstone shrines symmetrically surround the central sanctuary and then the unique architectural form has become the original of Angkor temple-mountain's evolution.

In the first half of the tenth century AD, King Jayavarman IV moved his capital from Yasodharapura to Koh Ker for unknown reasons, and he also continued to establish Prasat Thom as centre of the capital, which was surrounding by various temples. In the middle of the tenth century AD, King Rajendravarman II, reigned AD 941 – 944AD, subsequently returned to Yasodharapura and built temple-mountains which is artificial stepped pyramid with quincuncial Prasats, as well as both Eastern Mebon and Pre Rup is the most typical style among them. The prasats of Ta Keo temple-mountain, with its quincuncial configuration, reflected a mixture of the religious symbolism of Hinduism and local Khmer culture. The central prasat was consecrated for the Linga merging both the Shiva worship and local Khmer phallicism, hence being regarded as the embodiment of the king, or more precisely, the king-god. Besides, the linga symbolized strong fertility bringing prosperity to the empire and consolidating its domain. The prasats in four corners offered sacrifice to the ancestors, which was a fundamental tradition in Khmer culture.

As the new element of the Khmer architectural evolution, there has been a special concern that long-hall begun to play a significant role among temple-mountain group. Comparing with the layout plan of Eastern Mebon, Pre Rup and Phimanakas, it can be seen that 14 long-halls are symmetrically surrounding the first terrace of the Eastern Mebon, as well as 14 long-halls are also symmetrically surrounding both the first and the second terraces of Pre Rup. These long-halls are mainly constructed by sandstone, and gradually becoming the most remarkable characteristics of temple-mountain layout. However, as an architectural style, long-hall is gradually vanishing from sight in the temple-mountain's layout that is mainly changed in the following aspects: a huge artificial stepped pyramid is surrounded by successive galleries covered with masonry corbel-arch, and quincuncial prasats with equal-arm crossing plan.

Compared with the layout of Ta Keo temple-mountain, based on the features of long-halls, enclosed galleries and quincuncial prasats, hypothetically, it can be inferred the dating of Ta Keo temple-mountain is no earlier than Eastern Mebon and Pre Rup. Subsequently, during the evolution between Baphuon and Angkor Wat, which established in the 11th century AD, the enclosed galleries and quincuncial prasats turned into the most remarkable symbol of temple-mountain.

Among the above chronology, the dating of Ta Keo could be roughly between Per Rup and Baphuon. While in the above sequence of architectural style, as a transitional form, the ruins of brick-roofs, enclosed galleries and vanishing of long-hall all belong to the outstanding features of Ta Keo site. Therefore, for the evolution progress of temple-mountain, the above-mentioned ruins can show the precious aspects of the special stage in the Angkorian period.

Generally speaking, the emergence of Ta Keo temple-mountain marked the mature formation of the temple-mountain. Ta Keo temple-mountain is a milestone during the evolution of temple-mountains, along with its anterior Pre Rup, its contemporary Phimeanakas, and its and its successors Baphuon and Angkor Wat. For the first time, Ta Keo temple-mountain combined the pyramid with the enclosed gallery, which subsequently became a typical feature of the Angkorian architecture. Meanwhile, the enclosure wall was replaced, and the long-hall found in its predecessors lost its popularity.

3. Inscriptions of Ta Keo Temple-mountain

There are totally six inscriptions that numbered as K275, K276, K277, K278, K535, K536 at Ta Keo site. The inscriptions are mainly engraved on the jambs of easterngopuras. However, it has always been contentious diverse viewpoints for the content and dating of these inscriptions.

Beyond all doubt, for the Interpretation and research of Ta Keo inscriptions, G. Cœdès is the most famous scholar on behalf of the significant studies. He had firstly proposed Ta Keo inscriptions are engraved in ancient Khmer rather than in Sanskrit, as well as linked with the inscriptions (K382) of King Suryavarman in Preah Vihear temple and it could be much later than the dating of King Yasovarman I.

In accordance with the studies on the Ta Keo inscriptions such as K275, K276, K277, K278, G. Cœdès firstly suggested that these four inscriptions were dedicated toKing Suryavarman I, the word " ici" of K275 and K278 inscriptions maybe related to a fire caused by lightning in the construction process, which also described an atonement ceremony was held to eliminate the unknown knell, and after the ceremony the supervisors brought some new stone and elephants to continue construction process. Thus G. Cœdès deduced that Ta Keo temple-mountain was built not later than 1007AD.

4. Chronology of Ta Keo Temple-mountain

In the light of a multi-disciplinary research project in 1934, the construction project of Ta Keo temple-mountain was believedthat have been initiated in the late tenth century AD and terminated in an unfinished state in the early eleventh century AD. The construction processing was carried out mainly during the reign of King Jayavarman V and that of King Suryavarman I.

4.1 Pre-historical context of Ta Keo temple-mountain

It seemed to be a tradition in the Angkorian period that each ruler would establish one or more new Vnam Kantals, acting as the state temple, and facilitating worship of his own Devaraja, a symbol of the combination of the ruler and the Siva. King Rajendravarman (who reigned 944 – 968 AD) for instance, moved his capital back to Yasodharapura (probably located where Angkor Thom stands today) from Kor Ker (about sixty kilometers north-east of Angkor, the second capital at Angkor which was built by King Rajendravarman in the 960s AD), abandoned Phnom Bakheng, and then consecrated East Mebon (952 AD) on an artificial island in the centre of the Eastern Baray, Pre Pup (961 AD), and Phimeanakas (968 AD). The latter of these two are believed, by the modern scholarship, to be the state temple or to be the state shrine for worship of Devaraja.

4.2 Ta Keo project during King Jayavarman V's Reign (968 –1000 AD)

King Jayavarman V succeeded to the throne in 968 AD, built a new capital on the west side of the East Baray, which he called Jayendranagari (city of Indra the conqueror) and utilized Phimeanakas as Vnam Kantal early in his reign. In 975 AD, he started to construct his own Vnam Kantal, the state temple, Ta Keo temple-mountain; which was dedicated to Siva sometime around 1000 AD.

4.3 Suspension of the project

During the short reigns of the successors King Udayadityavarman I (Reign 1001 – 1002 AD) and King Jayaviravarman (Reign 1002 – 1011 AD), the construction of Ta Keo was suspended as a result of an unstable political situation and unceasing civil wars, which persisted until King Suryavarman I came to the throne.

4.4 Resumption of consturction during King Suryavarman I's Reign (1002 –1049 AD)

King Suryavarman I enthroned himself after victory in a civil war that had lasted for ten years. The dynastic changes brought not only new social reforms but also the new architectural or artistic forms imported from outside, and then it also boosted the maturation of Angkor architecture. Whereas Buddhist power expanded rapidly in the reign of the former King, King Suryavarman I utilized the opportunity to reform religion so as to re-obtain support from Brahmanism groups. Brahmanism groups were responsible for the formidable fortifications around his Royal Palace and state temple, the Phimeanakas, and also for the construction of the great Western Baray, extending over an area of eight kilometers by 2.5 kilometers. In 1050AD, his successor created a new and more impressive state temple, the Baphuon, to the north of the temple. Sivacarya enthroned the Brahmanism bishop with the patronage ofKing Suryavarman I. To solidify his power, King Suryavarman I married his daughter to his guru Yogiçvarapandita, who then was in charge of resuming the construction of Ta Keo temple-mountain and made use of the lower parts of the temple.

4.5 Terminating the project

The project had never been accomplished when it was terminated probably by reason of a lightning strike disaster. According to G. Cœdès studies on the ancient Khmer inscription with its number K277, an expiatory ceremony was held for conjuring away the evil presage, and in spite of this the construction was still terminated.

4.6 Later Utilization

In the temple were found lots of sculptures of Angkor Wat and Bayon Style as well as the inscription in the period of King Jayavarman VII, which indicates that Ta Keo temple-mountain was not abandoned in spite of its unfinished state.

5. Research and Conservation History on Ta Keo Temple-mountain

Several brief introductions of Ta Keo temple-mountain have emerged in historical documents since its initial discovery in the mid-nineteenth century. For instance, in 1863, French Representative in Cambodia, simultaneous the chief of investigation team of Mekong River, Doudart de Lagrée, marked the rough location of Ta Keo temple-mountain in the map of Angkor relics, and this was the first reference to the rediscovery or description of this temple in the modern historical documents.

After a decennary investigation following the same survey mission, F. Garnier, a member of Doudart de Lagrée's expedition included a simple description of Ta Keo temple-mountain in the 1873 publication entitled Voyage d'Exploration en Indochine. It was the first published description of Ta Keo and its main features: five rectangular platforms, an enclosed gallery dominating the second level, and five towers surmounting the upper level.

In 1880, L. Delaporte, another staff member of Doudart de Lagrée investigation team, published the first layout plan of Ta Keo temple-mountain, which was actually drawn by Ratte in 1873, in Voyage au Cambodge-L'Architecture khmère. This plan is almost useless because the monument had not yet been excavated from the covering tropical vegetation. And later, in 1883, J. Moura published some dimensions of Ta Keo temple-mountain with descriptions of the individual monuments and more complete documentation. In the preliminary study on the meaning of the temple's name, he suggested some elucidative introductions and presumptions, and he extolled admiration for the architectural values of Ta Keo temple-mountain. Meanwhile, he also noticed an offset phenomenon of east-west axis of the temple. However, J. Moura checked and modified the dimension of axis offset in his perception so much as to induce some errors of dimension.

After ten years, in 1890, Fournereau and Porcher cited the brief description by J. Moura in the publication of a research report, but did not correct any mistake of J. Moura's publication. In 1904, another comprehensive plan of Ta Keo temple-mountain was drawn by E. Aymonier. In the meantime, E. Aymonier also undertook a study on the inscription in Sanskrit at Ta Keo temple-mountain. In 1911, E. Lunet de Lajonquière had finished the publication entitled Inventaire descriptif des monuments du Cambodge in which Ta Keo temple-mountain site was registered with its serial number 533. The tropical forest covered Ta Keo temple-mountain at that time, and therefore his description of location was not quite accurate. In this publication about Ta Keo temple-mountain, another plan and section was also included, but the dimension still had many errors. By making modifications to

———————— Synopsis Ⅲ Historical Context of Ta Keo Temple-mountain

the above-mentioned plan and with reference to Ratte's drawing, L. Delaporte finally finished the best of all the historical Ta Keo plans in 1920. He also attached an elevation of a reconstruction in the publication, which similarly relied on the investigation report of Ratte and Loedrich.

Once the former Cambodian territories, including the Angkor region, were returned by the France-Siam Treaty of 1907, the first director of de Conservation des Monuments du Groupe d'Angkor, J. Commaille, started to implement a project to clear the tropical vegetation which covered Angkor temples. Actually, before setting up the base of operations at Angkor, the École Française d'Extrême-Orient (EFEO) already began the process of eradicating undergrowth and felling trees in 1901. The process of removing and cataloging the fallen stones that had piled up on site was also begun in 1911. The restoration work at this stage consisted in resetting the stones in their original positions. As a unique feature of typical temple-mountains, it is certain that Ta Keotemple-mountain was included in Commaille's grand plan of excavation and conservation.

Unfortunately, J. Commaille was murdered in 1916, so leadership was taken up by H. Marchal in cooperation with C. Batteur's team, de Conservation des Monuments du Groupe d'Angkor, and together they undertook the excavation missions at Ta Keo temple-mountain and continued the field work until the end of 1923. During the excavation process, much bulky humus-soil had to be transported away and then the fallen stone and scattered components were piled up with tidy cataloging. Meanwhile, during excavation, a tower was detected in the southern area of Ta Keo temple-mountain, which was very similar to the central prasat of Ta Keo, and this tower was given a registered number of No. 54 in either H. Marchal's plan or G. Trouvé's plan. In April of 1929, H. Marchal had also found a terrace located in not only the western embankment of the East Baray, but also in the main axis of the temple. In the northeast corner of Ta Keo, other smaller relics were detected as late as 1936, but hereafter no further relics or new features of importance were discovered. H. Parmentier undertook another documentation mission for Ta Keo in 1927. He did not issue any formal publication although he did produce a great deal of sketches. Fortunately, the detail of the above-mentioned activities was recorded in the EFEO's journal entitled Rapports d'Angkor.

Deserving acknowledgment, Gilberte de Coral Remusat, V. Goloubew, and G. Coedès launched a chronological study of Ta Keo inscriptions in Sanskrit and ascertained its construction time in 1934. Subsequently, G. Coedès continued to study eight inscriptions in Sanskrit at Ta Keo temple-mountain, and translated the full text of the inscriptions into the French language in 1952. Subsequently, either Glaize Maurice or H. Marchal had issued a brief introduction of Ta Keo temple-mountain in their publications. They had modified the errors of Ratte's plan, and re-drawn a new plan of Ta Keo, which preserved the offset of east-west axis and corrected the location of associated building in the outer enclosure. Finally, B. Groslier finished the axial-view drawing of Ta Keo temple-mountain layout without the associated building in the inner enclosure base.

At the end of the 1960s, J. Filliozat and B. P. Groslier entrusted Jacques Dumarçay with the investigation, survey and documentation of Ta Keo temple-mountain and missions were undertaken in 1967 and in 1969. Consequently, Ta Kèv: étude Architecturale du Temple was published in 1970 providing not only the accurate dimensions but also detailed descriptions of Ta Keo temple-mountain, with abundant measurement data and survey drawings. Prior to the CACH and CSA conservation and restoration project, Jacques Dumarçay's publication was the most perfect documentation work for Ta Keo temple-mountain.

As mentioned above, a lot of study has already been done to identify issues and research of the early of twentieth century. Owing to the leadership of H. Marchal and the cooperation with C. Batteur's team, de Conservation des Monuments du Groupe d'Angkor, had always undertaken the excavation missions and conservation projects that included Ta Keo temple-mountain in Angkor site until 1923. The excavation mission of Ta Keo temple-mountain was recorded in the staff member's detailed journals. The fallen stones and scattered components were transported away from the gaps in the temple's northern and southern sides, and were then piled up inside an area of low land around the temple site with tidy cataloging. In April of 1929, the excavation team had found a terrace that was located in not only western embankment of the Eastern Baray but also the main axis of the temple. In the northeast portion of Ta Keo temple-mountain, other smaller relics were detected as late as 1936 but hereafter, no further relics or features of importance were found at Ta Keo temple-mountain site.

As a result of this excavation team's careful work almost all of the observable fallen stones and scattered components have been found in conformity with records of the excavation work, and their former locations can be ascertained. Meanwhile, the excavation team had also undertaken another mission that was for consolidation works, mainly in dangerous structural parts, adopting two techniques. One was to erect concrete pillars for the sake of supporting collapsed structures or the declined walls, and taking up the task of removing the fallen stones from the upper terrace; another was to install iron hoops to strengthen the stonework and reinforce cracked stone components in order to prevent further deformation or decay. Inasmuch as the work was limited by the technology available at that time, these temporary consolidation techniques could not provide the ultimate solutions for the structural damage that presented a danger, these consolidation methodologies produced a positive effect in maintaining the structural stability at Ta Keo temple-mountain, and their contributions still have considerable value from today's viewpoint.

No major project was undertaken from the end of the 1940s to the 1960s, besides the clearing of vegetation and small-scale consolidation or reinforcement of structures. Later, during the dark years of the Khmer Rouge, when Cambodia became isolated from the rest of the world, many archaeologists were killed and all research and conservation activities in the country ceased.

Since the 1990's, once Angkor was inscribed on the World Heritage List, it was necessary to establish working mechanisms to promote national and international collaboration. Some excavation and consolidation works were resumed at Ta Keo temple-mountain. For example, brickworks were used to provide support in the gaps created by collapses on the four corners of second foundational terrace in order to reinforce the corner tower against structural deformation or deterioration. Many imminent danger zones were shored up with wooden frameworks in order to prevent the further declination of walls and collapse of structures. Through the implementation of these considerable consolidation works, a positive effect was made in preventing further deterioration of Ta Keo temple-mountain.

第四章 茶胶寺庙山的建筑形制

一、茶胶寺庙山的整体布局

根据调查及实测资料推测，茶胶寺庙山极有可能坐落于一组规模庞大建筑组群的中心位置[1]，这座规划严整且气势恢弘的建筑组群一直向东延伸约500米直至东池西侧的堤岸之上。毗邻东池的一座仅存角砾岩砌石基址的建筑遗迹恰好位于茶胶寺庙山建筑的东西向轴线上，并通过一段两侧列布界石的神道与茶胶寺庙山建筑主体部分连接起来。由此可见，无论是位于东池堤岸之上的建筑遗址，还是使其与庙山主体相连的神道残迹，皆应是构成茶胶寺庙山整体格局的重要组成部分。[2]（图4-1，图4-2）

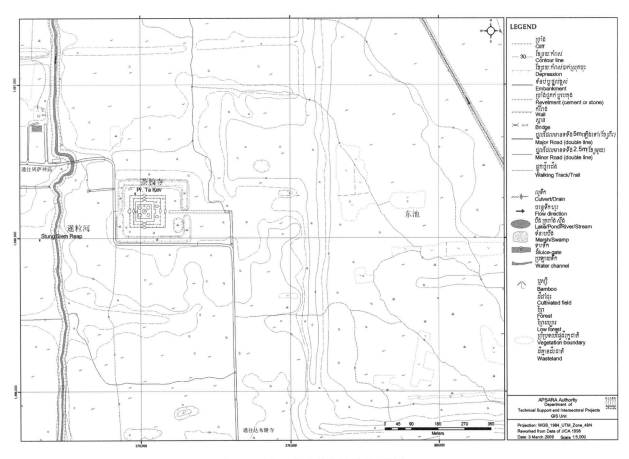

图4-1 茶胶寺庙山整体格局平面图之一

---

[1] Jacques Dumarçay. *TA KEV：ETUDE ARCHITECTURALE DU TEMPLE*, Paris：EFEO, 1971, pp. 10–15.
[2] Claude Jacques. *Angkor：Cites and Temples*. River Books Co. Ltd：BANGKOK, 2002, p. 120.

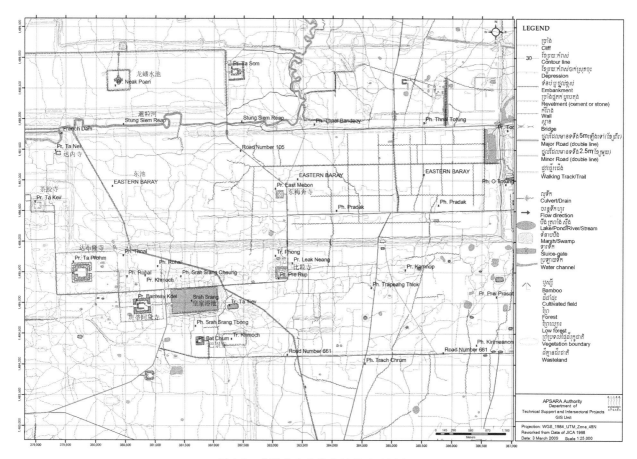

图 4-2　茶胶寺庙山整体格局平面图之二

自茶胶寺庙山主体向其北侧推移大约 500 米，可以抵达位于暹粒河与东池之间的一处土堤，今已被热带丛林所覆盖，推测或为茶胶寺建筑组群北部边界的遗迹。而在茶胶寺庙山南侧方向的茂密热带丛林之中，曾经发现有人工挖掘水渠的痕迹与密林中的填土小路并行，亦是在距离茶胶寺庙山大约 500 米处戛然而止，据此推测或为茶胶寺建筑组群的南部边界的遗迹。唯茶胶寺庙山西侧由于暹粒河所形成的天然屏障的阻隔，在大约 1000 米 × 780 米的长方形区域之内，庙山建筑主体部分略偏向此区域的西侧。虽然庙山建筑主体的中心位置是两条正交轴线的交汇之处，但却并非严格中心对称，庙山整体由此略微偏向东南。

在以茶胶寺庙山为中心面积约为 100 公顷的范围以内，尚存有诸多疑似与茶胶寺庙山相关的遗址或遗迹。例如，东池西堤之上的两座石砌平台遗址、庙山东南方向的一座砂岩砌筑的塔殿遗址、庙山西侧的一座建于阇耶跋摩七世（Jayavarman VII）时期的慈善医院遗址、庙山壕沟东北外侧的建筑遗迹以及庙山南北壕沟外侧丛林之中的各种凹坑等等。[1]（图 4-3，图 4-4，图 4-5，图 4-6，图 4-7）

---

[1] 茶胶寺庙山所在区域地形基本平坦，属于冲洪积地貌。主要地层的层位分布比较稳定，建筑基础底面位于同一地质单元、同一成因年代的土层上，地基持力层土层分布均匀。茶胶寺庙山地质勘查深度范围内的地层上部为填土、人工回填土，下部为第四纪沉积土，主要为粉土质砂及黏土质砂。地下水位埋深 6.6—8.9 米，标高 83.14—85.27 米，属于浅水。场地土类型为中硬土，建筑场地类别为 II 级，场区内的底层不存在地震液化问题。参见中国文化遗产研究院、北京特种工程设计研究院：《茶胶寺岩土工程勘察报告》，2010 年，第 4—8 页。

图 4-3　利用激光雷达（LiDAR）获取的茶胶寺庙山整体格局影像图

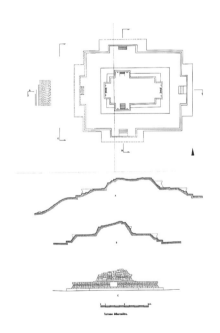

图 4-4　东池西侧的建筑遗址及实测图

东池与茶胶寺庙山之间，通过长达 500 余米的神道相连。由该神道西端的一座十字形平台的遗迹推测，这应是进入茶胶寺庙山建筑组群中心区域的重要入口之一。这座十字形平台由三层角砾岩石块砌筑而成，砌石的表面初步雕刻了线脚。根据现存平台遗迹推测其上原先应该覆有木制框架支撑的屋顶结构。从二十世纪二十年代拍摄的历史照片来看，这座十字形平台的两侧跺台上曾经置有石

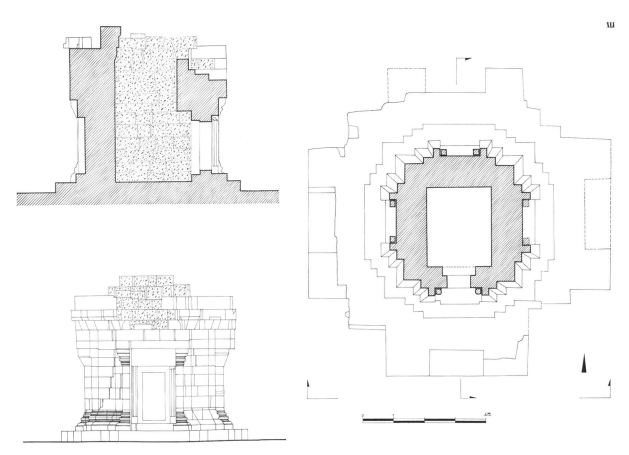

图 4-5 茶胶寺庙山东南角的建筑遗迹现状实测图

图 4-6 茶胶寺庙山东南角的建筑遗迹现状

图 4-7 茶胶寺庙山西北侧阇耶跋摩七世时期的建筑遗址

图 4-8　1923 年法国远东学院发掘清理茶胶寺庙山神道情况　© EFEO Archives

狮两尊，现皆已不存。与此十字形平台相接的神道宽度约为 10 米，除散布界石之外，沿着神道两侧部分角砾岩石砌筑的墙基尚隐约可见，一直延伸至东外塔门基座的外侧（类似的矮墙还出现在巴方寺的神道两侧）。十字形平台与东外塔门之间虽以神道相连，而沿神道两侧边缘由角砾岩砌筑的矮墙与东外塔门之间围合而成半封闭空间，此为神道原始格局抑或是后期所改建，目前尚不得而知。

对于茶胶寺庙山东侧神道部分的遗迹现象及其清理情况，二十世纪六十年代末，雅克·杜马西（Jacques Dumarçay）在其调查报告中曾经有如下描述："我们发现，庙宇东边的通道是从东湖而来的一条稍有填高的小路，它在跨越壕沟之前被打断，打断它的是一处带线脚的角砾岩基础的残构，最多有三层，它无疑应支撑着一座木制塔门。这个可能加建的小型构筑物与一段台阶结合在一起，台阶设有三级踏步，两侧夹以垛台，其上承托石雕狮像。这一台阶通至铺砌石板的小径（神道）标高，小径（神道）以有线脚的界石夹道。经由连续三次转折，至两个水池标高小径（神道）变得更宽，接近 8 米。通过三级踏步，神道与水池沿着侧面衔接起来，踏步沿着锥形体转向并留下一段长约 9 米的土堤。这些踏步在清理茶胶寺庙山东立面时已被清理畅通，在南北立面被清理出大约 10 米的距离。建筑组群的布局或已被完全改变，两个水池的填土不仅填至石栏的高度，而且超过并达到了第一层基台的第三级踏步高度。局部建造在填土之上的一段角砾岩砌筑的墙体，包括了第一层基台东外塔门基座以及入口门道的界石。此墙曾覆有大量填土，而靠近基座的填土更是高达 2 米。1922 年，当法国远东学院及其吴哥古迹保护处（CA）清理茶胶寺庙山的时候，这些含有大量破碎的砖瓦及界石的残片的填土被清理运出。庙山北立面的入口仅由跨越壕沟的通道标示出来，亦有可能是由第一层基台或基座较小的前突部分标示出此通道的位置。"[1]

另外，茶胶寺庙山的整体布局还应包括：坐落于神道南、北两侧的两座水池以及环绕庙山的壕沟。其中，环壕内侧及神道南、北两侧各设水池一座，由于淤土覆盖严重之故，今仅存大致轮廓依稀可辨[2]。环濠东西方向长度约为225米，略大于其南北方向的长度（195米），其南北两侧壕沟与庙山建筑第一层基台基底外侧的距离大约为33米，而西侧壕沟和东侧壕沟与庙山建筑第一层基台基底外侧的距离分别是15米和58米。壕沟驳岸以砂岩及角砾岩砌筑成阶梯状，大部圮不存，尤其是西侧及南侧的壕沟。东侧和北侧现有横跨壕沟通道的遗迹尚可辨识。（图4-8）

2012年3月至4月，中国文化遗产研究院在茶胶寺庙山壕沟北侧通道的东半部分进行考古发掘，布设5米×5米探方8个，实际发掘面积192平方米，发掘埋藏深达2.5米。通过此次发掘清理，其西、北、南三面驳岸的砌石阶轮廓清晰，结构完整。整体砌筑工艺表现为表层平铺一层砂岩砌石，其下大部分为规格大小不等的角砾岩错缝平砌而成，驳岸护岸坡度匀缓。具体而言，西坡驳岸由砂岩和角砾岩错缝平砌为16层阶梯状；南、北两侧驳岸上部保存较差，砂岩石滑落及侵蚀风化严重，至第7层以下的保存状况较为完整，推测应与西驳岸同样砌筑为16层阶梯状；壕沟底部宽度为2.5米[3]，自下而上逐层增宽，至顶层宽度增为10.6米。[4]（图4-9，图4-10，图4-11，图4-12）

表4-1 茶胶寺庙山建筑材料分布简表

材料名称	建筑位置	所占比例（体积）
长石砂岩	塔门、长厅、藏经阁、回廊、围墙、角楼、须弥台	c.58%
硬砂岩	中央五塔	c.20%
角砾岩	第一层基台、第二层基台及其院落地面的铺砌	c.20%
砖	塔门、藏经阁及回廊的屋顶部分	c.2%

作为一种主要的建筑材料，砂岩在吴哥时代的寺庙建筑中曾被广泛使用；而角砾岩历来是高棉寺庙建筑的基础用材，大量用于池基、围墙、道路和桥梁，庙山基台内部挡土墙也多以角砾岩砌筑。角砾岩的硬度较低，岩石内部的孔隙率决定了其强度。茶胶寺庙山大量运用砂岩作为主要建筑材料，体现出吴哥时代在采石技术、材料运输、石作工艺等方面的长足进步。茶胶寺庙山大约80%的建筑部分皆以砂岩砌筑并构成庙山建筑的主体部分，由角砾岩砌筑的第一、二层基台则为辅助部分，主要是为增加须弥台及中央主塔的地坪标高，以此烘托彰显出神之居所——"须弥山"的高峻挺拔与神圣庄严。

---

[1] 同上注。
[2] 1922–1923年，法国远东学院的亨利·马绍尔（H. Marchal）曾经对这两座水池进行考古清理，清理出的水池深度为4.1米，基底尺寸为4.8米，水池堤岸以12层角砾岩梯形砌筑而成，并在南侧水池之北堤和西堤发现了保存较为完整的砂岩压顶石。马绍尔认为其他各面的石块已经脱落，而杜马西则认为在建造庙山东侧神道角砾岩矮墙的时候，两个水池就曾经进行有意的回填，因为神道伸向南南侧分岔时稍有侵越至水池填土之上的痕迹。参见 Jacques Dumarcay. *TA KEV: ETUDE ARCHITECTURALE DU TEMPLE*, Paris: EFEO, 1971, p.11.
[3] 通过考古发掘，发现在壕沟最底层角砾岩砌石的外表面（壕沟内侧迎水面）使用了一种白膏泥填料进行封护，其宽度约为10–12厘米。根据分析白膏泥色泽白中透青，质地细腻，黏性强，在可以起到较好的密封作用的同时又具有很强的吸湿性，推测这是为防止角砾岩由于长期渗水而导致的风化，或为阻止由于渗水而引起的台阶下部或里侧的掏蚀而采取的专门工艺。（著者注）
[4] 此次考古发掘共计出土器物标本90件，其中，陶器13件，釉陶器4件，瓦54件，瓷器12件，建筑构件7件，以弧瓦、拱瓦居多，其次为陶瓷器。（著者注）

第四章　茶胶寺庙山的建筑形制

图 4-9　2012 年茶胶寺庙山考古发掘所揭示的
　　　　环壕驳岸构造情况之一

图 4-10　2012 年茶胶寺庙山考古发掘所揭示的
　　　　 环壕驳岸构造情况之二

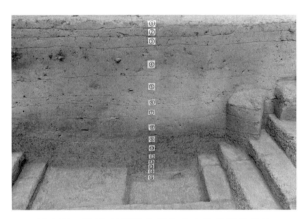

图 4-11　2012 年茶胶寺庙山考古发掘所揭示的
　　　　 环壕淤积土层情况

图 4-12　2012 年茶胶寺庙山考古发掘所揭示的
　　　　 环壕底部构造情况

二、茶胶寺庙山建筑形制概略[1]

茶胶寺庙山建筑主体部分自下而上主要包括：第一层基台（Outer Enclosure，亦称"庙山外院"）、第二层基台（Inner Enclosure，亦称"庙山内院"）、须弥台（Pyramids）以及坐落于须弥台之上的中央五塔（Prasat）[2]。其中，围墙和回廊（Gallery）分别环绕的第一层基台和第二层基台四周，在其正交轴线位置分别辟为八座塔门（Gopura）；另外，第一层基台的东侧南北对称布置有外长厅（Outer Long Hall）两座，第二层基台的东侧南北对称布置内长厅（Inner Long Hall）及藏经阁（Library）各两座。

第一、第二层基台及须弥台的东西方向长度皆大于其南北方向长度，第一层基台基底尺寸为121米×106米，第二层基台基底尺寸为94.7米×89.2米，第三层基台基底尺寸为60.7米×59.2米。第一层基台的高度约为1.73米（不包括围墙高度3.2米），第二层基台的高度约为6米，须弥台总高度13.89－14.39米，且整体呈现出西高东低的趋势（东北角13.89米，东南角14.08米，西南角14.39米，西北角14.33米），须弥台自下而上又可以分为三层（高度分别约为5.68、4.61、3.61米）。须弥台顶部五塔之

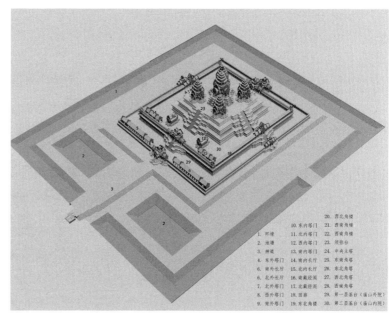

图4-13　茶胶寺庙山整体布局示意

中央主塔的现状高度约计21.24米（至须弥台的顶面），由此茶胶寺庙山自第一层基台的基底至中央主塔顶点的总高度约计43.3米。（图4-13，图4-14）

---

[1] 茶胶寺庙山建筑形制的测量数据均是在2008－2010年期间中国文化遗产研究院与天津大学合作应用三维激光扫描技术基于数据点云（point cloud）所获取。另外，本章对于雕饰线脚的描述则主要参考了雅克·杜马西于1971年出版的建筑调查报告中的实测记录（*TA KEV: ETUDE ARCHITECTURALE DU TEMPLE*, Paris, EFEO, 1971）。

[2] 通过对茶胶寺庙山建筑的地基与基础调查发现，自下而上第一层基台的基础坐落于由天然和人工回填组成的地基土之上，第一层基台外部四周石墙由角砾岩石块砌筑，内部充填细中砂（从密实度推断为夯筑），顶面铺砌1－2层角砾岩条石，同时内部夯筑的细中砂层为第二层基台的地基。第二层基台的基础结构与第一层基台基础结构类似，同时其夯筑砂层做为须弥台的地基。须弥台基础结构调查遇到一些问题，探槽开挖时发现其顶面铺砌角砾岩条石大于两层，在作业面小的情况下未能开挖成功，但从工程物探无损测试数据看，须弥台内部充填土的波速数值与其他两层基台近似，充填土应为夯筑的细中砂，即须弥台基础结构与第二层和第一层基台基础结构类似。明显的不同是须弥台石砌墙体材料除角砾岩外，墙体外侧还包砌一层砂岩块石，同时其夯筑的砂层作为顶部五塔的地基。参见中国文化遗产研究院、北京特种工程设计研究院：《茶胶寺岩土工程勘察报告》，2010年，第6页。

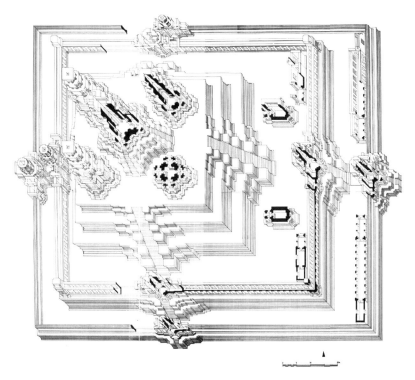

图 4-14　茶胶寺庙山建筑构成

茶胶寺庙山建筑结构的总体保存状况较为完整，但各层基台的局部和大多数单体建筑损毁严重，存在明显的结构安全隐患，主要表现为地基不均匀沉降、结构体坍塌、墙体倾斜、结构裂缝、构件错位或缺失等。从两层基台及须弥台的现状观察，损坏最严重部位均在四个转角处，特别是第二层基台及须弥台的转角全部坍塌，并导致其上部建筑单体存在严重险情；各层基台中部均保存较好，没有出现明显的结构变形，每层基台四边中部的台阶总体保存较好，局部出现石块移位或缺失现象；各层基台顶部的地面均有局部沉陷和铺地石缺失现象，造成地面出现多处坑洼，雨季会导致严重积水，并下渗至基台内侧，从而对基台边墙产生破坏压力。

茶胶寺庙山各单体建筑，包括中央五塔、各层的塔门、藏经阁、长厅、角楼、回廊和围墙，均出现不同程度的损坏，其中多处结构存在严重险情。在所有单体建筑中，中央五塔保存较好，除局部出现裂缝和构件错位之外，整体稳定性较好。其余建筑所存在的共同破坏状况是屋顶完全坍塌、墙体出现倾斜或开裂现象、构件错位或缺失，部分建筑存在地基不均匀沉降，特别是二层基台四个角楼因基台转角坍塌而处于危险状态。[1]

---

[1] 通过现场勘察并进行整体结构三维有限元分析，茶胶寺庙山结构稳定性的结论如下：（1）茶胶寺各单体结构在现存残损的情况下，各应力（拉、压、剪应力）均小于砂岩的破坏应力，且其相对于破坏应力非常保守，故在现存情况下处于稳定状态；（2）由于砂岩石材抗拉强度极低，茶胶寺部分结构出现的局部拉应力相对砂岩抗拉强度所余不多，有出现拉裂破坏的危险；（3）茶胶寺各单体结构在现存状态下的破坏风险基本源于基台继续破坏所造成相应部位石块的沉降；沉降破坏的发生，导致石块塌落或其他严重的破坏；（4）通过各单体结构的数值计算结果可以得出，茶胶寺单体建筑的拱顶结构主要通过四个角将重力荷载向下传递，易造成石块向两侧方向的分离；（5）对比数值分析结果与现状可以发现，茶胶寺庙山基台与各单体建筑的应力较大部位通常破坏较为严重。现状勘察与分析研究表明，茶胶寺建筑结构稳定性对于不均匀沉降非常敏感，失稳可能性随着沉降量的增加而加剧，而且由于砌筑建筑的石块之间无任何黏结材料，石块间以摩擦力来维持结构的稳定，因此石材表面越粗糙，结构越稳定。参见中国文化遗产研究院、湖南大学：《茶胶寺典型单体建筑结构危险性评估与加固技术研究》，2011年版。

1. 第一层基台（庙山外院）

第一层基台，亦称为"庙山外院"（Outer Enclosure），其中包括四面四座塔门、两座外长厅以及环绕基台四周的围墙。第一层基台的平面略呈方形，东西通长约为121米，南北通长约为106米，高度约为1.73米。基台的结构实为由9层（外观7层，底部2层被沙土覆盖）粗凿角砾岩砌筑的挡土墙，内部以回填沙土夯筑而成[1]。第一层基台的顶部四周边缘环绕以砂岩石块砌筑的围墙，墙身通高约3米，除四个角部以及局部塌落开裂之外，围墙大部保存较完整。墙身外侧立面雕刻线脚，内侧表面多为粗凿，似仍保留着未完工时的状态。砌石墙体的顶部为略呈弧形的压顶石，脊饰皆已不存。另外，每侧围墙的根部皆开凿有排水口[2]。雅克·杜马西曾对第一层基台雕刻线脚的形制进行详细的描述如兹："……勒脚，逐层凹进的两条平条线脚，凹圆线，束腰，凹进的两条平条线脚，反枭线，凸出很厚的平条线脚。檐口包括：斜面，凹条，两条平条线脚，半圆线，凹进的两条平条线脚，凸出的两条平条线脚，枭线，束腰，凸出的两条平条线脚，宽束腰，接着是盖顶及其支撑块石，正好嵌入脊尖装饰。"[3]

第一层基台的整体保存状况较好，各个角部及围墙局部有残损。基台南侧多处呈现不规则沉陷，局部有石块缺失或断裂，东南角因上部围墙倒塌缺失一角。基台北侧总体保存较好，东北角缺失部分石块，上部墙体转角倒塌，基台东侧保存较好，东南转角处构件缺失，并有明显沉陷。

环绕第一层基台围墙的四面中央沿庙山正交轴线分别辟为外塔门（Outer Gopura）四座。其中，东、西二座外塔门为十字形平面，面阔五间，进深一间，且分别于明间与两外侧室被辟为门道；与东、西两座外塔门相较，南、北两座外塔门的规模尺度略小，形制亦为面阔五间，进深一间，唯其进深尺度较小且仅于明间辟为门道。另外，在第一层基台的东北角和东南角沿庙山东西轴线分别对称设置两座外长厅（Outer Long Hall）[4]。

通过对今存遗迹现象的分析来看，第一层基台及围墙所围合而成院落的地面铺装似乎尚未最终完成，因此仅在四座外塔门与连接第二层基台踏道之间的通道部分施以砂岩石块铺砌，其余各处地面皆为角砾岩石块铺砌。然而，在这些角砾岩石块铺砌的地面之上，尤其是在基台内侧西外塔门和南外塔门两侧的地面上，却发现开凿有数量颇多的圆形或半圆形孔洞，其直径大约在330厘米至640厘米不等，而这些孔洞的布列方式沿第一层基台底部在面阔方向上表现出某种规律性，颇为引人注目。

对于上述遗迹现象，目前的研究或解释存疑较多，还不能完全理解或判断其功能与性质。但是，正如前文所述，在吴哥时代庙山建筑形制及年代序列之中，作为重要的过渡形式之一，茶胶寺庙山建筑组群中所出现的砖砌屋顶、石砌叠涩拱回廊以及长厅的逐渐式微等形制特征皆是非常令人瞩目的类

---

[1] "由下至上第一层基台基础坐落于天然和人工回填组成的地基土至上，基台外部四周挡土墙由红色角砾岩砌筑，内部填充细中砂（从密实度推断为夯筑），顶面铺砌1-2层角砾岩条石，同时内部夯筑的细中砂层作为第二层基台的基础。"参见中国文化遗产研究院、北京特种工程设计研究院：《茶胶寺岩土工程勘察报告》，2010年，第6页。
[2] 茶胶寺庙山原有排水系统应为有组织排水与无组织渗水相结合，部分雨水沿砌石缝隙自然渗流，其余大部则沿排水系统排出。各层基台皆设计排水坡度，每面角部有排水孔，雨水汇集后从排水口排往下层基台，再汇集排出庙山建筑后排向庙山四周场地之中。庙山周围场地为无组织排水，雨水大多汇流至壕沟内。参见中国文化遗产研究院、北京特种工程设计研究院：《茶胶寺保护修复工程总体设计方案－排水工程设计方案》，2011年，第25页。
[3] Jacques Dumarcay. *TA KEV：ETUDE ARCHITECTURALE DU TEMPLE*, Paris：EFEO，1971，p. 16.
[4] 外塔门、外长厅的建筑形制皆在下文详述，不再赘述。

型学或年代学的标志。鉴于此，观照在茶胶寺庙山第一层基台内所发现的诸多类似于建筑基址的遗存，或许也正是茶胶寺处于吴哥时期庙山建筑形制过渡与转型的一种体现。对此，雅克·杜马西认为："（茶胶寺）建筑的初始状况应该是，木构建筑曾完全占据了第一层基台的四周。多亏有这些桩穴，由此我们可以辨认出这些建筑的平面，尤其是在南面东侧建造的房屋平面，涉及一个从抱厦短边伸出的长厅。这样的平面与这一层台东北角和东南角用砂岩砌筑的长屋的平面是一样的。H. 帕蒙蒂埃曾经认为这是一座简单的临时工棚的遗迹，在我们看来，这有些大材小用了，而且那时的工程施工或许早已超过主体工程的三分之一，而就当时的进展而言，搭设这样一座工棚实在是太晚了。我们认为，这些房屋至迟在人们建造未曾事先考虑的围墙之时就已经拆毁了，而且显然在原始的设计意图中其平面是没有设置外塔门的。若是围墙被固定到外塔门山墙上的地方显出围墙厚出少许，则可将之解释为一个假门的保留部分，而在这种情况下，围墙就应是在山墙基础上加建的。但实际情况恰恰与之相反，墙身与塔门结合为一体，这样的话就应是同时代的，属于同一时期的皇宫塔门在其山墙上有相似的布置。"[1]（图4-15，图4-16）

图4-15　茶胶寺庙山一层台东北角的"建筑基址"　　图4-16　茶胶寺庙山一层台东南角的"建筑基址"

2. 第二层基台（庙山内院）

第二层基台，亦称为"庙山内院"（Inner Enclosure），坐落于第一层基台（庙山外院）之上，平面亦大致呈方形，东西通长约为94.7米，南北通长约为89.2米，通高约为6米。第二层基台的砌体构造实为以14层角砾岩石块砌筑而成的巨大挡土墙[2]，其内部亦为干砌角砾岩砌石构成的挡土墙，

---

[1] Jacques Dumarçay. *TA KEV：ETUDE ARCHITECTURALE DU TEMPLE*，Paris：EFEO，1971，p. 16.

[2] 针对第二层基台的线脚雕刻及其形制，雅克·杜马西也曾进行过详细描述："线脚元素很有可能是围绕着一个扁平束腰翻转的，它并不在墙的中间，而是随着塔门基础下半截的顶部束腰延伸出来。由此，所有的上半截线脚应是按更小的比例细刻的。这显出对施工维护的不利，因为束腰比勒脚凹进，而流淌的水会落在墙基砌筑层之上。挡土墙在轴线上中断之处乃是砂岩造的塔门基础。基础可分为逐层凹进的两半截，它通过接连六个梯形墙以达到其最大外凸，只有一段阶梯将其打断，而东、西内塔门有三个门。东边基础的线脚已作细刻，并在除上半截向外突起以外的部分作了精雕。在上半截，构成阶梯梯台的突起呈现出一些轻微变形：居中线脚不是缀有叶漩涡饰的扁平束腰，而是带花饰的半圆线。下枭线被翻转，并像上枭线那样缀有叶漩饰。这些线脚元素是围绕半圆线成轴线对称的，只有那些直的枭线除外。在下半截，只有下枭线精雕了莲瓣。在上半截的精雕已经完工，它包含了与下半截相同的线脚，但缩小了比例。线脚元素由低到高包括：勒脚，逐层凹进的两条平条线脚，缀有莲瓣的直枭线，逐层凹进的两条平条线脚，绷得很紧的凹弧线，平条线脚，扁平束腰，平条线脚，凹条，逐层凸出的两条平条线脚，斜面，以及缀有叶漩涡饰的束腰或缀有菱形内接花饰的居中半圆线，之上我们可看到同样的线脚和装饰。"参见 Jacques Dumarçay. *TA KEV：ETUDE ARCHITECTURALE DU TEMPLE*，Paris：EFEO，1971，pp. 19–20.

墙体的内部施以回填砂土夯实，挡土墙最宽处的厚度约为4米。[1]

茶胶寺庙山第二层基台的顶面边缘四周施以砂岩砌筑的回廊（Gallery）环绕，在回廊四个转角处分别设置角楼（Corner tower）一座。回廊四面中央沿庙山轴线分别辟为内塔门（Inner Gopura）四座，各座内塔门皆为面阔五间，进深一间。其中东、西两侧内塔门的明间与次间辟为三间门道，南、北两侧的内塔门仅明间辟为门道。东内塔门两侧南北对称设置藏经阁（Library）各一座。回廊东南角及东北角的内侧沿轴线对称分别设置内长厅各一座（Inner Long Hall）[2]。第二层基台地面与第一层基台类似，地面亦施以角砾岩石块铺砌。

第二层基台的东南角崩塌引发上部东南角楼局部坍塌，暂以红砖砌体进行临时支护；西南角坍塌严重，亦引发上部角楼失稳，墙体倾斜变形；西北角上半部坍塌，引发上部角楼的西北角倒塌，同时角楼北墙出现倾斜变形，以红砖砌体填充缺失的角砾岩部分，并以木结构临时性支撑支护行将倒塌的西北角边墙；东北角上半部倒塌，致使上部角楼倒塌变形分割成三个独立部分，且东侧残留结构处于危险状态中。

3. 须弥台

作为茶胶寺庙山建筑的核心部分，须弥台坐落于第二层基台（庙山内院）中央略偏向西侧的位置，系中央五塔的基座。须弥台自上而下共分为三层，总高度在13.89－14.39米且整体呈现出西高东低的趋势（东北角13.89米，东南角14.08米，西南角14.39米，西北角14.32米）。须弥台的各层构造皆是由外侧砂岩砌石与内衬角砾岩砌石共同构成挡土墙，墙身厚度约为4米，挡土墙内部亦以回填砂土夯实。[3]

须弥台的各层平面皆近似于正方形，尺度逐层向内收进（第一层60.7米×59.2米，第二层52.6米×50.6米，第三层46.2米×43.8米），在其四面中心位置分别设置有陡峻的踏道以连接须弥台的顶层与庙山内院。各面的踏道皆随须弥台之分层亦各自可以分为三级梯段，踏道两侧的每一级梯段又再可分出标高不同的两层跺台，踏道及跺台的表面皆刻有雕饰。各面踏道皆划分为34级台阶且在竖向标高上保持一致，唯其坡度却不尽相同：东侧踏道坡度为44°，南北侧踏道坡度皆为46°，西侧踏道坡度则为57°。

值得注意的是，须弥台四面踏道的宽度皆在其顶部收窄，似与各级梯段的高差产生某种关联。譬如，须弥台的东侧踏道总高度为13.84米，共分为5.73、4.46、3.65米等三级梯段，而此三级梯段的每一级又通过跺台分别划分为3.31米和2.42米（上下段之比为0.73）、2.56米和1.88米（上下段之比为0.73）、2.11米和1.54米（上下段之比为0.73）共6段；又如，须弥台西侧踏道总高度为14.39米，共分为5.69、4.82、3.88米等三级梯段，而此三级梯段的每一级又通过梯台分别划分为3.05米和2.64米（上下段之比为0.86）、2.61米和2.21米（上下段之比为0.85）、2.10米和1.78米（上下

---

[1] 参见中国文化遗产研究院、北京特种工程设计研究院：《茶胶寺岩土工程勘察报告》，2010年版，第9页。
[2] 回廊、内塔门、藏经阁、内长厅的建筑形制皆在下文详述，不再赘述。
[3] 须弥台东南角三层台均存在局部坍塌、现存结构出现多条宽10－60毫米不等的裂缝，上层台整体向外侧倾斜，部分石块松动、有随时塌落的危险。上层台塌落石块堆积在中层台顶部，中层台塌落石块堆积在下层台顶部，下层台塌落构件被摆放在距台不远处的地面上。台外侧砂岩石块多数破碎，表面风化严重；内衬角砾岩石块靠外侧部分破损严重。除角部之外，须弥台主体较为完整、结构稳定。（著者注）

段之比为 0.85）共 6 段。须弥台东侧踏道坡度为 44°，每级梯段的上下两端段之比约为 0.73；须弥台西侧踏道坡度为 57°，每级梯段的上下两端段之比约为 0.85，通过各层的高度数值及其视线分析可以判断，上述数值的内在关联显然并非偶然为之，借此或可了解到古代高棉匠师通过缩小各级梯段的相继高度而且增大上下梯台之间比率数值的方法，进一步强化透视灭点增加目视高度以衬托庙山建筑高峻挺拔的设计意匠。[1]

须弥台的东侧砂岩表面雕饰有精美的纹样，而其南、西和北三面则仅有局部线脚雕饰。须弥台踏道两侧垛台上下两部分雕刻线脚和装饰纹样且与须弥台的雕饰大致相仿，唯其某些线脚略有出入且尺寸略小。由于风化剥蚀非常严重，须弥台残存的雕刻皆已岌岌可危。现将须弥台各层装饰纹样略述如兹。

第一层须弥台表面的雕饰线脚及纹样从下至上包括：带有方块的内接花饰（由出自同一点的四簇叶束饰组成）的圭脚，无纹饰厚皮条线，带叶束饰的反枭线，逐层凹进的两层皮条线，带叶束饰的凹圆线，逐层凸出的两层皮条线，缀有莲瓣的枭线，其上是精雕花蕊的凹圆线、皮条线、带吊坠饰的凹弧线，细扁平束腰，缀有花饰（花饰内接于菱形，由出自一个方块的四个母题构成）；其上为凹条和细扁平束腰，花饰与另一母题（由两个对顶三角形叶饰构成，三角形交接点被花饰中央方块里所含的同一母题遮住）交替出现；逐层凸出的两层无纹饰皮条线，很宽的半圆线；这条半圆线上缀有花饰，花枝由三条蛇神（Naga）的蛇身构成，蛇首转化为植物枝蔓。除开带莲瓣的枭线之外，在此条半圆线以上可以看到同样翻转的线脚，而顶部束腰上则缀有一整条吊坠叶饰。第一级踏道梯段垛台的下半截雕饰包含有与须弥台相同的线脚，但在居中半圆线上下的凹条和凹圆线是无纹饰的，居中半圆线则缀有四瓣花饰；其上半截也包含有相同的线脚，但居中半圆线被缀有叶漩涡饰的扁平束腰所取代，凹条和凹圆线都是无纹饰的。第一层须弥台雕刻因长期风化和雨水侵蚀，石材表面已严重损坏，基台四转角处倒塌，各台阶两侧的边台不同程度缺失构件，均在转角处。顶部砂岩石地面已高低不平，局部易积聚雨水顺着基台边缘漫流，易对石材产生侵蚀而造成风化。基台边缘铺满从上一层基台跌落的石构件。

第二层须弥台表面雕饰线脚及纹样由低到高包括：有花饰与对顶三角形叶束饰交替出现的圭脚，无纹饰厚皮条线，缀有叶束饰的反枭线，逐层凹进的两层皮条线，缀有吊坠饰的凹圆线，逐层凸出的两层皮条线，带莲瓣的枭线，其上是精雕花蕊的凹圆线，无纹饰皮条线和凹圆线，带花饰的厚皮条线，带四瓣花饰的凹条，无纹饰束腰，逐层凸出的两层皮条线，有菱形内接花饰与对顶三角形叶束饰交替出现的宽束腰。除代莲瓣的枭线之外，在此束腰以上可以看到带有相同精细雕刻的翻转线脚，顶部束腰同样缀有以叶束饰环绕的吊坠饰。第二级踏道梯段垛台下半截雕饰包含了同样的线脚，但其比例缩小。那些凹圆线、凹条及所有平条线脚都是无纹饰的，居中束腰缀有连续的叶漩涡饰。上半截包含了与下半截相同的线脚和相同的精雕，依然缩小了比例。第二层须弥台仅局部完成雕刻，四个转角皆已坍塌，其余部分结构稳定性较好。经长期风化和雨水侵蚀，石材表面受损严重，基台边缘有不同程度构件塌落，主要集中在转角处。顶部砂岩石地面已高低不平，局部区域易积水。

第三层须弥台表面雕饰线脚及纹样元素由低到高包括：带方块内接花饰（由出自同一点的四簇叶束饰组成）的圭脚，皮条线，缀有叶束饰的反枭线，逐层凹进的两层皮条线，带吊坠饰的凹圆线，皮条线，带四瓣母题的半圆线，逐层凹进的两层皮条线，缀有菱形内接花饰的细束腰，逐层凸出的两层

---

[1] 有的学者认为，这种强化透视灭点的设计方法或源自印度，至迟在南印度地区跋罗婆王朝（Pallava）即已流行。参见 Jacques Dumarcay. *ARCHITECTURE AND ITS MODELS IN SOUTH – EAST ASIA*, *Bangkok*: *Orchid Press*, 2002, p. 64.

皮条线，细斜面，缀有莲瓣的厚枭线，之上是带花蕊的凹圆线。在此线脚之上依然可以看到翻转的同样的线脚元素，不过顶部束腰显然比圭脚要厚（43厘米/36厘米），并包含缀有以叶束饰环绕的吊坠饰。第三级踏道梯段垛台的下半截的雕饰：缀有与梯段勒脚相同母题的勒脚，厚平条线脚，缀有莲瓣的直枭线（花蕊精雕在一个细凹条上），无纹饰凹弧线，缀有四瓣母题的半圆线，平条线脚，凸出的无纹饰束腰，逐层凸出的两条平条线脚，斜面，缀有四瓣母题的半圆线。其上，可以看到同样的线脚，顶部束腰缀有以叶束饰环绕的一整条吊坠饰，上半截包含有相同的线脚和相同的装饰，但缩小了比例。[1]

第三层须弥台残损及结构破坏形式同第一、二层须弥台，亦为转角部位坍塌，其余部分结构稳定性较好，经长期风化和雨水侵蚀，石材表面受损严重，基台边缘有不同程度的构件塌落。

作为未完成的工程，茶胶寺庙山遗存的构造特征与施工方法，对研究吴哥时代庙山建筑的施工技术与流程具有重要的学术价值。根据茶胶寺庙山须弥台砌石的不同形态，可以将对砂岩石块加工过程分为"毛石——粗凿——精雕"等三个阶段。其中毛石阶段砂岩石块大多呈长方体形态，表面粗糙，部分石块有简单的凹凸处理，粗凿后的石块已经过简单打磨，粗略勾勒出线脚凹凸。精雕过程主要是在砂岩石块表面雕刻装饰纹样，线脚的凹进和凸出更加丰富，弧线更加饱满。三层须弥台基座的每层都在大致相同的位置分布有这三种阶段的石块，但进度略有不同，以下层加工进度最慢。由此可以推测，茶胶寺庙山建造工程可能在以角砾岩和砂岩将巨大的三层须弥台砌筑完成之后戛然而止，当时对砌石的粗凿加工接近尾声，雕饰线脚纹样的精雕也已经开始一段时间。由此或还推测，上、中、下三层须弥台的建造基本上同时开始，开始的位置和雕刻移动的方向也大致相同，但因为下层台基石块较多，因而在工作中断时，尚未开始对南面砌石进行精雕。

另外，在三层须弥台砂岩立面砌石上存留下一些排列无序、近于圆形的孔洞，可分为两类，一类孔洞较大，数量较少，其直径约6厘米，深8-9厘米；较小的孔洞直径约4厘米，深4.5厘米，数量较多。这些孔洞是建造寺庙时为插入用于抬起石块的木质工具而专门雕凿的。比较而言，毛石上存留的孔洞较多，精雕之后的线脚上可见的孔洞最少，粗凿阶段的砌石上可见的孔洞数量居中。在从毛石到粗凿再至精雕过程中，不断将小石块从原石上切下，完工之后的砌石比毛石阶段体积减小很多，精雕之后的砌石将呈现出完整的纹饰图像，砌石表面的孔洞痕迹也逐渐随之隐去。（图4-17，图4-18，图4-19）

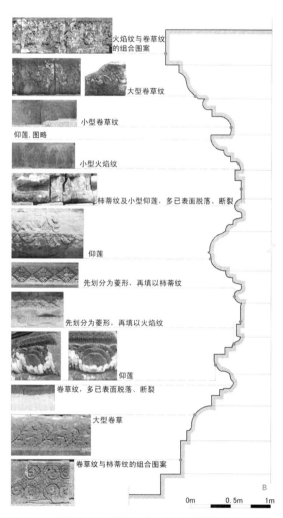

图4-17 茶胶寺庙山须弥台第一层雕刻纹样分布示意

---

[1] Jacques Dumarcay. *TA KEV: ETUDE ARCHITECTURALE DU TEMPLE*, Paris: EFEO, 1971, p. 31.

第四章 茶胶寺庙山的建筑形制

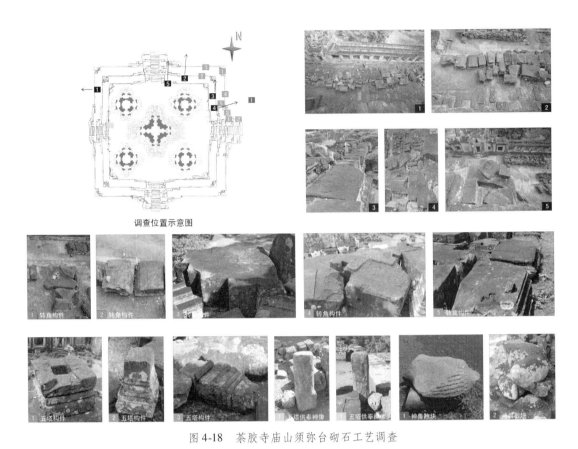

图 4-18 茶胶寺庙山须弥台砌石工艺调查

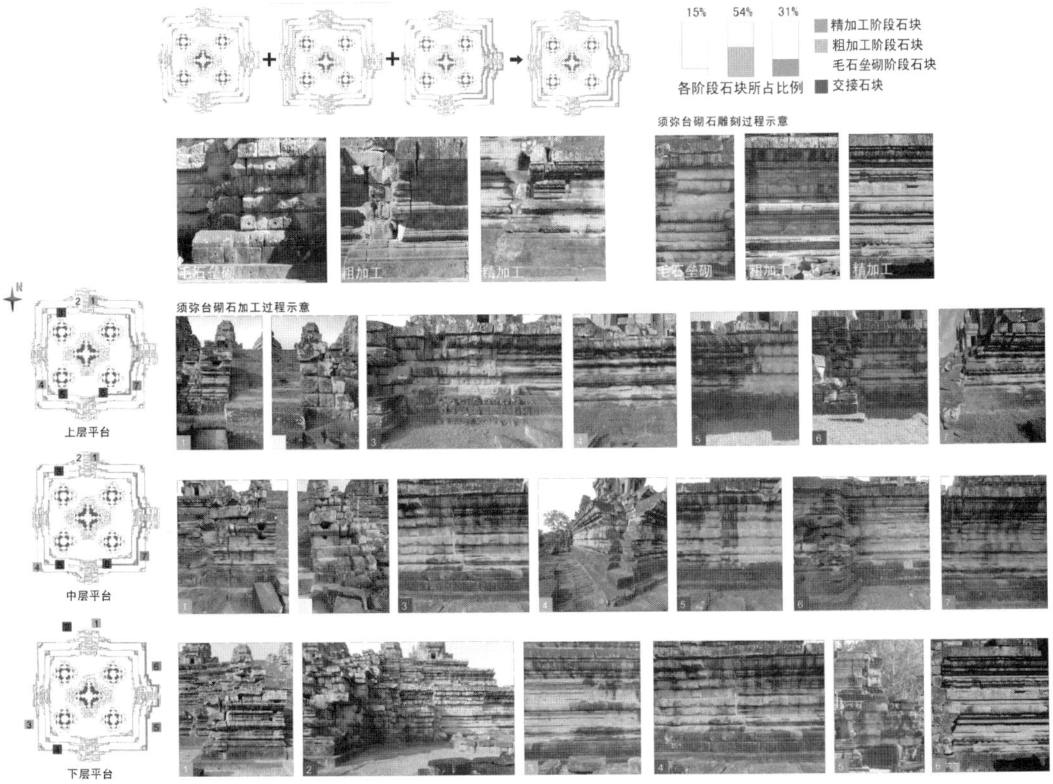

图 4-19 茶胶寺庙山须弥台砌石工艺流程示意

4. 中央五塔

茶胶寺庙山的中央五塔坐落于须弥台顶面之上，其构造完全以砂岩石块砌筑而成。在中央五塔的整体布局之中，中央主塔踞于中心位置，而在规模尺度方面略逊于中央主塔的四座角塔则分别踞于须弥台之四隅。五塔的平面形式皆为四面出抱厦的十字形平面，较之四座角塔，中央主塔在其抱厦和中厅之间增出过厅一间。五塔的建筑形制大致相近，唯因四座角塔的规模较小，因而中央主塔在平面、塔基、山花的构造方面显得更为繁复[1]。

通过不同砂岩石材之间较为清晰的分界线则可以分辨出[2]，除底部的塔基部分使用普通长石砂岩外，砌筑庙山五塔所使用的砂岩主要是莫氏硬度超过5.5的硬质砂岩[3]。或许由于这种硬质砂岩的加工切削较为困难，除角塔内部檐口施以雕刻之外，在绝大部分的硬砂岩砌石表面仅仅完成了初步的雕凿或削平。硬质砂岩的雕凿痕迹致密均匀，致使其表面呈现出一种特殊的凿毛粗糙质感。因为庙山五塔的立面砌石仅为粗略雕凿轮廓且未作任何雕饰，或可佐证五塔的建造工程应没有最终完成。[4]另外，在四座角塔的西门，东北、西北角塔的北门，西北、东南角塔的南门皆施以红色角砾岩砌石封闭，由此亦可推测茶胶寺庙山五塔的形制及其建造过程在历史上可能曾经发生过某些改变，但其中的具体细节尚不清楚。

图 4-20 茶胶寺庙山内发现的尚未完成雕刻的构件

茶胶寺庙山五塔的外观立面自下而上可以划分为塔基、塔身、假层、塔顶及塔刹等五个部分，现将其各部分的形制、装饰及其内部空间构成分述如下：(图5-20，图5-21，图5-22，图5-23，图5-24，图5-25)

（1）塔基

庙山五塔砌筑于砂岩塔基之上，中央主塔与四座角塔的塔基形制差别显著。

角塔的塔基立面可以划分为两段，每段系由4层砂岩叠砌而成。塔基四面的踏道两侧分别设有单

---

[1] "人们经常把高棉建造者的才能归结为一种从木建筑到石建筑的简单转化。我们认为这里情况并非如此，如果说某些做法，比如那些应用到门框上的做法并不十分适合于石材，那么茶胶寺的建造者已显现出对石材切割术非常实际的认识，并在方案设计和宗教象征的限度内直接以石材来构思寺庙，其构思显然很精密。" 参见 Jacques Dumarcay. *TA KEV：ETUDE ARCHITECTURALE DU TEMPLE*, Paris：EFEO, 1971, p. 33.

[2] 砂岩（Arkose and Graywacke）是一种致密的黏性沉积岩，其主要成分是直径在0.1毫米至1毫米的石英颗粒。吴哥地区的砂岩品种大致分为三类：灰黄色长石砂岩、红色石英砂岩和灰绿色长石玄武岩（硬砂岩）。其中以长石砂岩应用得最为广泛，其强度在三者之中最低，系由中小粒径的颗粒组成；石英砂岩组成颗粒的粒径亦较小，但其强度较高；硬砂岩的内部结构不同于其他两种砂岩，空隙率较小，硬度相对较高。（著者注）

[3] 通过中央主塔东南角的某些出现残损的部位亦可看到作为填充砌石的长石砂岩的存在。

[4] 茶胶寺庙山建造工程并未最终完成，通过遗存至今的不同雕饰阶段的痕迹，为研究施工流程提供可能。总体而言，施工流程应是在完成砌筑庙山各部轮廓之后，再对其进行雕饰和装饰。须弥台顶部五塔系硬质砂岩砌筑，未作精细雕刻，而在其他各座单体建筑之中，塔门的完成程度较高，其雕刻工序可以大致分为粗凿、初雕和精雕等三个阶段。精雕部分的山花与假层大多位于建筑的上部；建筑下部多为初雕与粗凿阶段，由此可见施工流程中的雕刻应是自上至下完成的。花柱系预制的石雕构件，在其安装之前贴近壁面的纹饰皆已完成雕刻，而其他饰面则为初雕状态，待花柱安装完成再行进一步的精雕。（著者注）

层垛台。中央主塔的塔基立面则可划分为三段,体量庞大且构造复杂,在其塔基四面每段踏道的两侧亦设有两层垛台。此外,中央主塔东立面的塔基与踏道顶部之间隔有一个小的阶梯平台,而在其南立面与北立面上,塔基则与踏道直接相连。值得注意的是,中央主塔四面踏道的上端宽度较下端略为收窄,踏道坡度约为47°,其上下两端水平方向的夹角约为10°。这种尤其在中央主塔东、西立面的塔基顶端通过收窄踏道宽度而产生强烈的透视效果的方法,显然是有意为之,由此更加彰显突出中央主塔高峻挺拔的庙山核心地位。

（2）塔身

从整体立面上来看,可将庙山五塔的塔身分为基座（散水）、塔身、檐口等三部分,而依其平面位置的差异又可以分为中厅、过厅和抱厦等。基座位于石砌塔基之上,是塔基与塔身之间的连接部分并可承担着建筑散水的功用。塔身及其檐口部分皆以尺寸不等的砂岩石块砌筑,各层及砌石之间未施以任何粘接材料,砌石水平分缝的贯穿位置略有差异,而竖向分缝则无甚规律可循。[1]上述砂岩砌石之间的水平分缝并非以一条直线贯穿始终,通常的做法是在位于下层砌石表面上凿出凹槽或凹面,从而使其上的砌石置入凹槽（凹面）内即可阻止产生沿水平方向的滑移;又因其凹槽或凹面的位置及方向并不固定,借此亦可阻挡横向侧移以增强砂岩砌体的稳定性[2]。五塔的檐口部分及其山花的标高随塔身各部分的高差而转折起伏,以中厅最高,过厅次之,抱厦最低。

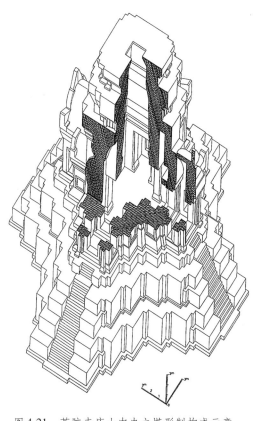

图4-21 茶胶寺庙山中央主塔形制构成示意

（3）假层

假层坐落于中厅的檐部之上,其平面呈逐级向内收分缩进的方形,亦可视为中厅平面外缘向内逐层缩进。假层的形制与塔身主体相类似,四面皆辟为假门,假门的两侧设有壁柱,其上部施以门楣及山花。五塔的假层部分残损状况甚为严重,其中,四座角塔的假层皆现存三层,中央主塔的假层仅余两层。

（4）塔顶

塔顶即为五塔的中厅、过厅及抱厦之上的屋盖部分,其构造皆施以叠涩拱逐级向内收进[3],拱身通高约为2.2米。五塔现存的塔顶部分残损甚多,唯中央主塔的过厅部分保留尚算完整。

---

[1] 砂岩石块砌层通过角尺校正,砌缝显出向内稍陷,通常是小于1厘米。在这一凹陷内对接着上部砌层相应的稍凸部分,这避免了石块向前滑动。在第一层围墙与第二层回廊外墙上,每一砌层的砌缝都凿有一个很长的榫孔,上部砌层相应的榫头正好插入其中。另外,墙体转角处的石块砌筑也是经过精心设计的。转角处多以"L"形石块为主,普通的条形石块为辅,其中"L"形石块起到了重要的加固作用。

[2] 砌石抗滑移系数越大（即石块表面越粗糙）,结构中的应力和滑移都会相应地降低,结果趋于稳定。参见中国文化遗产研究院、湖南大学:《茶胶寺典型单体建筑结构危险性评估与加固技术研究》,2011年6月。

[3] "尽管处于完全的暗部之中,依然对其进行了细致的雕凿,其剖断面几近完美的半椭圆形。叠涩拱的根部起点以轻薄的承托构件作为标志,托架在山墙上也未转向。"参见 Jacques Dumarcay. *TA KEV: ETUDE ARCHITECTURALE DU TEMPLE*, Paris: EFEO, 1971, pp.31-32.

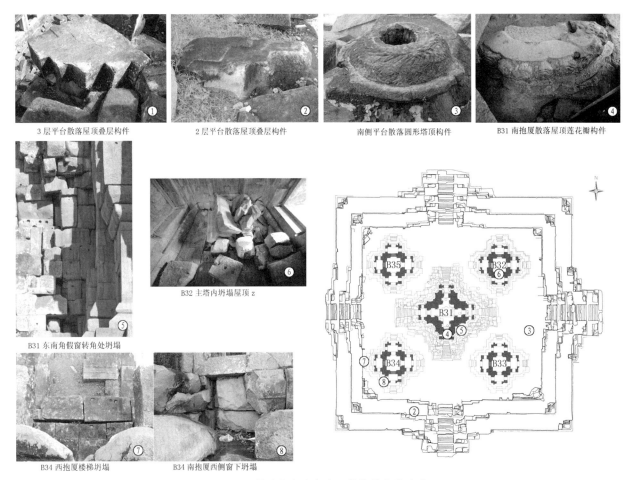

图4-22 茶胶寺庙山中央五塔散落构件分布

（5）塔刹

由于残损严重或未完成的缘故，五塔的塔刹皆已不存，在须弥台顶部地面仅存有类似塔刹残件的石构件数块。[1]

（6）山花

山花是五塔形制之中最为突出醒目的重要建筑构件[2]。虽然五塔的山花皆未作雕刻及装饰，仅粗略雕凿出简略的轮廓，但依其所处位置及形制差别仍可将五塔的山花分为如下三类：

---

[1] 帕蒙蒂埃（H. Parmentier）曾在中央主塔的北侧抱厦内发现一块石质构件残件，推测可能是塔刹中的某个部件。它已完成精雕，其线脚装饰包括一条缀有翻转莲瓣的反枭线，由此构成了另一条枭线的基底，该枭线支撑着莲瓣，石块在其中心被凿穿了一个直径6厘米的圆柱孔槽，整个组件应有约70厘米直径。

[2] "抱厦在立面上包括两根成列壁柱以支撑山花，并以两根花柱支撑门楣。在抱厦的侧立面上，各座窗框仅做修削。因此在立面上，连通抱厦的梯形墙石是光秃的，墙基与檐口之上是半个山花。半山花之上，墙面重新继续，饰有一道檐口，并以此支撑山花斜面的根部作为起点。此山花标高本应高出抱厦山花，此种布置仅在东立面上存在。抱厦侧立面曾设有七根窗棂（皆已不存）的假窗。主体建筑轮廓则是通过尺度琐细的梯形墙呈现出来，而梯形墙的走向却在抱厦檐口标高处停顿，在檐口之上又重新继续。主体建筑上方有一道檐口，檐口在平面上仅显现为单独的凸起，确切地说正是在这一平面上矗立起了塔，现在只留下逐层收分凹进的两个假层。在各个立面，假层在主体建筑的突出部分设为假门，其门扉石多经过粗凿。两根花柱支撑门楣，两根壁角柱支撑山花。主体建筑墙面之上是一道檐口，其顶部束腰在塔的当前情况下，要高出假门的山花。"参见 Jacques Dumarcay. TA KEV: ETUDE ARCHITECTURALE DU TEMPLE, Paris: EFEO, 1971, p.32.

正面山花：主要是指五塔抱厦与过厅之上的山花。此类山花按其形制又可分为单山花与双山花两类。其中中央主塔抱厦之上的山花为双山花（中央主塔的东立面与西立面可以看出明显的雕刻痕迹），其余位置皆为单山花。单山花砂岩砌石分为三层，而双山花砂岩砌石则分为四层。

侧翼山花：主要是指位于中央主塔过厅屋盖之侧面的山花，系由两层砂岩石块砌筑而成。

假层山花：主要是指位于每级假层的假门之上的山花，其构造方式实为假层砌石尺寸加大并略向外凸出大约10-15厘米。假层体量逐级向内收分，因而山花尺度也逐级缩小。其中，四座角塔假层的山花皆构为两层，尺寸较厚的一层采用方形砌石，而置于顶端的一层尺寸相对较薄，皆采用平砌方式。

（7）室内

各座角塔中厅室内檐部皆保存有雕饰线脚，题材多以浅浮雕莲瓣为主。除各塔在门框表面处施以雕饰线脚之外，其余皆无精细的雕刻与装饰。五塔中厅室内檐口的上沿发现有安装天花板的凹槽痕迹，至凹槽以上的塔心砂岩砌石皆经过精细砍削处理，以致每一处收分转折之处都以一个精细雕凿的斜面为标志（共分为5层）。中央主塔的中厅室内地面为砂岩石铺砌，自中心向中厅四周放出缓坡，坡度约为1/20。中厅的旧有格局已面目全非，由当地土著散乱摆放的各类雕像及建筑构件堆砌在中厅中央。四座角塔室内格局及雕像皆已毁失不存。东北、西北、东南三座角塔内，仅余承托双座雕像的基座残迹且断裂成碎块；西南角塔内的基座仅可承载单座雕像，亦已散落断裂。各座雕像基座下方应为与基座连为一体的方墩（或被凿出槽穴以容纳奉祀祭品）。基座直接安置于地面之上，其线脚部分被置入凹槽的石构件遮挡。

图4-23　茶胶寺中央五塔砌石石材类型示意

（8）门窗

五塔的窗依各部分形制差别可分为真窗和假窗两类，以上两类门窗抱框构件皆以八字插榫进行连接。五塔的真窗尺寸与相对应的门洞大致接近且参差不齐，其高度约为1.8米（高宽比约为2:1）。假窗则设于中央主塔过厅的两侧，假窗抱框表面饰有粗雕的线脚，其内侧尺寸高约为0.94米，宽约为0.88米；虽然窗框内侧壁面安装窗棂的上下榫孔仍清晰可见，但现场却未发现散落的窗棂构件。通向抱厦的门洞尺寸约为高2.25米，宽1.25米，门框之上为粗凿门楣雕饰，门框上方施以叠涩拱[1]（抱厦侧向窗上部门楣并未施以叠涩拱支撑），其凹槽或以木板内外相隔（叠涩拱朝向抱厦的开口原为固定在横向支架上的木板所遮盖，安装横向支架的凹槽已经凿至拱顶内皮）；另外，门洞两侧上下均有榫槽，疑为旧时安装木门所致。门框内外两侧亦为粗雕线脚，门洞两侧仍保存有安装花柱所用榫槽的残迹，但现场并未发现形制完整的花柱遗存，仅有部分未完成的花柱残迹。

---

[1] 这种做法目的是将上部荷载分散至门框两侧的墙体上，从而起到保护门楣的作用。

图 4-24 茶胶寺庙山中央五塔顶部塌落的构件

中央主塔的整体稳定性较好，三层基座局部转角构件缺失，但不影响其稳定性。塔身稳定部分区域出现结构裂缝，部分梁体受压断裂，石材有较严重的风化、破损现象。东北角塔基座基本稳定，南侧与东侧因基台坍塌而出现下沉现象，东抱厦、南抱厦分别与塔身开裂且向外倾斜（约 3 厘米）。西抱厦壁柱向外错动距离接近 10 厘米，塔身部分构件被压裂。东南及西北角塔整体稳定性，顶部大量石构建塌落，基座上的石构件缺失较多，个别构件压裂。

5. 塔门

如前所述，在茶胶寺的第一层基台（庙山外院）围墙的四面中央沿庙山轴线分别辟为外塔门四座，分别称作：东外塔门、南外塔门、西外塔门、北外塔门；第二层基台（庙山内院）回廊四面中央沿庙山轴线分别辟为内塔门四座，分别称作：东内塔门、南内塔门、西内塔门、北内塔门。

以上八座塔门根据其平面构成、门道数目、山花排布等不同情况可以分为 I 型（东塔外门和西外塔门)、Ⅱ型（南外塔门和北外塔门）、Ⅲ型（东内塔门）、Ⅳ型（南内塔门和北内塔门）、Ⅴ型（西内塔门）等五种类型；若以塔门平面布局形式划分，则可将其分为"一"字形、"丁"字形、"十"字形等三种形式。其中，塔门的平面布局皆可分为五种基本空间单元：主厅、内侧室、外侧室、过厅、抱厦，这些空间单元进行组合而构成形制各异的塔门。建造塔门的建筑材料主要以砂岩为主，另外还应包括局部用于砌筑塔门屋顶的红砖、塔门基座内部填充的角砾岩以及制作门扇或室内装修的木材等。

作为进出庙山的通道，各座塔门亦构成重要的祭祀空间，其主厅内部通常奉祀林伽或神像[1]。八座塔门之中，较之西侧塔门，东侧塔门坐落于庙山正面，地位显赫且等级最高，因而建筑形制亦最为复杂。虽然各座塔门的等级形式并非对称一致，但整体而言，外塔门的体量皆大于内塔门。例如，东外塔门（Ⅰ型）为十字形平面，面阔五间为14.93米，进深为9.09米，[2]前后各出抱厦一间[3]，中厅与南北两间内侧室相连，内侧室的两侧则为南北外侧室各一。[4]而东内塔门（Ⅲ型）仅在其内侧出抱厦一间，其外侧仅以双层山花叠置取代抱厦；至于南、北两侧的塔门，Ⅳ型的内塔门正面皆施以双层山花叠置，而外塔门（Ⅱ型）的正面山花则为单层山花。由此可见，内塔门的规制等级似乎略高于外塔门。[5]

构成各座塔门的核心是主厅及其上部空间，逐层收进方形平面的假层直接砌筑于中厅四周墙体之上，主厅入口门楣施以两层叠涩拱支撑。由于叠涩拱的外侧被门楣所遮挡，根据现存榫孔推测叠涩拱内侧原应为木板覆盖。另根据主厅入口门洞抱框表面残存的榫孔痕迹，亦可推测塔门原先应设有木制门扇。塔门的抱厦、内侧室及外侧室的作用虽为次要，但也皆施以叠涩拱顶覆盖。内侧室的叠涩拱并未直接坐落于墙体之上，而是由从墙体檐口收进的鼓座承托。外侧室山墙则以壁柱支撑其上部的山花及假门，其内侧以花柱及粗凿门楣作为装饰。

---

[1] 东外塔门中厅门框及其西抱厦门框发现有明显的砍凿痕迹，凿痕距室内地坪约为60-70厘米。同样的情况也出现在东内塔门中厅西侧的门框上，雅克·杜马西认为古代高棉庙山建筑的建造顺序一般是先定下东西方向与南北方向的两条轴线，两条轴线在多数情况下都被镌刻于奠基石上并在位置上供奉林伽，然后再建造容纳林伽的塔门。因此这些被砍凿的门框原因可能是由于宗教信仰改变时尺度大于门框的林伽或神像基座被搬移出塔门的缘故。法国远东学院的亨利·马绍尔（H. Marchal）在二十世纪二十年代对茶胶寺庙山进行清理时，曾经发现并挖掘了位于东外塔门中厅内的轴线基石。参见 Jacques Dumarçay & Pascal Royére：Cambodian Architecture：Eighth to Thirteenth Centuries, Leiden；Boston；Köln：Brill, 2001. p. 137.
[2] 从东抱厦入口台阶外地面至现存最高点高度为11.62米。
[3] 东外塔门的东抱厦室内南北长2.25米，东西长1.63米，檐口至室外地面高约6.4米，南、北墙各开有一明窗。西抱厦室内南北长2.28米，东西长1.06米，檐口至室外地面高约4.35米，南、北墙各开有一明窗。
[4] 东外塔门南侧室室内东西长2.23米，南北长1.93米；北侧室室内空间与南侧室相同。南北外侧室西墙开有明窗，墙厚平均0.7米，窗间柱为5根；东墙为假窗，墙厚平均0.75米。窗间柱均7根。侧室檐口至围墙外地面高6.47米。
[5] 以南内塔门为例，其基座总高5.52米，其中下层基台高3.12米、上层基台高2.4米。中央辟为连接庙山第一层基台和第二层基台顶部的高台阶，台阶宽3.3米，每步台阶平均高40、宽22厘米，共14级。主厅和东西侧室基座高0.55米，东西外侧室基座高0.44米。南内塔门平面呈"一"字形，面阔五间15.8米，南北进深4.5米，残高7.87米。中厅南门与庙山第二层基台的高台阶相接，北门则进入庙山第二层基台的院落之中，门宽1、高2.2米。门框外侧两边分别立有直径28厘米的八边形花柱，花柱紧贴门框和外侧壁柱，花柱支撑长2、高0.64、厚0.3米的门楣，门楣上方为体量硕大的山花。中厅上部第一层假层墙体外边缘东西长4.7、南北长4.2、高1.42米，东西两侧设有假门；第二层假层残高0.8米，四面设有假门，南北两侧的假门上方是双山花形式。主厅平面呈方形，内部空间东西长2.7、南北长2.37米，从地面至已塌毁屋顶残高7.5米。中厅两侧有宽1.19米、高2.06米的门通往东西两侧室。东西侧室台座高0.43米、外墙东西长2.48米、南北宽3.02米，台座上皮至檐高3.69米，东西外侧室台座高0.36、外墙东西长3.19、南北宽2.6米，台座上皮至檐高3.15米。在每间房屋的南北墙上都有一扇窗，北侧为通透的真窗，南侧均为封堵的假窗，每扇窗上均有七根直径14厘米的窗柱，窗宽1.3、高1.42米。东侧室与东外侧室、西侧室与西外侧室的内部空间分别相通，均为一间房屋，东西长5、南北宽1.76米。

茶胶寺塔门形制或许是在其原有的十字形平面木构寺门基础上，将主室中央的屋顶部分以逐级收进的"假层"从而构成塔殿的立面形象，推测这种形制可能是茶胶寺庙山建筑的创制之一，及至茶胶寺庙山之后巴方寺的塔门，塔殿的形象更为典型完整，逐层收进的塔身完全以莲花宝顶作为收束。

上述塔门，虽建筑形制不尽相同，但自下而上又皆可将其建筑元素分为基座、墙体、假层、屋顶、山花等部分，现将其各部分形制概括如下：

（1）基座

塔门基座形制可以分为两类：

其一，或可称其为"大须弥座"，这是一种与庙山第一、二层基台结合建造的塔门建筑基座形式，即在塔门各座门道的相应位置设置阶梯状的须弥座，高度约为2.15米，其线脚元素由低到高包括：细平条线脚、反枭线、浅凹条、斜面、逐层凹进的两条平条线脚、平条线脚、平栱。"大须弥座"又可分为"整体式"和"逐层收进式"等两种形式。其中，第一层基台上的南外塔门与北外塔门（Ⅱ型）属于"整体式"须弥座形式，而位于同一层基台上等级规制更为显要的东外塔门与西外塔门（Ⅰ型）的基座采用"逐层收进式"须弥座形式[1]；第二层基台上的四座塔门基座则皆为"逐层收进式"须弥座形式。

其二，或可称其为"小须弥座"，实为在各座塔门大须弥座顶面上所设置的基座，高度为0.45米，其线脚连续包括：平条线脚、逐层凹进的两条平条线脚、初具莲瓣饰形的枭线、逐层凹进的三条平条线脚、并由斜面来靠上墙的光面或壁柱柱身。由于"小须弥座"直接置于基台顶面之上，根据其所处不同位置亦可分为主厅须弥座、内侧室（次间）须弥座、外侧室须弥座以及前后抱厦须弥座等几类。小须弥座的标高高于庭院地面，其中又以主厅须弥座的标高为最，内侧室须弥座次之，外侧室须弥座标高最低，而前后抱厦须弥座的标高则与内侧室须弥座的标高略同。

（2）墙体

塔门墙体自下至上皆可分为墙基、墙身、檐口三部分，而根据其位置则可区分为中厅墙体、内侧室墙体、外侧室墙体、前后抱厦墙体等。

墙基建于"小须弥座"之上，标高亦随之高低变化。不同部位墙体的檐口标高不尽相同，其中以主厅最高，内侧室次之，外侧室最低，前后抱厦的檐口标高则与内侧室相同，整体呈现出中间高两边低的形式。另外，内侧室及外侧室檐口之上原置有端头瓦，今皆已不存。

墙身由砂岩石块砌筑而成，根据构造方式及位置随形设计切割，例如外侧室转角将砌石切割成"囗"形截面作为壁柱。墙体砌石间的水平分缝多以贯通为主，水平分缝一直贯穿外侧室、内侧室、中厅以及前后抱厦，直至门窗处断折为止；其竖向分缝规律不甚明显，通缝与错缝的情况皆有。门道之处的墙体断开，设置门框、壁柱、花柱等。

塔门各部分檐口线脚雕饰风格类似，唯中厅的墙基和檐口部分雕饰更加丰富细腻，其线脚雕饰由低到高包括：反斜面、凹条、平条线脚、细刻莲心的板条、逐层凹进的两条平条线脚、细刻莲瓣的枭线、四条皮条线夹着凹圆线、刻有从角部分岔的叶束饰的枭线、平条线脚、精雕花饰且环绕以彼此相

---

[1] 以东外塔门为例，其中厅第一层基台高约1.34米，第二层基台高约1.03米。南、北入口第一层基台高约1.52米，第二层基台高约0.89米，设有踏步11级。

对并从同一点衍生出的双叶枝蔓装饰的束腰。

（3）假层

塔门主厅之上皆设有三层且逐层方形平面收分的假层，并与主厅的墙体檐口部分相接。每一假层结构类似，四面外凸两级形成假门，假门由两层抱框组成，其最外一层抱框由壁柱支撑山花构成，内层抱框则由花柱支撑门楣构成。假层的内壁砌石打磨平整，逐渐向上收分，与外部结构完全没有对应关系。假层墙体砌石分为上下两层砌筑，较之使用内外两层方型石块的砌筑方式，假层通常采用长条砂岩砌石且"丁头冲外"以增强整体的稳定性。假层的基座平面为方形且在两翼皆有突起且有线脚雕饰，上假层基座即是置于下假层突起的边缘，其线脚元素由低到高包括：缀有从角部分岔的斜叶束饰的反枭线构成了檐顶线脚，四条皮条线夹着浅凹条，缀有莲瓣的枭线，逐层凹进的三条平条线脚，平条线脚，带菱形内接花饰的束腰。塔门假层四面皆设假门，假门两侧出壁柱以支撑假层山花（多与塔门山花构成双层山花）。下假层假门因其多被内侧室及前后抱厦屋顶遮盖，构造及工艺较为简单粗糙。

（4）屋顶

各座塔门屋顶多已全部坍塌，仅从山花构件遗存的拱形曲线等痕迹推测应为砖砌叠涩拱结构。

（5）门楣与山花

门楣与山花是塔门装饰的重要部位且情况较为复杂，不仅包括壁柱所支撑的主轴线门道上部和次轴线门道上部的门楣与山花，还应包括装饰性假层小壁柱所支撑的小山花。[1] 由于塔门山花的设置具有很大灵活性，与其他结构的交接情况亦不相同，因而出现了诸多类型的山花，山花尺度的不同则又进一步导致其装饰内容的变化。再者由于山花结构的不稳定性，大多已经塌落，通过现场调查情况来看，塔门山花多由三至四层砂岩砌石雕刻组合而成。山花底层砌石厚度较大，两端边缘部分雕刻蛇神那伽（Naga），蛇神顶端略凸出部分以承托山花第二层砌石，其厚度与第一层相近，边缘则为卷叶蔓草雕饰。山花顶层多以茎叶雕饰且高度较高，也最易坍塌损毁。根据其所处位置及作用，诸座塔门的山花可以大致分为三类：正面山花、侧翼山花、假层山花等三类。

正面山花位于塔门主轴线门道上部，以壁柱、花柱及门楣承托，是塔门雕饰的重要组成部分。

侧翼山花位于塔门次轴线门道上部，分为内侧室山花、外侧室山花等。侧翼山花背依屋顶，仅在正面施以雕饰。

假层山花属于装饰性假层雕饰之一，与假层构造融为一体，由于体量较小，假层山花实为以构筑假层的砌石外立面直接雕刻而成的。

（6）花柱

置于门道两侧的花柱亦为塔门重要的装饰部位之一，与各座塔门的内侧门框和外侧壁柱支撑山花的结构作用相比，塔门花柱诸多的雕饰细节都被赋予了极其丰富多样的装饰内涵，并且具有鲜明的风格演变的时代特征。一直以来，对于花柱的讨论和断代成为吴哥艺术史家最为关注的研究对象，取得的成果也颇为丰硕（详见本书第二章）。

---

[1] 茶胶寺保存完整的山花主要集中在塔门，东外塔门外放置的拼对山花、南外塔门及南内塔门的山花不仅雕刻完整，保存状况也相对较好。

表 4-2 茶胶寺庙山塔门山花形制简表

位置	层数	宽度（mm）	高度（mm）	说明
东外塔门抱厦山花	3	3200	2450	宽高比 1∶0.765
南外塔门内侧室山花	3	3050	2300	宽高比 1∶0.754
南外塔门屋顶山花	3	2400	1860	宽高比 1∶0.775
西外塔门内侧室山花	4	4480	残高 2300	现存三层
西内塔门内侧室山花	4	4050	残高 2050	现存三层
西内塔门外侧室山花	3	3750	2920	宽高比 1∶0.780
北内塔门内侧室山花	3	3755	2835	宽高比 1∶0.755

6. 长厅

茶胶寺庙山的四座长厅分别位于第一层基台（庙山外院）与第二层基台（庙山内院）的东北角及东南角。第一层基台上两座长厅依其位置分别称为南外长厅和北外长厅，第二层基台上的两座长厅则分别称为南内长厅和北内长厅。基台上的南北长厅皆沿庙山东西轴线对称布局，现将其形制叙述如兹：

南外长厅[1]坐落于第一层基台东南角围墙内侧，东外塔门之南侧[2]。建筑布局坐南朝北，其平面略呈"凸"字形，自北至南依次由抱厦、主厅、后室等三部分组成。南外长厅的南北通进深为 27 米，抱厦、主厅及后室的面阔依次为 2.4、3.6、3.5 米。[3]

抱厦面阔一间，进深三间，三面开敞。抱厦的基座高度为 0.34 米，抱厦外侧尺寸南北长为 7.7、东西宽为 3.3、残高约计 3.2 米。石柱边长 0.39、高 2.35 米，顶部支撑跨度 5.8 米（底高 4.9 米）的过梁，过梁以上的屋盖现皆不存。抱厦及中厅正门均为宽 1.5、高 2.2、门槛宽度为 0.6 米。抱厦正门两侧皆立有直径 24 厘米的八边形花柱一对，花柱上刻有浅浮雕纹饰；花柱及门框石支顶上部厚重的门楣，门楣之上立有山花。各柱头皆无雕刻，柱础仅以砂岩条石相连并刻有线脚。

主厅东、西两侧各设 9 扇明窗，连续排列构成一个长方形空间，窗框为单层砂岩结构，厚 0.8 米，窗框四边为八字形斜交，有榫槽相互插接，窗框上有雕刻的线脚。每窗设有圆柱形窗棂 7 根，窗棂为整块砂岩雕刻，两端有榫头（上下窗框雕有榫槽），并雕有复杂的分层线脚装饰。各间窗框上是雕有

---

[1] 北外长厅的形制与南外长厅相同，不再赘述。
[2] 南外长厅残损甚为严重，屋顶已经全部坍塌，残存的墙体变形开裂，尤其是中厅西墙整体向东侧倾斜，现以临时性木结构斜撑予以支护。
[3] 每座均包括一个大厅，它通过抱厦向东塔门敞开。在相反方向，从主厅延伸出较窄的一个小厅。大厅在每面纵向墙上通过九个窗来非常充分地采光，窗以九根窗棂来装饰。小厅在每面纵向墙上通过一个窗来采光，有七根窗棂。山墙是不透光的。这两座屋的门没有装门扇的槽。开向抱厦的门左右夹以花柱，花柱支撑着带装饰的门楣，其前立有两根壁角柱，它们应该只与抱厦的柱顶楣高度上的山花相接。柱顶楣落在壁角柱和四根支柱上，它撞在山墙上的位置也就是它落在两根壁柱上的位置。这面墙上开了一个门，门上包含了惯常的装饰。在室内，除了门窗框之外没有任何装饰。在室外，墙基是带线脚的。在抱厦，这一墙基构成了小栏墙，支柱和壁柱就嵌入其中。大厅和抱厦之间没有高差，只是小厅的墙基更低。这一墙基包括：厚勒脚，凹条，两条平条线脚夹着半圆线，凹圆线，两条平条线脚夹着板条，凹条，最后由斜面来攀上墙的光面。顶部的檐口翻转复制了相同的线脚，再上面是厚斜面与一排端头假瓦，其凿剪仍未完工。这两座建筑曾覆盖了由梁架支撑的瓦，多亏某些梁木的槽穴尚在原位，由此我们认为可局部复原这一屋架：这一屋架由两根 45°倾斜的骨架构成。在大厅，屋盖曾由八个落在窗的侧柱上的桁架来支撑。抱厦的屋顶则只落在两个桁架上，小厅想必也如此。

线脚的墙檐，墙檐上部的结构均已塌落。其中，砂岩砌石外墙南北长28.8、东西宽3.6、残高3.28米，每扇窗长1.62、高1.8、厚0.45米，窗间距平均为0.74米。中厅进入后室的门宽1.5、高2.2、门槛宽0.6米。主厅正门门框两侧设有壁柱，门框上方的门楣及山花皆无雕饰。

后室进深及面阔皆为一间，砂岩砌石基座高0.38米，外墙南北进深7.7米，东西面阔3.3米，残高3.35米，墙厚0.57米；与中厅连接入口处设有门框，未设花柱；东、西两侧墙中央各辟有明窗1扇，窗洞宽1.5、高1.55米，7根窗棂形制与主厅相近但尺寸更小。后室的高度比抱厦和主厅略低。

北内长厅[1]坐落于第二层基台东北角回廊内侧，东内塔门之北侧。其建筑布局与外长厅大致相同，坐北朝南，其平面亦略呈"凸"字形，自南至北依次由抱厦、主厅、后室等三部分组成。北内长厅南北通进深（含墙身厚度）为16.88米，抱厦、主厅及后室的面阔（含墙身厚度）依次为3.14、3.7、3.24米。根据北内长厅山墙上残存排列有序的槽孔痕迹推测，内长厅可能采用木结构屋架支撑其屋盖。

抱厦面阔一间，进深两间，三面开敞。抱厦的基座高度为0.24米，抱厦外侧尺寸南北长度为5.49、东西宽为3.14、残高约计3.4米。砌石方柱共有两对，柱径为0.6米，高约2.92米，柱头亦无雕刻，柱础相连并有线脚雕刻，入口处有门框，门框两侧有八角形壁柱，壁柱上有浅浮雕纹饰。柱头以上的屋盖现皆不存。

主厅入口设有壁柱及门框，门框上方门楣素面无雕刻。主厅墙壁施以单层砂岩石块砌筑，墙身分墙基、墙身和檐部三段，西侧墙壁中央辟为明窗一扇，窗框为单层砂岩结构，窗框四边为八字形斜交，有榫槽相互插接，窗框上有线脚雕刻。窗框内安装圆柱形雕饰窗棂7根，皆为整块砂岩雕刻而成，窗棂两端有榫头（窗框上有榫槽）并有复杂的分层线脚装饰。窗框上有带线脚雕刻的檐部。檐口上部已基本全部塌落，唯有后室入口墙壁之上残存的山花较为完整。山花由三层砂岩石块堆砌组合而成，其表面的雕刻仅为粗略的轮廓并未完成；根据山花内侧呈三角形排列的五个榫槽，或可推测其木结构屋架的式样。后室入口设有抱框且不设壁柱，西侧墙中央也辟为明窗一扇，形制与主厅窗相近但尺寸更小。后室和抱厦高度比主厅略低。

7. 藏经阁

藏经阁位于茶胶寺庙山第二层基台（庙山内院）的东侧，沿庙山东西向轴线南北两侧对称设置，按其位置分别称其为南藏经阁与北藏经阁。从整体风格来看，以上两座对称布局的藏经阁形制相同且保存状况类似[2]，其空间布局、使用功能及创建年代等理应大致相同。[3]

藏经阁平面略呈"凸"字形，其整体布局自西至东依次分为抱厦、主室、门头等三部分；自下至上则可划分为基座、墙体、拱顶、顶窗及屋盖等几部分。以北藏经阁为例，其基座东西向长度为12.5米，南北向长度为7.91米，其最高点至第二层基台为6.48米；抱厦东西向长度为1.53米，南北长度为1.89米；主室东西向长度4.91米，南北向长度3.01米，残高约5米。

---

[1] 南内长厅的形制与北内长厅相同，不再赘述。
[2] 南北两座藏经阁分别在其基座、墙体、檐口的转角部位皆有明显的开裂走闪。北藏经阁东门头山墙歪闪甚为严重，施以铁质构件临时加固。两座藏经阁的屋顶及山花部分皆已完存。
[3] 根据砂岩材质的分析，茶胶寺庙山中央五塔的建造年代应区别于其他建筑，而砌筑藏经阁主室墙体的砂岩石材为何在色彩、硬度及表面风化状况等方面皆明显地区别于同一建筑的其他部分，其中原因尚不得而知。

藏经阁基座以砂岩砌筑可大致分为三层，尺度逐层向内收进。基座之西侧以 5 级台阶与抱厦的西入口相接。藏经阁西侧抱厦是进出藏经阁的唯一通道，其面阔、进深皆为一间且尺度远逊于主室。主室南北两侧墙体顶部分别设有两层檐口，底部檐口为半筒拱形的曲线屋面，上层檐口之基座矗立于一层拱顶之上[1]，在其正中皆辟为横向高窗，窗框之间设以 9 根雕饰线脚的圆柱形窗棂[2]，其上部的各级山花及顶层屋盖部分皆已不存。主室上檐口线脚雕饰自下而上包括：斜面，两条平条线脚镶边的板条，凹圆线，逐层凸出的两条平条线脚，枭线，平条线脚，束腰。主室下檐口的前凸突部分几乎完全支撑其上的上层檐口基座，上层檐口雕饰线脚自下而上包括：斜面、凹条、两条平条线脚镶边的板条、凹圆线、两条平条线脚镶边的半圆线、凹圆线、斜面、枭线、平条线脚、厚扁平束腰、平条线脚、细"S"形线、平条线脚以及墙身秃面。至于藏经阁的外部立面，抱厦立柱的柱脚（与其柱头线脚相同）以及主室墙基皆是未经精雕的线脚，与其对应的上檐口线脚雕饰由低到高包括：勒脚、反枭线、平条线脚、凹圆线、平条线脚、直枭线、平条线脚、浅凹条、两条平条线脚镶边的半圆线、细枭线、平条线脚、凹条、两条平条线脚镶边的板条、凹条以及斜面搭接墙身秃面，柱头上所有的枭线都是直的。[3]

藏经阁的东侧门头外侧实为假门雕饰，其立面线脚与雕饰仅是初具雏形，其上的门楣及各级山花皆已不存，仅余部分花柱残迹；东门头内壁凿有一个平整的壁龛，其边缘以三叶拱状雕饰与两条带有两条平条线脚镶边的"S"形线交汇，平条线脚并未超过壁龛之厚度。

另外，藏经阁主室入口门框上残存有可供安装双扇门的凹槽，门框之上砌为两层的斜撑拱顶，拱顶内部未设安装横向支撑的凹槽。抱厦内门两侧施以两根花柱支撑装饰性门楣。抱厦南、北侧窗框之上的檐口雕饰包括：两条平条线脚镶边的板条，凹圆线，逐层凸出的两条平条线脚，"S"形线，逐层凸出的两条平条线脚，扁平束腰等。抱厦的西侧入口并未预留供安装门扇的凹槽。抱厦及主室的地面皆以砂岩石块铺砌，抱厦地面标高略低于主室。

8. 回廊

回廊坐落于庙山第二层基台之上且环绕其四周，回廊内侧朝向庙山内院敞开，各面均不设进出回廊的专门通道或路径。回廊东、西两侧分为 26 间，东西内塔门的南北两侧分别设有 13 间；回廊南、北两侧分为 29 间，南北内塔门之东侧为 19 间、西侧为 10 间。

回廊在朝向庙山外院一侧的开间辟为假窗，每窗均施以 7 根圆柱形雕饰窗棂。若以外立面计，假窗的开间数目与其朝向庙山内院一侧并不相同，东立面与西立面在相应内塔门的南、北两侧皆分为 15 间，南立面与北立面在相应的内塔门东侧分为 21 间，西侧则分为 12 间。回廊的窗框构件，施以八字形插榫进行拼接[4]，或可看出这种连接方式来自木结构技术的源流。[5] 在回廊的四个转角处分别设置角楼一座，形制较为特殊，从中尚可看出早期砖砌塔殿样式的源流，惜各座角楼的残损甚为严重，

---

[1] 两个半筒拱实际上一部分呈现为侧墙的厚度，另一部分呈现为内檐口的出挑，内檐口形成了支撑中殿拱顶的鼓座托架。
[2] 拱顶上端与横向高窗底部的线脚雕饰包括：勒脚、直枭线、逐层凹进的两条平条线脚。
[3] Jacques Dumarcay. *TA KEV：ETUDE ARCHITECTURALE DU TEMPLE*, Paris：EFEO, 1971, p.25.
[4] 回廊角楼的水平层砌块之间多施以类似于中国古建筑石作的"银锭榫"、"燕尾榫"等铁质构件进行拉结，中国文化遗产研究院在茶胶寺保护修复工程西南角楼解体过程中曾经发现多处使用铁质构件拉结的痕迹。
[5] 插榫连接方式并不适用于石材，此种做法极大地削弱了构件的抗压性能。

目前仅存大致轮廓及其残迹。[1]

回廊的构造方式与长厅多有相似之处，依其立面可以分为基座、墙身、屋顶等三部分。回廊因其平面简洁，所以基座仅由规格不一的角砾岩石块砌筑堆叠而成；墙身则可以分为墙基、墙体、窗及檐口等四部分，檐口之上的端头瓦仍有少量存放于原位，是研究回廊形制源流与变迁弥足珍贵的实物遗存。由于回廊的屋顶部分皆已不存，其形制究竟是砖砌叠涩（西外塔门与西北侧回廊交接处的檐口之上残存少量红砖遗迹），还是砂岩叠涩拱尚且无法进行准确辨别。

须指出的是，在第二层基台四周发现的 5 处建筑基址残迹似与回廊的整体格局有关，因为其平面布局不仅与茶胶寺庙山现存的四座长厅类似，亦与比粒寺、东梅奔寺内的长厅格局大致相当，因此可以推测这些建筑基址残迹应属于某种类型的长厅，或许这些长厅建筑在回廊及塔门的建造之前即已存在。如前所述，观照吴哥时代庙山建筑形制的源流与变迁，长厅的式微与回廊的出现皆是吴哥时代庙山建筑形制革故鼎新的重要标志。在茶胶寺庙山回廊与疑似长厅的建筑基址残迹之间，似乎存在着某种颇为耐人寻味的关联。或许可以进行如下推测：苏利耶跋摩一世在位时期的茶胶寺庙山业已初具规模，可能在当时国王所倡导的革新风潮之中，在尚未完工的茶胶寺庙山第二层基台上增建了回廊与四座内塔门，由此将原有设计中的多座长厅废弃，从而使得茶胶寺庙山的总体格局更加趋于完整与统一。然而，茶胶寺庙山的回廊与四座内塔门、角楼连为一体，却未设置任何通向回廊内部的入口，因此其使用功能存在重大缺陷。这似乎也从另一个侧面表明，苏利耶跋摩一世时期茶胶寺庙山建筑形制的变革或许仅是源自对某种形式和象征的追求，却没有对其实用功能进行充分的考虑和理解。而茶胶寺庙山以降，在吴哥时代的庙山建筑中，回廊形制在诸如巴方寺、吴哥窟、巴戎寺等重要寺庙中皆得到了大规模的运用，并且一改茶胶寺庙山时期回廊的简朴粗陋的作风，构成庙山建筑中最重要的空间元素及装饰部位。

---

[1] 关于回廊的保存状况，略述如兹：西南回廊屋顶全部倒塌，南内塔门西侧 9 米回廊全部倒塌，仅在外侧保存 7 扇盲窗；西北回廊保存状况同西南回廊；北西回廊外墙残存 8 扇窗，靠北内塔门一侧窗严重变形，以钢筋混凝土临时支撑，内部地面及内窗靠北内塔门一侧全部塌毁，内窗只保留西侧的 2 扇；东北回廊屋顶全部坍塌，外侧 10 扇窗、内侧 8 扇窗全部倒塌，另外 2 扇内窗倾斜变形处于危险状态；东北回廊外墙北侧靠角楼处塌毁 1 扇窗，余者较好，内侧窗中有 3 扇倒塌，基座亦损毁；东南回廊外墙靠东南角楼的 2 扇窗和内侧 2 扇窗倒塌，其余保存较好；南东回廊屋顶全部坍塌，除靠东南角楼第 3 扇窗塌毁一半，其余屋檐以下、包括基座部分均保存较好。

三、茶胶寺庙山的保存现状及建筑勘察简述

茶胶寺庙山包括庙山建筑本体和周围环境（庙院）两大部分。茶胶寺庙山建筑的主体部分又分为基台和上部建筑两大类。

第一类基台平面布置及基本形状保持较为完整，一层、二层和须弥台（第三层台、第四层台、第五层台）基台外部砌石挡墙基础未出现明显的下沉和变形，除各基台角部出现明显塌落缺失外，其余墙体保持稳定，无明显残损现象，未发现不良地质作用，场地稳定。各层基台局部由于场地渗水时有沙土析出，从而导致地面有局部下沉，但尚未对整体安全造成影响。但一、二层基台及须弥台角部缺失影响基台安全，且对建筑整体及上部建筑的完整影响较大。

第二类为各层基台上的各类单体建筑，包括各类塔殿、塔门、长厅、藏经阁、回廊等，均为石砌结构，其保存现状大致分为三类：其一，结构主体保存较好，建筑形制大部分完整，屋顶等局部有塌落、开裂等病害（中央五塔）；其二，结构大部分保存，基本形状尚存，但残缺较为严重，结构存在较大倾斜、塌落、开裂等严重隐患，建筑构件缺失严重，屋顶塌落（内外塔门，藏经阁）；其三，结构大部塌落，建筑形体残缺不全，构件大部遗失，仅有部分墙体、梁、柱尚在原位（内外长厅，二层基台上的回廊及各座角楼）。

茶胶寺庙山建筑保存状况与残损勘察将病害类型分为砌体裂缝（墙体、基座等结构因建筑体变形出现开裂现象）、构件破碎（石构件受外力作用或长期风化结构产生破损）、构件断裂（石构件受外力作用造成剪切破坏而出现的断裂）、构件错位（建筑局部结构或部分构件明显偏离原位）、结构倾斜（因基础产生不均匀沉陷造成建筑局部结构或石构件产生一定角度的倾斜）和基础沉陷（因地基不均匀沉陷，造成建筑局部结构随地基下沉），主要表现为：

其一，建筑形象缺失，结构稳定性受损，散落构件长期堆放不利保护；其二，砌体结构存在倾斜、开裂、塌落的危险，结构安全存在较大隐患；其三，大部分基台的角部由于应力荷载集中、材料性能劣化以及外界扰动等原因，皆出现碎裂、塌落等现象，造成较大的结构隐患；其四，部分石质构件（石柱、门楣、山花、门窗）的残损直接影响建筑外观及结构安全；其五，部分重要石刻表面风化、剥蚀严重，其中表层剥落和起鼓病害的分布具有明显的特征。其中，须弥台表面雕刻剥落起鼓病害十分严重。根据统计，约计20%的表面雕刻已经剥落消失，30%－40%表面剥落病害正在发生，随时都有剥落灭失的危险。另外，茶胶寺庙山砂岩表面附着生长大量的地衣、苔藓等微生物，对石构件表面产生侵蚀和污染。[1]

另外，茶胶寺庙山建筑各层台基排水不畅，导致场地积水，从而威胁基台下夯土安全。庙山的原有排水系统因年久失修，加之基台局部沉降、地面铺石丢失等因素，已无组织排水功能。在雨季形成基台多处积水，且大部分水以下渗方式排走，对基台及建筑物安全造成隐患。在各建筑物的室

---

[1] 详见《中国政府援助柬埔寨吴哥古迹保护二期茶胶寺保护修复工程总体设计方案（茶胶寺石刻表面保护试验与研究）》，2011年。

内，因大多数建筑屋顶基本坍塌而无防雨功能，雨水直接落入室内，而由于室内四壁的遮挡而形成积水池，这已普遍对建筑物的隐患造成影响，因此大的排水系统及小的建筑局部排水均存在较大问题。[1]

鉴于茶胶寺庙山建筑残损勘察是保护修复工程设计所涉及的内容，本章不再进行详细叙述，现谨将其保存现状及残损勘察基本情况择要简略叙述如兹。

1. 基台部分

茶胶寺庙山所有单体建筑皆坐落在各层基台之上，其基台的变形破坏必然造成其上部各单体建筑的破坏。例如，二层基台四个角部的坍塌，致使其上部的角楼处于极度危险之中。而基台内部沙土的流失造成地基承载力下降，使建筑的基础产生不均匀沉降，导致整体建筑出现结构变形、墙体倾斜、构件破损，以致建筑局部坍塌。

(1) 第一层基台（庙山外院）

整体保存较好，但各角部及局部有残损问题。南侧已呈多处不规则沉陷，局部有石块缺失或断裂，东南角因上部围墙倒塌缺失一角。北侧总体保存较好，东北角缺失部分石块，上部墙体转角倒塌，东侧基台保存较好，东南转角处构件缺失，并有明显沉陷。

(2) 第二层基台（庙山内院）

东南角倒塌引发上部东南角楼局部垮坍，以红砖砌体进行临时性支护；西南角垮塌，引发上部角楼失稳，墙体倾斜变形；西北角上半部坍塌，引发上部角楼的西北角倒塌，同时角楼北墙出现倾斜变形，以红砖砌体填充缺失的红石部分，并以木结构临时性支撑支护行将倒塌的西北角边墙；东北角上半部倒塌，致使上部角楼倒塌变形，分割成三个独立部分，且东侧残留结构处于危险状态中。部分角部的石构件，不同程度地被上面跌落构件砸断。

(3) 第三层基台（须弥台之第一层）

尚未完成雕刻，线脚分明，但因长期风化和雨水侵蚀，石材表面已严重损坏，基台四转角处倒塌，各台阶两侧的边台不同程度缺失构件，均在转角处。顶部砂岩石地面已高低不平，局部易积水，水顺基台边缘漫流，易对石材产生侵蚀，造成风化。基台边缘铺满从上一层基台跌落的石构件。台阶踏道保护状况较好。

(4) 第四层基台（须弥台之第二层）

局部完成雕刻。四个转角坍塌，其余部分结构稳定性较好。经长期风化和雨水侵蚀，石材表面受损严重，边台也有不同程度构件塌落，主要集中在边台转角处。顶部砂岩石地面已高低不平，局部区域易积水。台阶踏道保存较好。

(5) 第五层基台（须弥台之第三层）

残损及结构破坏形式同第三、四层台，四个转角坍塌，其余部分结构稳定性较好，经长期风化和雨水侵蚀，石材表面受损严重，边台也有不同程度的构件塌落，主要集中在边台转角处。各面台阶踏道保存较好。

---

[1] 详见《中国政府援助柬埔寨吴哥古迹保护二期茶胶寺保护修复工程总体设计方案（茶胶寺排水工程设计方案）》，2011年。

2. 单体建筑部分

（1）第一层基台上的单体建筑

①南外塔门

屋顶全部坍塌，南北两侧门楣山花全部塌落无存；东西侧墙损毁一半，两侧室的墙体分别向东、西两个方向倾斜，墙体呈现多处裂缝，东侧窗构件错位，二十世纪二十年代法国远东学院（EFEO）及吴哥古迹保护处（CA）曾以钢筋混凝土柱进行简单的临时支护。南侧边台阶石构件有缺失，台阶两侧边台呈多处裂缝。

②西外塔门

建筑顶部全部坍塌，西外廊门楣和屋顶全部塌落，南北两侧门大部倒塌。特别是南侧门只保留南墙，并以钢混结构支护，南、北两侧结构分别向两侧倾斜。中厅残留的结构较为稳定，两侧台基及台座缺失构件严重，北侧与墙体连接位置明显下沉。

③北外塔门

整体结构相对稳定，屋顶全部塌落无存，局部结构有下沉错位现象，中厅顶部出现多处裂缝，部分构件处于行将掉落的危险状态，东侧室东南窗整体扭曲。北侧台阶及边台砌筑后没有雕刻。

④东外塔门

屋顶全部塌落，东西廊门楣和屋顶全部掉落，基础有不均匀沉降，侧门各间墙体向内侧倾斜，南、北两面墙以钢筋混凝土结构支护，墙体多处开裂，东侧台阶及边台保存较好，只有部分构件缺失。

⑤南外长厅

屋顶全部坍塌。前廊各柱皆存，而柱上部大梁只保留两条，包括前门过梁在内的其他大梁以上结构全部倒塌。正厅西边窗整体向东倾斜，并从北侧第二扇窗至第四扇窗完全倒塌，倾斜的窗体以木结构斜撑支护。多处窗框被压劈裂，后室东墙与南墙向内侧扭曲倾斜。

⑥北外长厅

前廊柱全部倒塌前门东框倒向一层基台边墙，唯西门框尚存。正厅屋顶全部塌毁，现保留的结构较为完整，窗柱和窗框均有不同程度损坏。后室西墙与北墙完全倒塌，保留的东墙向北侧倾斜。

⑦第一层基台围墙

南墙：连接南门西侧长度约18米的墙体倒塌，东南转角坍塌成一豁口，其余部分保存较好；西墙：连接西门南侧长度约3米的墙体倒塌；北墙：连接北门西侧长度约16米的墙体倒塌，剩余部分约有长度12米的墙顶结构缺失；东北角墙体转角处坍塌形成一处豁口，残留结构随第一层基台转角处有明显下沉现象。

（2）二层基台上的单体建筑

①东内塔门

中厅屋顶塌落，但三层结构总体保存较好，西廊、南北侧室及两侧门顶部屋面及山花全部倒塌，南北侧室上部残留后期维修的砖砌体屋面。

②南内塔门

南内塔门高台基本稳定，但每层台上部石块缺失较多，并有错动现象。中厅及东侧室墙体基石稳定，无明显裂缝。中厅保存第二层塔身，第三层塔身只残留几块石构件，顶部不存。西侧室西南墙和西墙严重扭曲变形，整个墙体连带南侧假窗向东倾斜，西侧假窗明显与次间墙体脱离，墙体裂缝达20厘米，顶部石块严重错位。屋顶完全坍塌，构件无存。

③西内塔门

整体结构稳定性较好，除屋顶坍塌外，其余结构保存完整。

④北内塔门

残留部分结构基本稳定，中厅保留二层塔身，顶部全部倒塌，中厅西门楣及山花全部倒塌。西侧室屋顶残存有后期砌筑的砖屋面。西次间与梢间之间顶部山花保存完整，余者或全部倒塌或只保留一部分结构。东侧室北墙向内侧倾斜。

⑤南藏经阁

前廊屋顶和山花全部塌毁，前廊整体向西倾斜。正厅二层以上屋面及东侧前山花坍塌，正厅东、西两端有较宽裂缝，使结构体向东、西两个方向倾斜，建筑西南角以钢筋混凝土柱支护。

⑥北藏经阁

结构残损状况同南藏经阁，且比其更危险。主厅东假门上部构件之间严重错位，行将倒塌，且墙体多处开裂。石构件多处受剪力破坏。

⑦南内长厅

屋顶全部塌落，墙体有多处开裂，且墙体和柱子有多处出现不均匀沉降和结构体倾斜，但整体稳定性尚好。中厅南北两侧门上部的山花各保留一半结构，其余山花全部倒塌。

⑧北内长厅

结构残损状况与南内长厅相同，前廊只保留南门，中间柱和梁及以上结构全部坍塌，除中厅与后厅之间上部的山花保留，其余山花全部坍塌，后室的窗上以钢混柱支撑断裂的窗梁。

⑨角楼

四处角楼未建成或未完成雕刻，目前都只残留一层结构，其上部应还有结构体存在，现已无存。角楼内侧面积1.4米×1.4米，角楼与回廊相通，另外两侧朝外的墙设有假门。受第二层基台四个转角坍塌的影响，均出现不同程度的倾斜、变形和局部结构倒塌现象，且多处于危险状态中。

⑩回廊

南西回廊屋顶全部倒塌，南内门向侧9米回廊全倒塌，外侧保存七扇窗，南侧剩余七扇窗；西南回廊屋顶全部倒塌，其余保存较好；西北回廊保存状况同西南回廊；北西回廊外墙残存八扇窗，靠北内门一侧窗严重变形，以钢筋混凝土支撑，内部地面及内窗靠北内门一侧全部塌毁，内窗只保留西侧的二扇；北东回廊屋顶全部坍塌，外侧十扇窗、内侧八扇窗全部倒塌，有两扇内窗倾斜变形，处于危险状态。基座保存尚好。东北回廊外墙北侧靠角楼处塌毁一扇窗，余者较好，内侧窗中有三扇倒塌、基座亦损毁；东南回廊外墙靠东南角楼的两扇窗和内侧两扇窗倒塌，其余保存较好。南东回廊屋顶全部坍塌，除靠东南角楼第三扇窗塌毁一半，其余屋檐以下、包括基座部分均保存较好。

（3）须弥台上的单体建筑

①中央主塔

整体稳定性较好，三层基座局部转角构件缺失，但不影响其稳定性。塔身稳定部分区域出现结构裂缝，部分梁体受压断裂，石材有较严重的风化、破损现象。

②东南角塔

基座基本稳定，南侧与东侧因基台坍塌而出现下沉现象，东廊和南廊分别与塔身脱离约3厘米并向外倾斜。西外廊壁柱向外错动10厘米，塔身部分构件被压裂。

③西南角塔

整体稳定，顶部大量石构建塌落，基座上的石构件缺失较多，塔身整体稳定性较好，个别构件被压裂。

④东北角塔

结构残损状况同东南角塔。

⑤西北角塔

结构残损状况同东南角塔。

（4）各单体建筑残损原因分析

其一，从建筑地基下生长出的高大植被，破坏了基础结构并造成地基承载力下降，这与基台因内部植被的生长导致的破坏是一致的。

其二，由于建筑变形导致各结构体受力关系发生变化，多数构件受集中应力作用造成剪切破坏，使其失去承载力，如门窗的过梁和部分墙体构件；建筑变形也导致一些纵向受力构件产生偏心受压，使中部断裂，如门窗的边框。

其三，部分较大型构件与下部构件的搭接面过小，如门楣和门过梁的搭接面只占门楣底部面积的1/4，另外3/4从梁探出，造成结构的不稳定，下部结构若出现变形，易使该构件以上结构整体倾覆。

其四，由于水和生物的侵蚀而出现石材的风化，易导致构件承载力下降，造成上部结构处于危险之中。

3. 周围遗迹部分

（1）神道

神道为茶胶寺的主轴线，现尚存部分路基，高出周边地面，两侧散落数件残损石构件。

（2）散水平台

尚可分辨边缘轮廓，部分地面有石铺面，大部分已损失殆尽，周边地面因废弃有植物生长沉积物堆积而不完整，场地标高有待调整。

（3）环壕及水池

四周壕沟轮廓尚完整，断面深度及宽度不一致，但基本保持原规模，雨季仍可作为排水通道，其北侧与东侧护壕上有桥涵遗址，但未完全发掘。环壕原在内侧砌有石护岸，现基本毁坏，局部保存有散落石块，一些部分为原位置及原砌法。

茶胶寺庙山东侧的南、北两水池保存有基本轮廓，但边缘不清晰，原池周边护岸基本无存。雨季仍可汇集雨水，旱季基本干涸。

4. 茶胶寺庙山建筑勘察的初步结论

其一，茶胶寺庙山各层基台主体完整，但均存在不同程度的倒塌现象，主要集中于角部，并已严重危及各自上部建筑的安全。

其二，茶胶寺各层基台上建筑的位置、规模尚存，大部分结构完整。但各建筑均存在严重的倒塌、破裂、构件缺失现象，大部分建筑保存都已不完整，且相邻建筑由于结构上相互牵连而相互影响，破坏隐患较多。

其三，茶胶寺庙山保存现状的致损因素包括风化作用与结构性破坏。

其四，柬埔寨的热季季风性气候导致的温差变化及多雨，是造成茶胶寺庙山石材的风化破坏的主要原因。结构性破坏则源于基础（台座）的不均匀沉降。其建筑各单体结构是由大量石块堆砌而成，其间无黏接材料，且各局部结构之间联系不紧密，各单体结构缺乏整体性。

Synopsis Ⅳ The Architectural Composition of Ta Keo Temple-mountain

1. The Site Plan

As the embodiment of the most unique feature of temple-mountains, Ta Keo temple-mountain is located in the central zone of the Angkor Site. Leaving eastward from the Victory Gate of Angkor Thom and crossing the Siem Reap River, one can reach Ta Keo temple-mountain within a 900-meter distance. The temple is located slightly to the west of the centre of its rectangular surrounding of 1000 × 780 meters. The east boundary of the surrounding is the west embankment of the Eastern Baray (reservoir), where a dock terrace is exactly in the axis of the principal entry of Ta Keo temple-mountain. Another dock terrace is 200 meters northward, located in the axis of Phimeanakas-Victory Gate of Angkor Thom. Furthermore, the west embankment is also noteworthy as crucial part of this territorial approach.

To the west of the surrounding lies a series of relics, including a hospital built in the period of Jayavarman VII. To the south-east there is a tower with an appearance quite similar to that of Ta Keo temple-mountain, and there are many other important relics to the South (Ta Prohm), North (Ta Nei and Neak Pean), and West (ChauSay Toveda and Thommanon) of Ta Keo temple-mountain's surroundings.

Such a temple-mountain is a symbol of the sacred Meru Mountain of Hindu cosmology and Ta Keo was constructed in the end of tenth century AD as the state temple. The human royal beings were the subject of an apotheosis and the whole temple became a residence for them and other gods: it represented a cosmic mountain, a vertical world axis and symbolic mythological centre of the physical, metaphysical and spiritual universe. With these explanations in mind, it may readily be understood not only that the spatial, architectural, stylistic and ornamental principle of the temple is as important as its architectural form, but that the form and its connotative meaning are two sides of the same coin. The space, building, and ornamental style should be understood as a codified text or associative implication of significant religious, historical, ethical and normative mythological concepts. The magic function of the relief served as a royal and godly invocation and brought inanimate building materials to mythological life.

Beyond all doubt is the fact that Ta Keo temple-mountain is a typical example, representing the spatial, architectural, stylistic principle of temple-mountain layout and composition, with all its characteristics. Moreover, the unfinished state of Ta Keo temple-mountain has outstanding value for explaining how the ancient Angkorian construction came into being; it is the first realization in sandstone of such a structure (generally dedicated to deify nobility) after the Phnom Bakheng that crowned a natural hill serving as its core. Ta Keo temple-mountain is constructed with much more care in the systematic cutting and placing of enormous stone blocks, the arrangement of which can be viewed easily due to the absence of almost any molding or ornamentation; and the architectural ornamentation in the temple plays a critical role in its identity. Finally the religious and cultural symbolism of Ta Keo temple-mountain can be perceived as a synthesized reflection of Hinduism and local Khmer culture.

Originally, the access to the Ta Keo temple-mountain was from the east across a moat by means of a paved causeway, proceeded by lions in the style of the Bayon and lined with boundary stones. About 500 meters further to the east is the bank of the Eastern Baray. The external enclosure wall forms a rectangle of 120 x 100 meters and is in sandstone on a laterite base. The second terrace dominates the first with an imposing molded laterite base and four axial sandstonegopuras. From the courtyard, standing in front of the three tiers that form the fourteen meter-high central pyramids, one is left with a powerful impression. The upper platform is square and almost entirely occupied by the quincunx-shaped towers in their unfinished state. These open to the four cardinal points by projecting vestibules. The corner towers are set on plinths and are dominated by the central tower set on an elevated base, with the development of its porticoes and frontons adding to its grandeur. Fragments of pedestals and of Lingas are found both in and around the towers.

Leaving eastward from the Victory Gate of Angkor Thom and crossing the Siem Reap River, one could reach Ta Keo temple-mountain within a 900-meter distance. As a result of academic studies and archaeological survey, it is currently accepted that Ta Keo temple-mountain is the result of different historical components. Perhaps it had been the central area of a specifically large-scale urban planning of the Khmer Empire's capital, notwithstanding the integrity of its layout has completely vanished in the tropical forest. In the inaugural stage of construction, it is presumed that Ta Keo temple-mountain probably extended directly to the west embankment of the East Baray, where the dock terrace remains spanning the whole territory of the temple, along an west-east axis. An over 500-meter-long causeway leads to the East Outer Gopura of Ta Keo temple-mountain from the embankment of the East Baray. The main existing construction in the causeway is the remains of a cruciform terrace, made of three levels of molded laterite stones. The little construction that should have been built with a wooden structure pavilion upon the terrace was probably connected with a three-step stair framed by string walls that supported stone lions. The lions were recorded and photographed during the excavation mission in the 1920's but are no longer in situ. The stair attains its largest width (approximately ten meters) at the level of the causeway. Along the causeway there are some scattered boundary stones, as well as the debris of the laterite parapet. Archaeological survey finds the parapet seeming to stretch until the foundation of the Eastern Gopura of the first level of the pyramid, hence forming a semi-enclosed space regardless of the unknown height of the parapet. The scale of the parapet can be extrapolated from that in the Baphuon temple site.

To the north of Ta Keo temple-mountain, approximately 500 meters, a steep embankment was well connected between the East Baray and the SiemReap River, and presumably, it was the north boundary of Ta Keo temple-mountain at some time in the past. In the west and in the south it was not necessary to excavate to find similar landmarks of the boundaries of the earlier extent of the temple. Exceptionally, the Siem Reap River is the natural boundary in the west of Ta Keo temple-mountain. In the south direction, along the axis of the temple, a soil-paved road, 500-meters-long, remains. As a result of considering these features it is concluded that Ta Keo temple-mountain is located in the center of rectangular area (1000 × 780 meters) with a little offset to west side.

In addition, Ta Keo temple-mountain is surrounded by a moat with a laterite embankment. The embankment of the moat in the west and south sides is seriously damaged, inadequate for us to deduce whether some paths once existed or not. What appears recognizable is the path across the east and north sides of the moat and from the measurement of these surroundings, Ta Keo temple-mountain's site area can be calculated to be approximately 4.7 hectares. Its foundation shape is rectangular within the size of 106 × 121 meters, and its height from the top of central prasat to the ground is approximately 22 meters.

From March to April 2012, Chinese Academy of Cultural Heritage-CACH carried out archaeological excavation mission at northern causeway and northern moat of Ta Keo temple-mountain site. During the excavation, 90 artifacts are found including 13 clay potteries, 4 glazed potteries, 54 tiles, 12 ceramics, and 7 architectural components. Among the findings, the amounts of architectural components are maximum. Secondly, the findings are the pieces of pottery and ceramics. At the last layer (16th layer) of the excavation, as the cushion layer of the embankment, its material is a special clay plaster which outline is 10 – 12 cm wide. The special clay plaster, known as microcrystalline kaolin, white-green color, fine texture, plasticity, strong stickiness, small seepage, good sealing, and it has strong moisture absorption to prevent decay or weathering caused by long-term seepage water. The view from the site excavation, the soil is very dense and hard solid, difficult to scrape. It can also protect the fresh soil foundation of the embankments due to prevent the seepage water. The further details of material analysis is undergoing in laboratory.

After excavation, the structure of the west, the north and the south embankments is entirely clearance. The construction materials of the embankment are sandstone and laterite stone. And the architectural form and masonry techniques can be known and surveyed. The structure of embankment has sixteen layers (In the survey of Jacques Dumarçay, it is eight layers). The condition assessment of the upper parts of northern or southern embankments is not good with serious weathering and collapsed, but it is integrated under the seventh layers. In the site plan for the function and form of Ta Keo temple-mountain, two causeways had played very important role. As we known, the east causeway is linked between the temple-mountain and the eastern baray, but as another important pass way of temple, where the north causeway was connected or extended? Many scholars suggest that north area of Ta Keo temple-mountain maybe the royal palace area or the urban centre of King Jayavarman V. It is actually very little evidence for the north area of Ta Keo temple-mountain. Therefore, the excavation of north causeway is just very small step of the archaeological campaign of Ta Keo temple-mountain.

2. Architecture Inventory

The emergence of Ta Keo marked the mature form of the temple-mountain. Ta Keo is a milestone during the evolution of temple-mountains, along with the earlier Pre Rup, its contemporary Phimeanakas, and its successors Baphuon and Angkor Wat. For the first time, Ta Keo combined the pyramid with the enclosed gallery, which became a typical feature of the Angkorian architecture. Meanwhile, the enclosure wall was replaced, and the long hall, found in its predecessors, lost its popularity.

In Ta Keo temple, the integrity of the monument was largely strengthened in design by combining the gallery together with the four corner towers in the second floor as well as thegopuras in both the first and second floors. The central prasat (tower and sanctuary combined) on the upper level, became the core space of the temple-mountain architecture through the development of its cruciform plan with porches and doors opening in four directions

The First Terrace of Pyramid

A series of regularly arranged holes can be observed on the first floor near the western and southern entries, and some of them are overlapped by the enclosure andgopuras of the first level. The diameter of these holes varies from 330 mm to 640 mm. Based on such observation, it could be inferred that there were once wooden long halls. The halls must have been removed before the enclosure and gopuras were erected. However, the initial function of the wooden long halls along with the alteration date remains unclear.

The Second Terrace of Pyramid

Some relics on the second floor were discovered forming a part of the foundation of constructions that relate intimately to the enclosed gallery. A possible conclusion is that there once existed long halls in the identical configuration of Pre Rup and Eastern Mebon. Furthermore, it seems reasonable to deduce that the constructions on the second floor firstly emerged in the form of long halls (in Jayavarman V's reign) and were later replaced by the enclosed gallery (in Suryavarman I's reign).

Prasat

Five prasats arranged in a quincunx-shape are made entirely of a kind of very hard sandstone, namely Graywacke. They had a cross-shaped plan and the cell was preceded by four porticoes. In the central tower, which stood on a double crucifix-shaped base platform, there are four vestibules between the cell and the porticoes. The roof, which consists of four gradually diminishing tiers, culminates 45 meters above ground level and is impressive and almost cyclopic in its bare, unfinished state. It can be observed that the remarkable material applied in the base part of the five prasats is different from that in the upper part: arkose for the former and greywacke for the latter, with a noticeable division line. The idea behind this choice of different materials is still unclear.

Gopura

There are a few doorframes in thegopuras that have been chiseled and partly cut away at the height of approximately fifty centimeters. After Dumarçay, this might be due to the religion conversion and consequently the displacement of original large-dimension sculptures (such as statues and Linga) and pedestals. Without more evidence, such hypothesis is somewhat far from being proved.

Library

Two libraries have an interesting structure with a single room inside, but from the outside, because of the two lowered half-barrel vaults resting on the perimeter walls, they appear to have a nave and two aisles.

Gallery

The second level rises to height of 5.8 meters and has an innovative feature, namely a rounding gallery measuring 81 ×74 meters with a false vaulted ceiling of corbelled bricks, and blind colonnaded windows on the inner side.

第五章 茶胶寺庙山建筑复原初探[1]

一、茶胶寺庙山总体格局的复原

茶胶寺庙山可能位于一组规划严整且规模宏大的建筑组群的中心，庙山东侧延伸大约500米直至东池西侧的堤岸之上，而堤岸之上仅存角砾岩砌石基址的建筑遗址及其与庙山相连的神道残迹，皆应是构成茶胶寺庙山整体格局的重要实物遗存。在占地面积约为1000米×780米的长方形区域之内，唯因茶胶寺庙山西侧由于暹粒河所形成天然屏障的阻隔，庙山建筑主体部分略偏向此区域的西侧[2]。（图5-1）

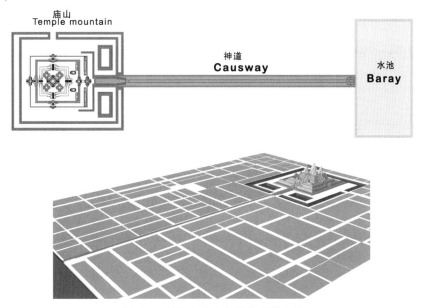

图5-1 茶胶寺庙山整体格局的复原

公元九世纪末叶以降，因陀罗跋摩一世、耶输跋摩一世等高棉帝国的早期统治者先后在其都城诃里诃拉洛耶、耶输陀诃罗补罗的中心修建起巴空寺、巴肯寺等大型庙山建筑，在此后大约四百余年的历史进程中，作为古代高棉建筑艺术、技术及其复杂信仰体系的集中呈现，诸如空中宫殿、比粒寺、茶胶寺、巴芳寺、吴哥寺、巴戎寺等宏伟壮观的庙山建筑成为吴哥时代建筑艺术风格最为显著的标志。

---

[1] 本章主要内容是在中国文化遗产研究院与天津大学建筑学院合作开展《吴哥古迹茶胶寺建筑形制与复原研究》的初步研究成果基础上完成的，课题负责人是温玉清、吴葱。本章所采用的复原图是由课题负责人指导的研究生及本科生伍沙、王祥、阎金强、任思捷、马庆阳等协助绘制的。（著者注）

[2] 详见本书第四章《茶胶寺庙山建筑形制简述》的相关内容。

茶胶寺庙山建筑研究

对于古代高棉人而言，庙山建筑似乎始终笼罩在深邃的宿命之中，高棉古典艺术始终追求着一种源自印度原型却持续不断的本土化演进，并且沿着在每一步连续的台阶上向着崭新的、富有创造力的轨迹发展。

吴哥时代的庙山建筑形制拥有一以贯之的空间图示与表现形式：逐层收进的砌石基台（须弥台）、高耸的塔殿、环绕须弥台四周的回廊及附属建筑（如藏经阁、长厅等）、庙山外围的神道、水池及壕沟等，共同构成了一种杂糅复杂宗教教义与象征内涵的建筑空间组合，其中又以坐落于须弥台之上的中央主塔作为统领庙山建筑组群的核心。庙山建筑的须弥台及其上的高耸塔殿是神之居所——须弥山，象征着宇宙的中心；环绕须弥台四周的回廊是象征诸神领地的连绵山脉，而庙山周围充溢流动的环壕之水是宇宙海洋的象征。庙山建筑的规划经营过程，则是古代高棉复杂信仰体系与王权政治的集中呈现。作为国家寺庙及其所崇奉的核心是凝聚"神王合一"的提婆罗阇（Devaraja）的化身，国家奉祀之神赋予高棉国王统治的王权，而国王则通过创建属于自己的庙山建筑以彰显其被赋予权力的合法性与正统性。

古代高棉匠师通过石构建筑所蕴涵的各种元素，熟稔地运用材料和工艺的技巧，创造出塔殿与神山作为基本的建筑元素以构筑庙山建筑的主体，并以塔殿为中心的十字对称平面布局。塔殿大多为方形平面且朝向东方，塔身结构逐层收分而上，塔内多供奉象征湿婆的林伽或毗湿奴像；神山则多以砌石须弥台承托顶部坐落着的五座呈"梅花状"（quincunx）布置的塔殿，并通过各级基台之间陡峻踏道或台阶的连接以突出庙山崇高神圣的象征性。（图5-2，图5-3）

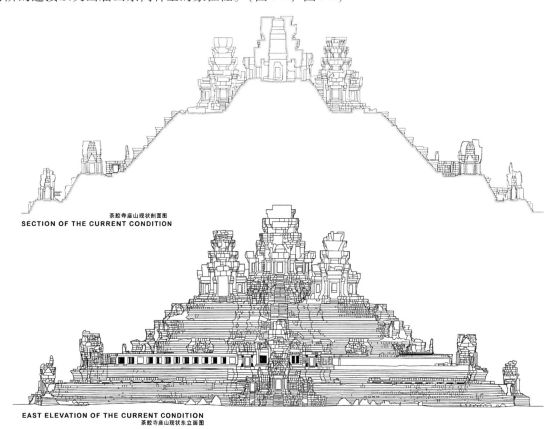

图5-2 茶胶寺庙山东立面及剖面现状图

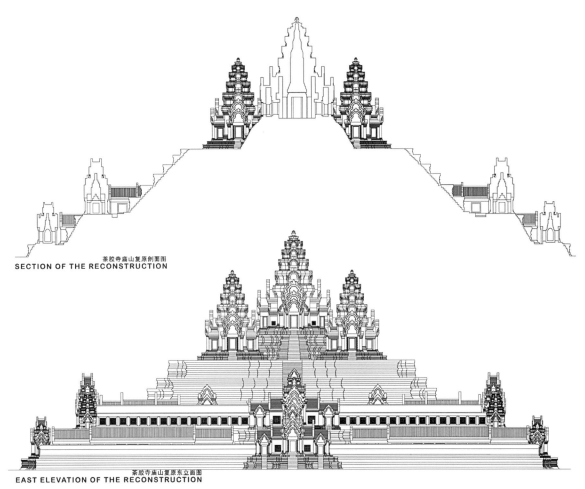

图 5-3　茶胶寺庙山立面及剖面复原图

作为吴哥时代庙山建筑构成的核心，塔殿的位置主要可以分为两类：其一，须弥台顶层之上；其二，分散布局于各层须弥台及其外围。而中心塔殿的布局方式则又可以分为两类：一种是单独塔殿形式，另一种是中央五塔形式。吴哥时代这种庙山建筑格局所表现出来的这种中心汇聚的秩序感，或与源自印度的曼荼罗空间图式有着千丝万缕的联系。曼荼罗，也称作"曼陀罗"、"曼拿罗"、"满拏啰"，系梵文 Mandala 的音译，其意为"坛"、"坛场"、"坛城"、"聚集"等。在古代印度的宗教传统中，曼荼罗被视为修持能量的中心，以"万象森列、融通内摄、圆融有序的布置"构成各类宗教信仰表述宇宙模式、或显现其宗教所见之真实宇宙的集大成者。为了使神秘旷奥的教旨变成看得见、摸得着的信仰模式，因此曼荼罗通常被赋予了极其丰富的宗教奥义，其涵括四极、强调中心与聚集的空间意义，遂成为古印度城市及神庙建筑的基本范型。曼荼罗"内聚外屏"的空间构成模式强调中心与边界，以中心为主导向外辐射，以边界为约束向心凝聚，其表现形式为一种包含了方与圆的形式的对称图形，其平面呈"十字轴对称"、"九宫分格"、"方圆相间"的结构，其中央耸峙的须弥山象征宇宙万物的中心，环绕须弥山的外围则象征世界的八个基本区域。（图 5-4，图 5-5）

正是源自印度文化传统的深刻影响，曼荼罗作为吴哥时代庙山建筑布局与空间图示的摹本，真实而具象地赋予了庙山极其丰富的宗教奥义。空间意义上的曼荼罗，即可理解为"聚集"，意味着

图 5-4　茶胶寺庙山复原图之一

图 5-5　茶胶寺庙山复原图之二

悉心布置的曼荼罗空间，可以聚集一切法相、法力、法器，以摒除修炼中各种魔障的干扰，随心所愿而达悉地；认识论意义上的曼荼罗，涵括"轮圆"、"俱足"之意，是指可以达到对教旨最为充分的领悟。因此，深受印度文化浸润滋养的古代高棉人认为，"须弥山"既是神的驻锡之所，也是宇宙的中心。[1]

---

[1] 作为信仰模式与建筑空间图示的互动表达，曼荼罗图示具有超越种族和地域的稳定性、延续性，几乎毫无变化地保存在诸多文化形态与时代风格演变的进程中，成为时空变幻中恒久不变、历久弥新的文化基因。诸如北京正觉寺、香山碧云寺、西黄寺的金刚宝座塔以及云南昆明官渡妙湛寺塔、内蒙古呼和浩特小召金刚宝座塔等，无论是藏传佛教，抑或历史更为久远的婆罗门教、湿婆教、毗湿奴教，还是曾经遍及东南亚地区的大乘佛教与南传上座部佛教；无论是繁缛精致的中华帝京金刚宝座塔，还是高峻威严的吴哥庙山建筑，跨越千年之间，相距万里之遥，纵有其不同历史文化、社会风俗的差异引发的建筑功能及其艺术形象的变迁与转型，但究其本质，建筑的空间图示与形态皆与曼荼罗如出一辙。（著者注）

纵观古代高棉建筑史，茶胶寺的经营与建设正处于庙山建筑形制的转型时期，无论其建筑形制还是整体布局皆有可能产生了重要的转变。首先，塔门在茶胶寺庙山中呈现出新的建筑样式，不仅与须弥台结合在一起，而且结合塔殿形制的假层逐层收进，空间跨度缩小，塔门从进出庙山的交通空间演变成为供奉祭祀神像之所，亦构成庙山重要的祭祀空间之一；其次，茶胶寺庙山出现了十字形四面皆开敞平面形式的塔殿，一改早期庙山塔殿方形平面、东侧单一门道的形制；再者，茶胶寺庙山首次出现回廊，回廊的出现更加凸显了庙山建筑的象征意义，同时规整了寺庙的整体布局，强调了组群的整体性。在茶胶寺庙山内首次出现回廊及回廊与角塔结合的建筑形式，经后期演变和发展成为古代高棉建筑的重要特征之一。始建年代早于茶胶寺庙山的东梅奔寺和比粒寺围绕须弥台建有周圈的长厅，外层环绕着围墙，是这一时期庙山建筑的重要特点。或许是苏利耶跋摩一世时代的革故鼎新，通过茶胶寺庙山与空中宫殿的回廊形制，可以看出这一时期庙山建筑的发展呈现出新的变化趋势：第一，回廊取代围墙，成为平面及立面构图中的重要因素。第二，回廊不仅表现在建筑形式上，而且在功能上也取代了围墙和长厅等建筑类型，尤其是长厅建筑类型，在茶胶寺之后的吴哥时代庙山建筑形制中逐渐式微并最终消失殆尽。

至迟在公元十世纪末或十一世纪初，吴哥时代庙山建筑以须弥台为核心的平面布局逐渐走向成熟，形制及其装饰也得以基本确定：即以砂岩砌筑的围合拱廊环绕作为庙山主体部分的须弥台与中央圣殿或中央主塔。

较之茶胶寺庙山建筑组群而言，位于第一层基台东侧南北对称布局两座石构长厅、第二层台东侧南北对称布局的两座石构长厅、第二层基台上的围合石构回廊以及砂岩砌筑的石塔完全取代砖塔等特征，可以推断茶胶寺的始建年代亦应晚于东梅奔寺和比粒寺。而始建于公元十一世纪中叶的巴方寺，其砌石构造的回廊不仅环绕第一层和第二层基台，而且以十字交叉型的回廊布局成为连接中央五塔的重要标志。至此，以吴哥窟为典范的吴哥时代庙山建筑形制已初现雏形。

在上述年代序列之中，茶胶寺庙山的建造年代应介于比粒寺和巴方寺之间。在上述庙山建筑形制序列之中，作为一种过渡形式，砖砌屋顶、石砌叠涩拱回廊的出现以及长厅的逐渐消亡也都是非常显著的标志[1]。因此，茶胶寺塔门的砖砌屋顶与回廊等遗迹愈显弥足珍贵，应是佐证茶胶寺庙山形制处于承上启下的特殊阶段的重要实物遗存。[2] 总而言之，吴哥窟所体现出庙山建筑的完美形制并非一日之功，古代高棉哲匠的灵感和创造在历经近三个世纪（公元九世纪至公元十二世纪）的传承和变迁之后，吴哥时代庙山建筑的完美形制才得以全方位的呈现。

观照吴哥时代庙山的形制源流与变迁，其中的基台或须弥台的演变或许另有规律可循。

例如，有可能始建于公元七世纪末期的阿约寺（Ak Yum），以围墙环绕的三层基台（须弥台）形制，应是吴哥时代庙山建筑最初的原型的意义；及至公元九世纪中叶以降，位于诃里诃罗洛耶的巴空寺庙山以及位于耶输陀罗补罗城内的巴肯寺庙山，皆是以五层基台（须弥台）作为庙山建筑组群的核心，并在基台四周或顶面之上对称布置数目众多的独立塔殿，而基台的最外围则由在建筑地坪砌筑的一圈围墙环绕四周；公元十世纪上半叶位于贡开地区的普兰寺庙山为七层须弥台，亦由一圈围墙环绕

---

[1] 柏威夏寺（Preah Vihear）和空中宫殿是石砌叠涩拱顶回廊的最早实例。（著者注）
[2] 维克多·格罗布维（V. Goloubew）根据巴塞特寺（Vat Baset）、埃克寺（Wat Ek）、吉索山（Phnom Gisór）以及大塔卡姆寺（Prasat Ta Kam Thom）等公元十一世纪的建筑遗构，推测茶胶寺的回廊及两座藏经阁应始建于苏利耶跋摩一世时期（Suryavarman I，公元 1002 – 1049 年在位）。

四周，围墙之内不设附属建筑，须弥台则被置于整个建筑组群轴线的末端。自此之后，以巴塞曾空隆寺和空中宫殿为例，其庙山主体皆为三层须弥台，周围地坪以一圈围墙环绕，这似乎进一步彰显出三层须弥台的原型意义。若对比巴塞曾空隆寺和空中宫殿时代大致相近的东梅奔寺和比粒寺，东梅奔寺的整体可以划分为三层基台（须弥台），比粒寺中央为三层须弥台主体，围墙环绕须弥台的布局得到进一步强化，围墙建立在须弥台周围升起的两层基台之上。

总体而言，茶胶寺庙山的整体布局更为紧凑峻拔，其整体布局或可划分由五层基台构成，第一层基台和第二层基台的四周设以围墙和回廊环绕，以及顶部三层的须弥台。第一、二层基台可以视为顶部须弥台的基座，而为突出中央五塔及须弥台高峻挺拔的设计意匠，古代高棉匠师第一、二层基台抬高，并在其上分别建造了围墙和回廊。围墙和回廊的布置不仅具有烘托须弥台主体地位的作用，而且极大地丰富了庙山建筑立面的层次。

表5-1　吴哥时代主要庙山统计简表

名称	层数	平面格局	建筑材料	中央塔殿数目	排布方式	基底尺度（m）	高度（m）
阿约寺 Ak Yum	3	至少1圈	砖	5	十字形	100×100	6.8
巴空寺 Bakong	5	1	角砾岩、砂岩	1	独塔	65×67	14
巴肯寺 Bakheng	5	1	砂岩	5	十字形	76×76	13
巴塞增空寺 Baksei Chamkrong	3	1	角砾岩	1	独塔	27×27	13
普兰寺 Prang	7	1	角砾岩、砂岩	1	独塔	62×62	35
东梅奔寺 Eastern Mebon	3	2	砖	5	十字形	110×110	≤12
比粒寺 Pre Rup	3	2	砖	5	十字形	50×50	12
空中宫殿 Phimanakas	3	2	角砾岩、砂岩	1	独塔	29×37	12
茶胶寺 Ta Keo	5	2	角砾岩、砂岩	5	十字形	120×100	38
巴方寺 Baphuon	3	3	角砾岩、砂岩	1	独塔	120×100	24
吴哥寺 Angkor Wat	3	2	角砾岩、砂岩	5	十字形	215×197	23

茶胶寺庙山虽与比粒寺具有大致相同的平面布局，可是最终达到的视觉效果却完全不一样。由表 5-1 可以看出，比粒寺须弥台的基底尺度 50 米 × 50 米，但是倘若与茶胶寺庙山平面设计尺度达成统一其基底面积达到 120 米 × 110 米（包括第一、二层基台）的话，其总高不过约为 12 米。而茶胶寺庙山基底面积为 120 米 × 100 米，总高却达到了吴哥时代庙山建筑基台（须弥台）的最大高度 38 米。

图 5-6　巴方寺庙山复原图

巴方寺可以视为在东梅奔寺的基础上吸收茶胶寺庙山新的形制特点发展而来，比如回廊的应用显示出更为娴熟的设计技巧。吴哥窟则由于建造技术的进步、建筑类型的丰富、建筑规模的扩张以及审美旨趣的改变等因素的影响，215 米 × 197 米的基底面积已使之不能如茶胶寺和巴方寺一样，去极力追求高耸挺拔的建筑形象，而是保留三层须弥台设计原型的基础上在水平方向展开整座寺庙的宏大格局。

二、茶胶寺庙山建筑的初步复原

1. 中央五塔的复原（以中央主塔与东北角塔为例）（图5-7，图5-8，图5-9，图5-10，图5-11，图5-12）

（1）假层

由于茶胶寺庙山中央五塔的假层及顶部的保存现状皆不完整，在雅克·杜马西关于茶胶寺庙山的研究中，推测庙山中央五塔之中角塔的假层应为四层，但未作详细说明。[1]

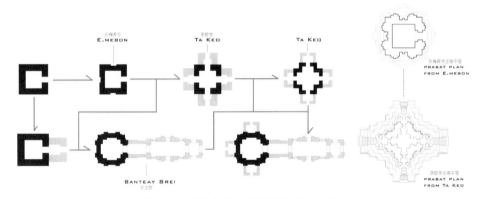

图5-7 茶胶寺庙山塔殿平面形式示意

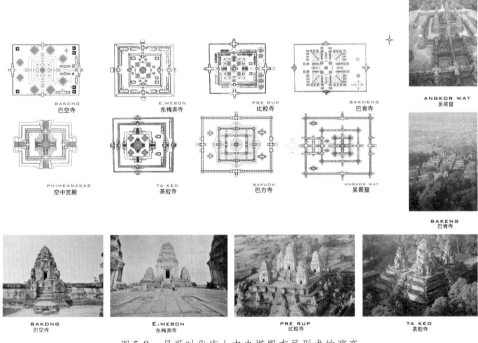

图5-8 吴哥时代庙山中央塔殿布局形式的演变

---

[1] Jacques Dumarçay. *TA KEV：ETUDE ARCHITECTURALE DU TEMPLE*, l'Ecole française d'Extrême–Orient EFEO：1971, pp. 35.

第五章 茶胶寺庙山建筑复原初探

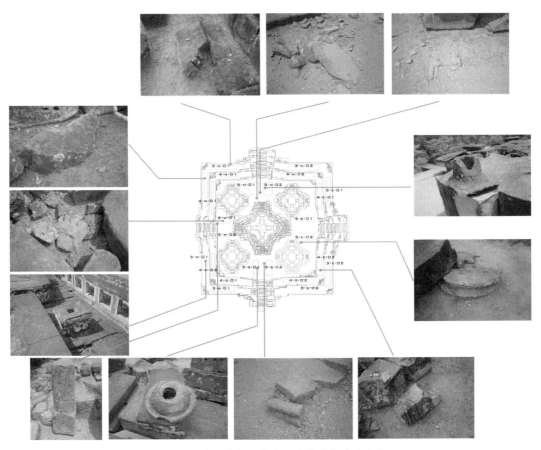

图 5-9　茶胶寺庙山中央五塔散落构件的分布

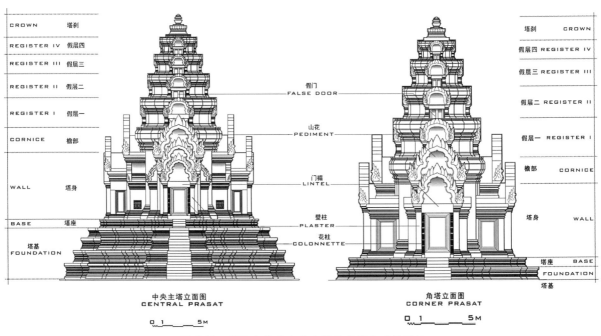

图 5-10　茶胶寺庙山中央主塔及角塔的建筑形制

149

茶胶寺庙山建筑研究

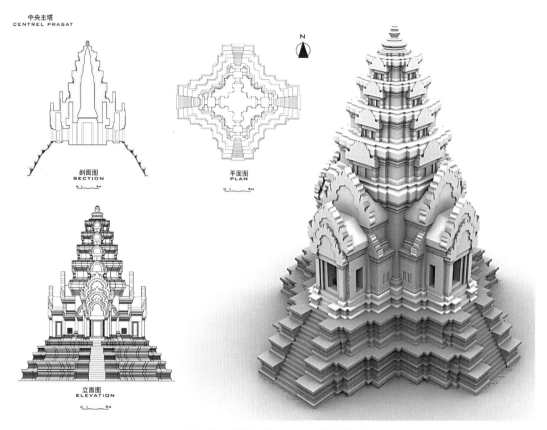

图 5-11　茶胶寺庙山中央主塔复原图

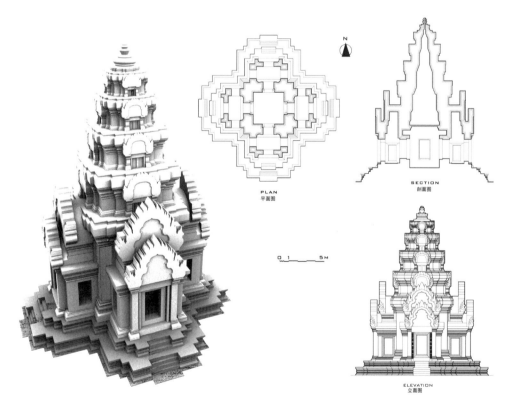

图 5-12　茶胶寺庙山角塔复原图

本章对于茶胶寺庙山中央五塔假层的复原或可从与其最上部假层相连的塔顶部分入手。虽然没有完整的塔顶遗存，但现场的散落构件中发现了一些疑似构成塔顶部分的砌石构件，这些构件为复原塔顶部分提供了重要的参考。通过推测塔殿最上部假层的尺寸，从而可以复原假层的数量。对发现的与塔顶相关的散落构件进行形制鉴别与尺寸测量可推断塔顶尺寸应大约为0.8-0.9米见方。

以假层及顶部保存较为完整的东北角塔为例，通过三维激光扫描实测其现存三层假层的面阔尺寸自下而上分别为6.1米、4.75米、3.85米。若以现存的最上部假层尺寸3.85米与散落构件中0.8-0.9米见方的塔顶构件进行匹配，这在尺度比例上很显然是极不协调的。若以第四层假层的推测高度约为3米计，似与散落的塔顶构件尺寸的匹配较为合适，由此或可推测茶胶寺中央五塔之角塔假层的层数为4层。较之坐落在须弥台四隅的角塔，其形制与等级更高的中央主塔假层的层数想必应不少于角塔，所以暂以中央主塔假层的层数亦为4层。参照现存假层平面逐层向内收进，以及与其对应的立面逐层收分的匹配拟合，可以复原茶胶寺庙山中央五塔的第四层假层。

（2）塔刹

吴哥时代的建筑遗存之中，塔殿顶部结构多为砌石（或砌砖）组成的叠涩拱屋面，或以塔刹构成塔殿顶部的主要形制（如女王宫）。

根据茶胶寺庙山中央五塔周围发现的诸多类似塔刹的散落构件，其中在须弥台南侧发现的散落构件可以拼对成直径约为0.92米，厚度约0.3米的砂岩圆盘，推测应为塔刹底座的组成部分之一，由此可以佐证茶胶寺中央五塔顶部形制应为塔刹。然而，由于吴哥时代建筑遗存中塔刹保存完整的实例并不多，加之茶胶寺庙山未完成以及对于散落构件的清理寻配工作尚未完善等条件所限，因此对于庙山中央五塔塔刹的复原较为困难。幸而与茶胶寺庙山建造时代与形制特征较为接近的女王宫、巴方寺（塔门）的塔刹遗存保存比较完整，成为茶胶寺庙山塔刹部分的复原的重要参照实例。

通过对茶胶寺庙山中央五塔周围散落构件所进行的寻配拼对来看，五塔的塔刹形制大致与巴方寺现存的塔刹实例比较接近，遂以巴方寺内现存塔刹的实例作为依据对茶胶寺庙山五塔的塔刹进行了复原设计。

（3）山花

茶胶寺庙山中央五塔之中，山花的保存状况皆不甚完整。以东北角塔为例，根据保存现状及周围散落构件的情况可以大致确定：正面山花共由三层砂岩砌石组成，自下而上砌石数目依次分别为3块、2块、1块，位于山花最上层的砌石呈三角形，山花各层砌石的水平分缝之间皆施以高约1-2厘米的凸起以增强结构稳定性。

较之正面山花，各级假层山花的砂岩砌筑方式与之有较大差别：砌石共分为两层，处于下部的一层由三块砌石构成且体量相对较大，而位于上部的砌石由长条形石块砌筑且将其一端伸入假层砌石内部，使之与假层结构紧密连接，此做法与女王宫塔殿假层山花十分相似，遂以女王宫作为参照实例对茶胶寺庙山五塔假层的山花进行复原。

中央主塔抱厦门楣之上正面山花的砌石层数为四层，此乃由于其形制为双层山花所致，并以雕饰方式实现其在立面上增加一层山花的效果。中央主塔的假层山花形制多与角塔类同，惟其在过厅的窗楣之上两侧亦有类似山花的砌石构件，根据位置与轮廓推测，其形制推测应该类似于女王宫中出现的翼形山花。

（4）线脚及雕饰

茶胶寺庙山中央五塔的线脚较之先前的做法似有较大的改变，特别是在塔基与塔身交接之处的线脚表现尤为明显。通常的做法是，塔殿墙基处的线脚在不同位置的标高都是相同的，因而墙身与基座之间的分界线甚为清晰。然而茶胶寺庙山五塔可能是由于塔殿平面形式的变化，构成塔身的建筑元素增加，其线脚标高亦随之进行相应的调整协调。现根据塔基与塔身连接之处砌石层数的不同，推算出线脚的设计标高，佐以参照茶胶寺庙山内各座塔门的线脚形制进行复原设计。

2. 塔门的复原（图5-13，图5-14，图5-15，图5-16，图5-17，图5-18，图5-19，图5-20，图5-21，图5-22）

（1）屋顶

根据实地调查发现，在茶胶寺庙山周围环壕与水池的填土中有数量不菲的砖瓦碎片；另外，二十世纪二十年代法国远东学院进行考古清理时拍摄的历史照片也显示当时的第一层与第二层基台上散布着大量的碎砖。另外，诸如西外塔门、东内塔门、南内塔门、西内塔门、北内塔门等五座塔门的屋顶部分尚局部残存砖砌叠涩屋顶的遗迹，而且在某些屋顶四周砌石表面残存着与砌砖尺度相吻合的刻槽，这些刻槽或可使砌石与砌砖屋面之间进行密实的连接以阻止雨水的渗入，由此推测其他塔门的屋顶结构亦应与此类似，所以茶胶寺庙山塔门屋顶部分皆以砌砖叠涩形制作为参照进行复原。

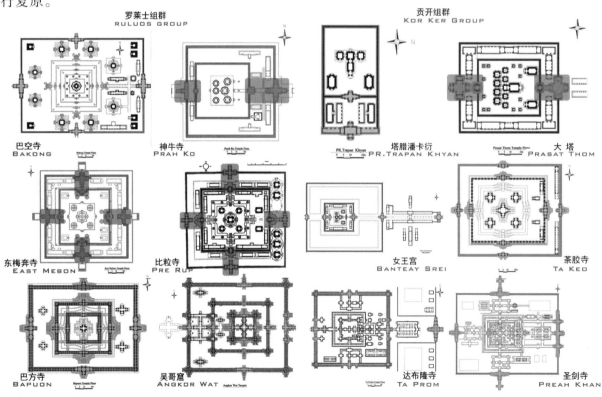

图5-13 吴哥时代庙山塔门布局形式的演变

第五章 茶胶寺庙山建筑复原初探

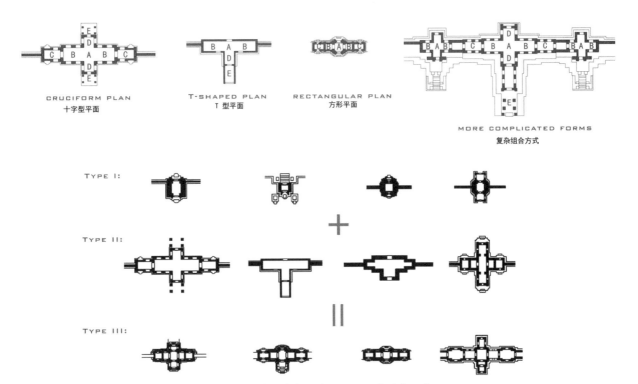

图 5-14 茶胶寺庙山塔门平面组合形式示意

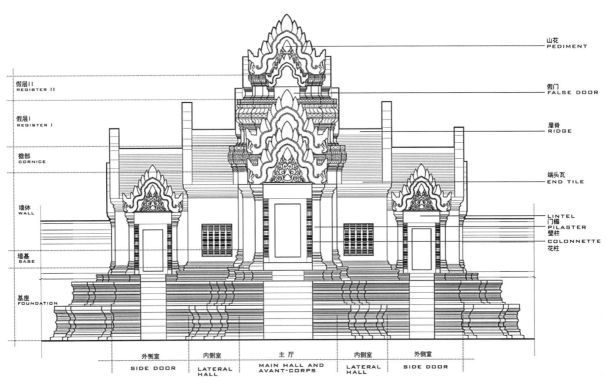

图 5-15 茶胶寺庙山塔门的建筑形制

153

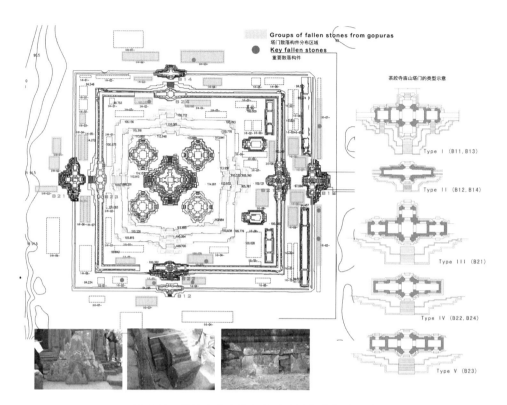

图5-16 茶胶寺庙山塔门类型及其散落构件分布

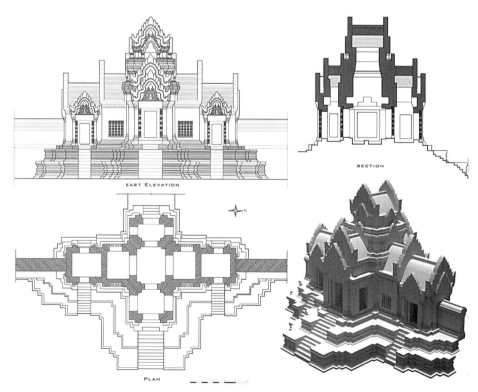

图5-17 茶胶寺庙山东外塔门建筑形制复原

第五章 茶胶寺庙山建筑复原初探

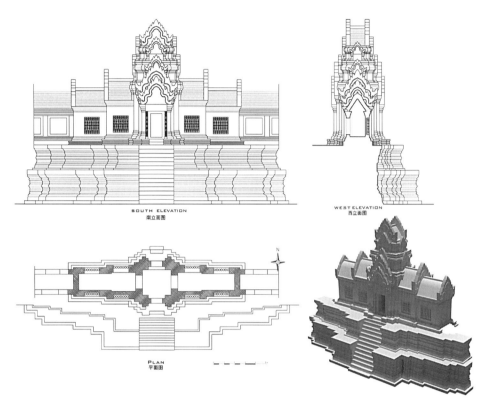

图 5-18 茶胶寺庙山南内塔门建筑形制复原

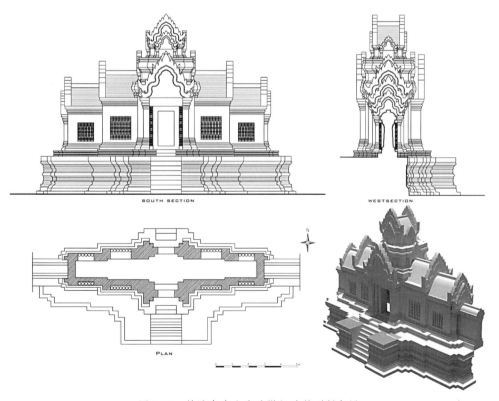

图 5-19 茶胶寺庙山南外塔门建筑形制复原

155

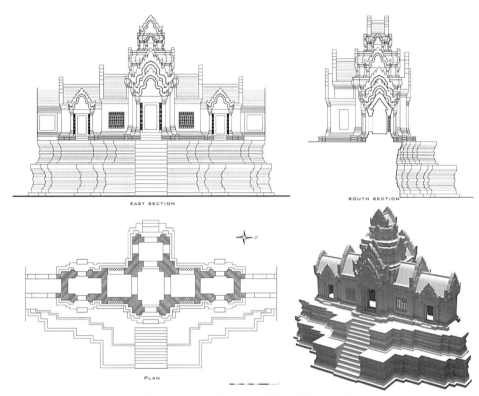

图 5-20　茶胶寺庙山东内塔门建筑形制复原

（2）端头瓦

在东内塔门内侧室檐口部分与主厅墙体相交之处，皆雕刻出与主厅墙体连为一体的小段端头瓦。在北外塔门外的基台上也残存有一块长约 1 米的断面呈"L"型的散落构件，其尺寸与檐口端头瓦的位置相匹配。另外，在各座塔门的内侧室与外侧室檐口顶部的边缘雕凿连续通长的凹槽，似为固定端头瓦之用。值得注意的是，茶胶寺庙山长厅和回廊的顶部亦有端头瓦遗存，但上述三种端头瓦的形制尺度皆不相同。

（3）山花

根据现场实测数据的分析，可知茶胶寺庙山各类塔门山花的高宽比基本相同，遂以雕刻完整、保存状况较好且易于测量的东外塔门山花作为复原各类塔门山花参照的原型。[1]

（4）东外塔门前后抱厦之山花

在东外塔门外有一组初步拼对完成且雕刻完整的山花，根据其底边宽度稍大于东外塔门东侧抱厦入口两侧壁柱及门楣的跨度，可以确定此山花是东外塔门东侧抱厦的最外层山花。东外塔门前后抱厦（东西两侧）檐口门楣上皆施以假层，假层外侧以山花封护前后抱厦屋顶端口，但两座山花之间却留有宽约 30 厘米的空隙，这种山花形制的实例在女王宫的假层山花中有所体现。另外，在东外塔门西侧抱厦假层的端部残存有砖块，似乎可推测上述假层山花背后的空隙或以砌砖作为填充。这种双层山花之间或在山花与主厅墙体之间留有空隙的做法，在茶胶寺庙山的其他塔门中亦有发现。

---

[1] 法国远东学院曾经于二十世纪六七十年代对茶胶寺庙山东外塔门的散落山花构件进行分类和拼对，并将初步完成拼对的山花分类放置于东外塔门东侧的第一层基台外侧。（著者注）

表 5-2　茶胶寺庙山塔门山花形制简表

名称	层数	宽度（mm）	高度（mm）	高宽比
东外塔门抱厦山花	3	3200	2450	1/0.765
南外塔门内侧室山花	3	3050	2300	1/0.754
南外塔门假层山花	3	2400	1860	1/0.775
西外塔门内侧室山花	4	4480	残高2300	现存三层
西内塔门内侧室山花	4	4050	残高2050	现存三层
西内塔门外侧室山花	3	3750	2920	1/0.780
北内塔门内侧室山花	3	3755	2835	1/0.755

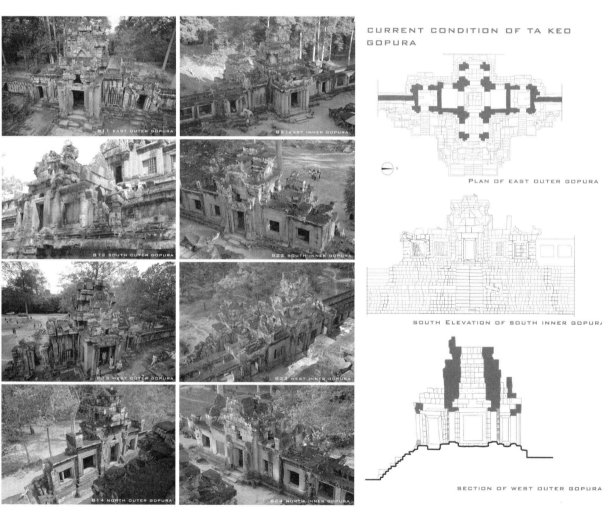

图 5-21　茶胶寺庙山各座塔门保存现状

（5）山墙的双层山花

茶胶寺庙山塔门的外侧室山墙辟为假门，通常形制做法应以壁柱支撑着门楣及山花，但茶胶寺塔门外侧室檐口上却无山花痕迹。其中缘由或可从以下三点说明之：其一，空中宫殿的塔门与茶胶寺庙

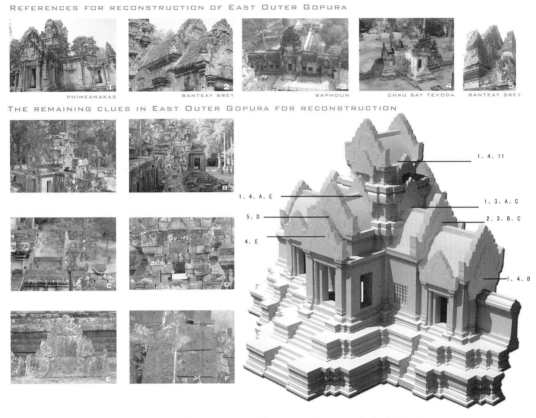

图 5-22　茶胶寺庙山塔门山花构件的分布及砌石构件形制分析

山塔门的建造年代大致相同，其山墙部位亦辟为假门且檐口以上施以双层山花；周萨神庙塔门或因假门可能被围墙覆盖遮挡之故，其山墙部位皆不设假门，但其山墙的檐部也施以双层山花。茶胶寺东外塔门外侧南北各有一块上部凸起、局部雕刻的散落山花构件且形状对称，推测应为东外塔门外侧室山墙之山花。另外，在第一层基台南外塔门处的散落构件中，亦发现双层山花且背后雕有凹槽，以凹槽尺寸可以确定其为南外塔门外侧室山墙处的山花遗迹。由此，茶胶寺庙山的各座塔门外侧室山墙上部皆以双层山花形制进行复原。

（6）南、北内塔门的假层

在第一层基台南内塔门的西侧，散落堆放着诸多尺寸规格较为一致的长条形石块，推测应与塔门假层"丁头冲外"的砌筑方式相吻合且其背面雕有与假层匹配的凹槽。

以上文提及实测各座外塔门第一级假层与第二级假层的比例相同为基准（约为0.49），推测内塔门各级假层之间的比例关系亦应相同，因此南北内塔门假层的高度得以确定。实测东内塔门、西内塔门现存的两级假层高度之比皆为0.7，从而佐证了各座内塔门假层比例相同的假设（实测南北两座外塔门现存的假层高度之比也大致相同），遂可以东内塔门、西内塔门两级假层高度之比对南北两座内塔门的假层进行复原。

（7）线脚

茶胶寺庙山各座塔门线脚雕饰的复原皆以保存较为完整的东内塔门线脚作为参照的原型，并根据其下部假层线脚与上部假层线脚之比例关系的适当调整进行复原。

3. 附属建筑的复原（长厅、藏经阁、回廊、角楼）（图5-23，图5-24，图5-25，图5-26，图5-27，图5-28，图5-29，图5-30，图5-31，图5-32，图5-33，图5-34，图5-35，图5-36，图5-37，图5-38）

（1）屋顶

主要参照茶胶寺庙山塔门屋顶部分施以砌砖叠涩对附属建筑的屋顶进行复原，砌砖叠涩的高度和出挑尺寸皆以东外塔门屋顶砖叠涩的实测数据为准。其中，内外长厅檐部残存的山花上凿有榫孔与凹槽，似与木结构屋架构造相关[1]；在西外塔门北侧与回廊相接处塌落的山花下残存少量砌砖的遗迹，可以大致推测回廊的屋顶应为砖叠涩；在第二层基台西北角塔外侧的第一层基台上发现有疑似角塔屋顶结构的散落构件，以此可以确定角塔顶层的形制与构造。

（2）端头瓦

由于茶胶寺庙山的长厅建筑有两种类型的端头瓦，遂在复原研究过程中曾经提出了多种设计模型，并通过其尺寸与形制特征点的比较匹配，得以确定两种类型端头瓦的形制。另外，由于尚有部分端头瓦残存于回廊的檐口之上，因此可以参照实测资料进行复原。但对于藏经阁而言，由于在现场的散落构件中未发现有类似端头瓦的实物遗存，故其形制特征及具体尺寸皆不得而知，因此主要参照其建造年代稍早于茶胶寺庙山的女王宫藏经阁的端头瓦进行复原。

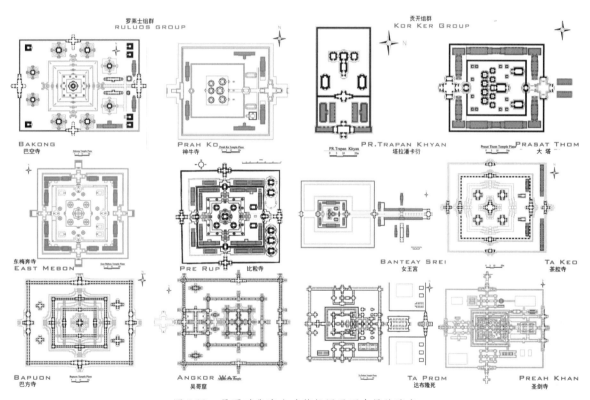

图5-23 吴哥时代庙山建筑长厅平面布局的演变

---

[1] 关于古代高棉的寺庙建筑遗迹中的木结构屋架，雅克·杜马西所著的《高棉的屋架与瓦》中有详细的论述。该书中通过现存石构建筑遗迹复原木结构屋架，从考古学调查中分析获得瓦片的铺装方式，并与印度、中国、越南等地区的木结构屋架与屋瓦进行比较。雅克·杜马西认为古代高棉木结构屋架以抬梁式（由梁与短柱构造而成）为主且所使用的技术也多有不同；木结构屋架自重很大，经常截面较大的水平横梁。

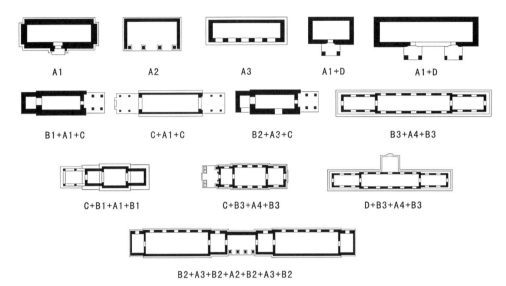

图 5-24　吴哥时代庙山中长厅平面布局的组合方式

图 5-25　茶胶寺庙山长厅遗存现状及其建筑形制

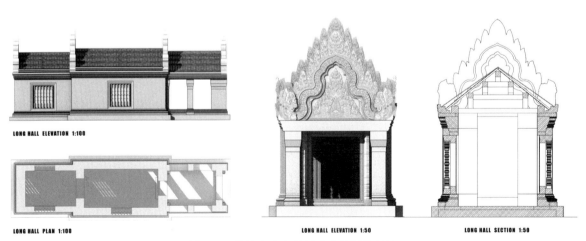

图 5-26　茶胶寺庙山长厅建筑形制的复原

第五章 茶胶寺庙山建筑复原初探

图 5-27 茶胶寺庙山建筑长厅形制复原

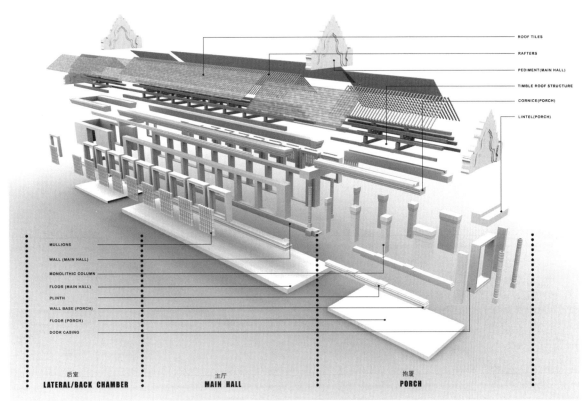

图 5-28 茶胶寺庙山长厅形制复原之构成示意

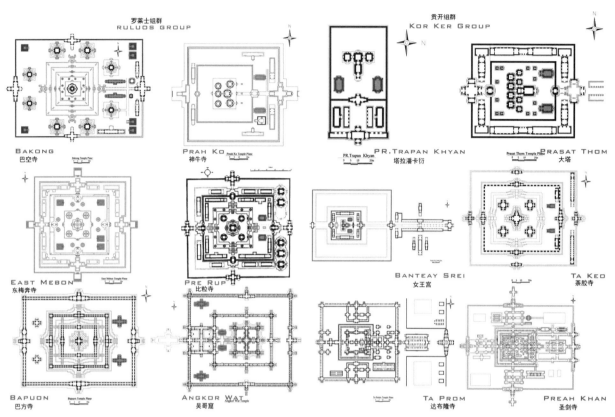

图 5-29　吴哥时代庙山建筑藏经阁平面布局的演变

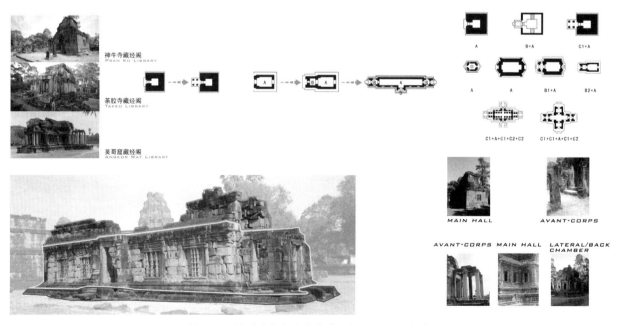

图 5-30　吴哥时代庙山建筑藏经阁平面组合方式

第五章 茶胶寺庙山建筑复原初探

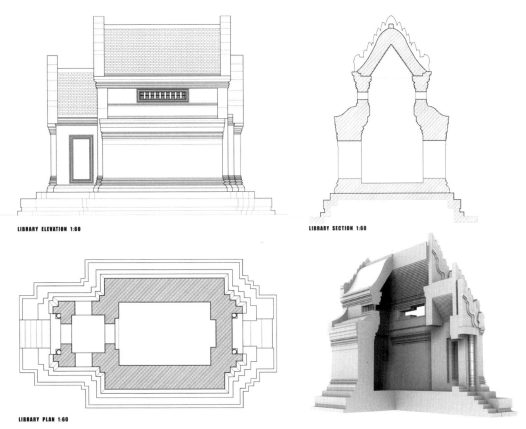

图 5-31 茶胶寺庙山藏经阁形制复原之一

图 5-32 茶胶寺庙山藏经阁形制复原之二

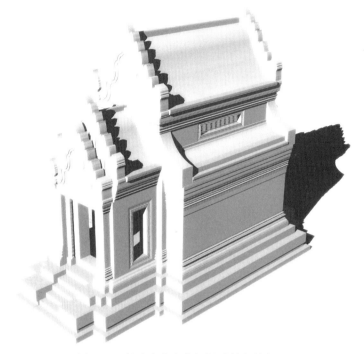

图 5-33 茶胶寺庙山藏经阁形制复原之三

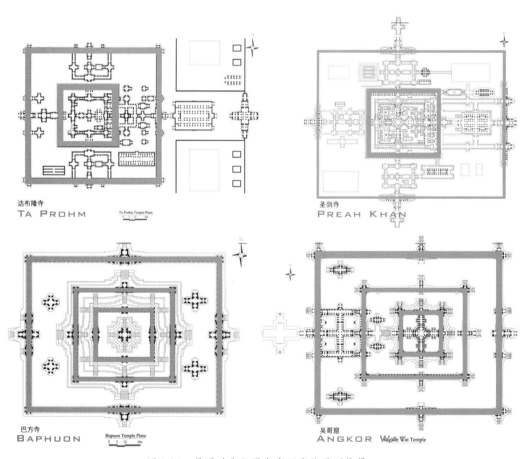

图 5-34 吴哥时代主要寺庙回廊的平面格局

第五章 茶胶寺庙山建筑复原初探

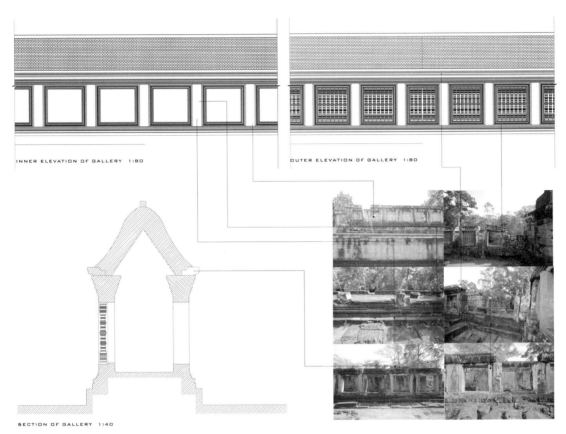

图 5-35 茶胶寺庙山回廊形制复原之一

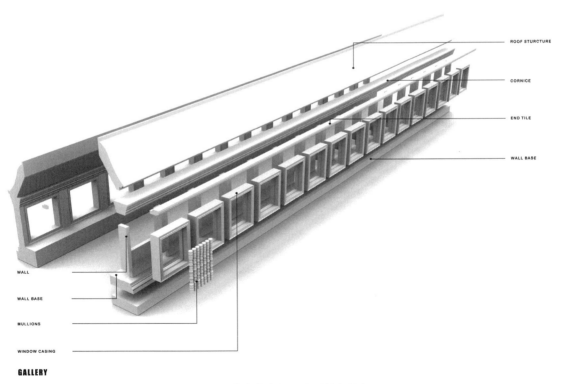

图 5-36 茶胶寺庙山回廊形制复原之二

图 5-37　茶胶寺庙山回廊角塔建筑形制复原之一

（3）山花

如前所述，根据现场实测数据的分析，可知茶胶寺庙山各类塔门山花的高宽比基本相同，遂以雕刻完整、保存状况较好且易于测量的东外塔门山花作为复原各类塔门山花参照的原型。而对于各类附属建筑的山花部分，亦是通过大量散落构件的实测数据，基本确定其山花形制的高宽比是一个固定的数值，进而参照东外塔门外侧初步拼对完成的各类山花形制进行复原。

（4）长厅的木构架

由于地处高温高湿的热带气候环境，古代高棉建筑中的完整的木结构屋架几乎没有存世。然而，雅克·杜马西在茶胶寺庙山清理过程及其调查研究报告中，对于清理现场所发现的屋顶散落瓦块和檩条尺寸皆有较为翔实的记录。因此，长厅木结构屋架的构件尺寸都是根据其记录进行复原的。

对于木构架的断面尺寸则是根据北内长厅现存山花上的榫孔尺寸进行确定的，并且参考了雅克·杜马西对于女王宫长厅木结构屋架的复原。

图 5-38　茶胶寺庙山回廊角塔形制复原之二

根据女王宫中的木结构榫孔遗迹，雅克·杜马西认为："垫梁坐在顶部砂岩内侧缺口部分，在墙壁顶端环绕一圈。梁没有架在墙壁顶端，而是在较低的墙壁上作插榫，梁与山墙使用楔子固定。梁上立有短柱，支撑着栋梁和檩条。其上面挂有椽子，椽子的下端插进垫梁。坐在墙壁顶端的砂岩，由于底部非常不稳定，用楔子与旁边的砂岩连接起来固定。从连续的砂岩材质的檐头瓦散落在建筑物内的情况看，椽子有可能是收于檐头瓦的里面。但是，由于垫梁和其他横梁的不稳定，因此在横梁一侧做暗榫，与山墙连接。顶棚大概是挂在室内的突出部分上。由于山墙的厚度不足以支撑系梁的重量，因此系梁的插榫痕迹在支撑顶棚的檐口内侧下方的厚墙部分。由于是用较薄的人字墙形成的，因此横梁的交圈部分做得不深，横梁采用榫接扒钉固定。"

对于端头瓦的构造方式，是根据其可能的形态提出多种样式的假设，最终以实测数据进行拟合拼对以确定其复原的形制。

（5）角楼

茶胶寺庙山第二层基台的顶面边缘四周施以砂岩砌筑的回廊环绕，在回廊四个转角处分别设置角楼一座。各座角楼的第一级假层以上部分皆已经不存，其假层的层数则是通过中央五塔的层数及其尺度比例关系的分析而得出的。通过对庙山五塔的复原研究，以确定各级假层高度和与其面阔宽度的收进缩减变化规律，因此角楼的第一级假层与其顶层的尺寸得以确定，角楼的层数亦随之确定。

4. 建筑装饰的复原[1]（图5-39，图5-40，图5-41，图5-42，图5-43，图5-44，图5-45，图5-46，图5-47，图5-48，图5-49，图5-50）

论及茶胶寺庙山的建筑装饰，或可将其分成以下情形：其一，建筑装饰保存状况较为清晰完整的部位；其二，雕饰刚刚开始或由于没有机会完工而草草收场的部位；其三，雕饰从来没有机会展开以至于不能确定其是否存在于古代高棉匠师最初的蓝图之中的部位。鉴于此，对于茶胶寺庙山建筑装饰的复原亦可以相应地进行如下的划分：首先对于较为清晰完整的装饰部位进行汇总，并在其残损的基础上进行复原，这部分工作可以为其他没有完成或尚未开始的建筑装饰的复原提供重要的参考依据，即从已经完成的部分来管窥或推测整体建筑装饰的风格；另外，对于尚未完成的建筑装饰部分的可能性探讨，以及对尚未进行的建筑装饰部分是否存在的探讨，则是基于实地调查收集数据资料和相关的分析进行的推断。纵观吴哥时代的各类建筑装饰部位，皆有其历史源流及风格演变的过程，以下主要针对茶胶寺庙山诸如花柱、山花、假层线脚、须弥台线脚等重要建筑装饰部位及其纹样题材进行复原。

（1）花柱的雕饰及纹样

装饰性花柱属于茶胶寺庙山建筑装饰中较为完整的部位。花柱遗存主要出现在各座塔门、内外长厅和藏经阁等建筑的入口处，中央五塔的入口处虽亦设有花柱，但却仅为粗凿未做任何装饰。上述花柱的比例尺度适宜进行细致入微的装饰性雕刻，并因其所处位置的特殊性且有较明显的风格变化特征，故被视为茶胶寺庙山建筑艺术风格及其年代推断的重要依据之一。

通过实地对八座塔门、四座内外长厅和两座藏经阁的花柱所进行的现状调查可知，虽然多数花柱

---

[1] 关于茶胶寺庙山建筑装饰的研究及背景可以参照本书第三章的相关论述，不再赘述。

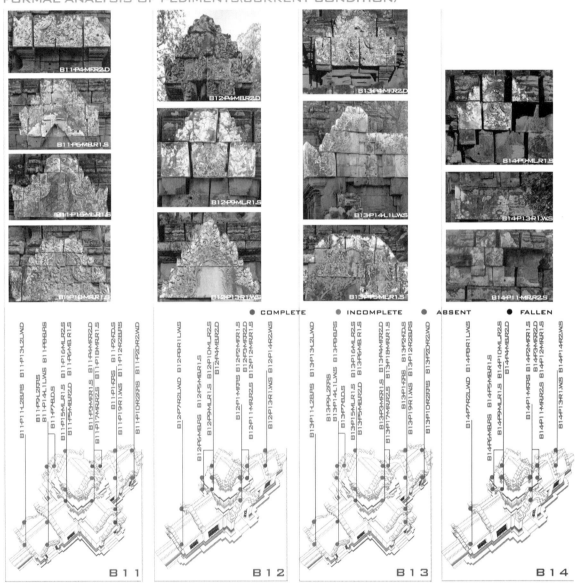

图 5-39　茶胶寺庙山塔门山花构件分布及其拼对之一

并未全部完成雕饰，但仅凭残存的纹样及其对称性，也可以推断其最初的设计构想，并且在实地调查测量中发现，茶胶寺庙山各座塔门、内外长厅和藏经阁的花柱在装饰题材上也十分接近，其装饰构成皆包括柱头、柱底、柱基以及柱身中间重复性的装饰单元等几部分，仅仅在于各自的尺寸和比例有所差异，遂针对较为典型的塔门花柱进行复原。

值得注意的是，设有外侧室侧门的塔门（东外塔门、西外塔门、西内塔门）的花柱可以分为两类，一类是主室入口两侧的花柱，另一类是外侧室入口两侧依附于门柱石并不单独用料的小型花柱。另外，花柱柱底部分并没有发现任何经过雕饰的痕迹且被单独留白，推测此留白区域并非设计初衷，那么最初的意图应是做成什么样式呢？

回顾法国高棉艺术史家简·布瑟利耶（J. Boisselier）对于吴哥时代建筑花柱演进的研究。初期的

第五章 茶胶寺庙山建筑复原初探

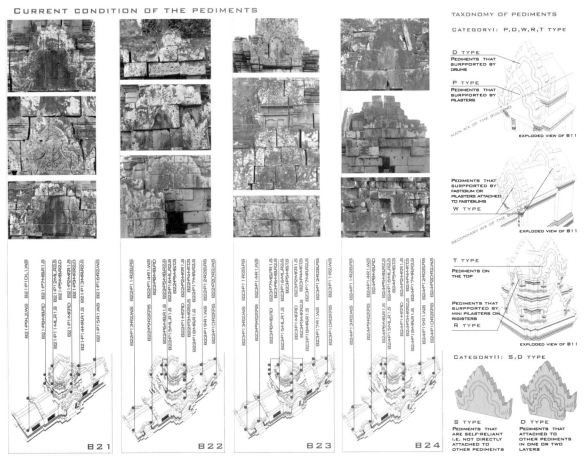

图 5-40 茶胶寺庙山塔门山花构件分布及其拼对之二

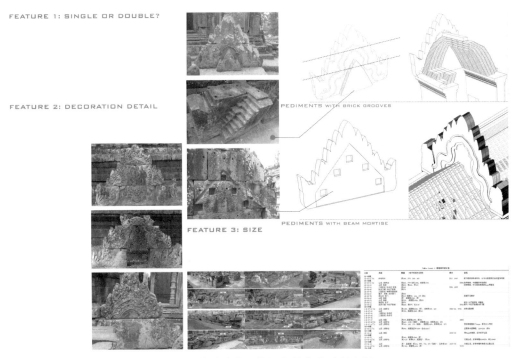

图 5-41 茶胶寺庙山塔门山花构件形制分析

图 5-42　茶胶寺庙山塔门山花构件雕刻纹饰的复原

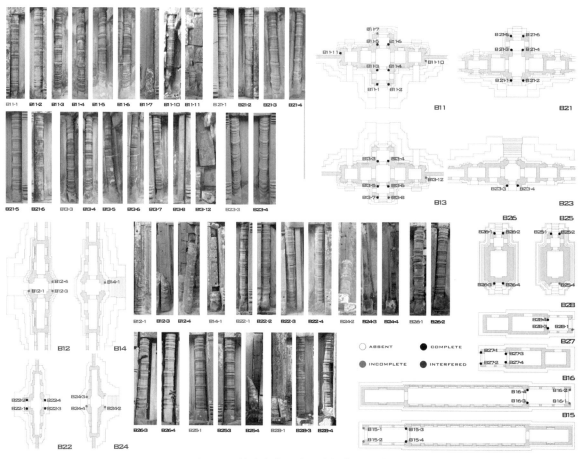

图 5-43　茶胶寺庙山塔门花柱构件分布

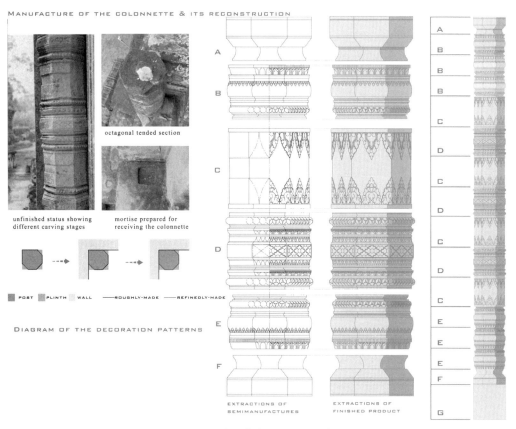

图 5-44　茶胶寺庙山塔门花柱复原

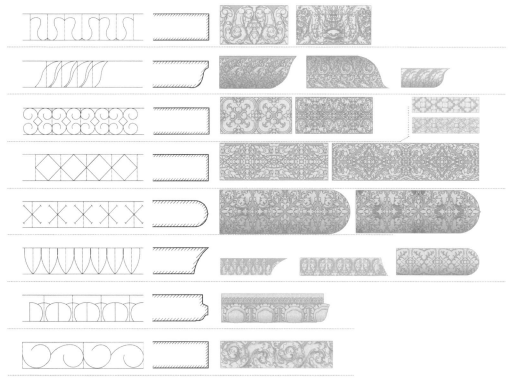

图 5-45　茶胶寺庙山须弥台雕刻纹饰组合方式

图 5-46 茶胶寺庙山须弥台雕刻纹饰复原之一

图 5-47 茶胶寺庙山须弥台雕刻纹饰复原之二

第五章 茶胶寺庙山建筑复原初探

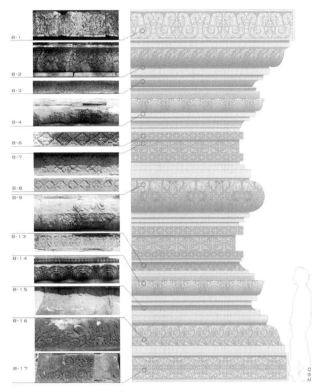

图 5-48 茶胶寺庙山须弥台雕刻纹饰复原之三

图 5-49 茶胶寺庙山须弥台雕刻纹饰复原之四

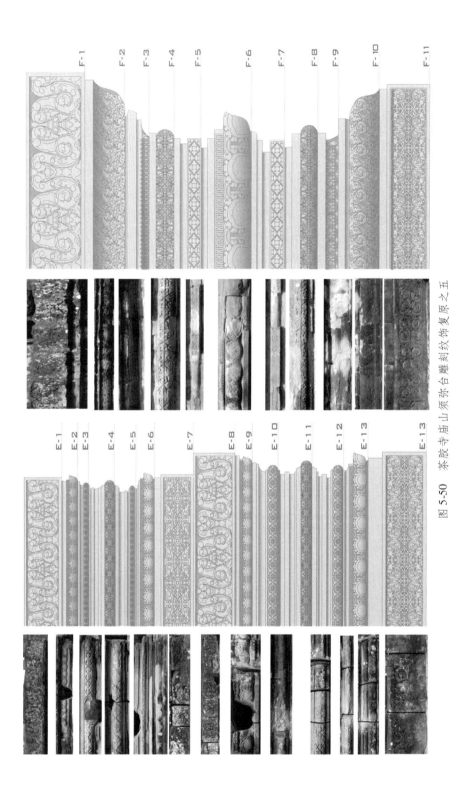

图 5-50 茶胶寺庙山须弥台雕刻纹饰复原之五

花柱柱底与柱身并无太大的区分，延续了柱身重复的线脚，并且这种情况往往在很大程度上与柱顶部分对称。但很显然茶胶寺庙山的情况却并非如此。首先，花柱的柱顶并没有留出类似有待雕饰的空白，因而其柱底与柱顶有可能并不对称。其次，如果花柱的柱底也延续类似柱身的线脚，那么上述留白区域显然是没有任何道理，而且为什么未在粗凿阶段即一并进行雕饰呢？参照北仓（Northern Kelang）的花柱，或可以推断，茶胶寺花柱的这部分留白区域的雕饰内容应在很大程度上区别于线脚的装饰纹样，抑或为造像？而茶胶寺花柱的这种留白做法也在某种程度上显示出工艺流程的问题，由于花柱安放位置所限，因而粗凿之后的花柱会被精雕加工出大致的线脚，并且在花柱贴近墙体或壁柱的两个面内进行精细的纹样雕刻，而在上述留白区域雕饰更为复杂纹样题材则或是由更富于经验的匠师统一进行的。

（2）山花雕饰及纹样

较之花柱而言，山花所在的位置乃是最佳观瞻视线的汇聚之所，因此成为古代高棉建筑最为重要的装饰部位。茶胶寺庙山内的山花装饰主要分布在塔门、内外长厅、藏经阁和中央五塔上，回廊角楼也应有与中央五塔类似的山花装饰。

由于庙山中央五塔的山花多为粗凿表面的砂岩砌石，四座回廊角楼也倾圮不存，所以暂仅对茶胶寺内的八座塔门、四座内外长厅和两座藏经阁的山花散落构件进行分类和调查。

在茶胶寺庙山各处构成山花的散落构件中，长厅山花的保存状况较为完好，遂采用三维激光扫描结合摄影测量的方法对其进行复原。塔门的山花情况较为复杂，不仅包括壁柱所支撑的主厅门楣上部和外侧室门楣上部的山花，还包括装饰性假层壁柱所支撑的小山花。而且根据平面布局与砌体构造的变化，各类塔门山花的设置又具有很大的灵活性，山花与砌石构造之间的交接情况不同，亦会出现不同类型的山花。而山花尺度各异则进一步导致其装饰纹样题材的变化。鉴于茶胶寺庙山内的各类山花大多已经塌落，所以针对现场山花散落构件的调查、识别和拼对显得非常必要。对山花进行的复原则是首先根据山花残件推定各类塔门的山花布置方式，并对山花进行分类与识别之后依次进行的。

（3）线脚雕饰

法国远东学院的雅克·杜马西在其调查报告中曾对茶胶寺庙山的各类雕饰线脚进行了详尽的叙述[1]，成为茶胶寺建筑装饰复原研究的重要参考资料。由于茶胶寺第一层基台和第二层基台的立面雕饰线脚基本完好却无雕饰纹样，而须弥台雕刻部分的线脚及雕饰纹样大都清晰可辨，于是复原工作是通过对须弥台线脚和纹样雕饰对应关系的分析，归纳出相对固定的装饰纹样母题，并根据分布位置的重要性及其比例尺度关系进行相应的变化，在此复原过程中订正了几处法国远东学院绘制的茶胶寺庙山测绘图中的几处疏漏。

另外，根据须弥台线脚和纹样的对应关系，可以用来推测须弥台第一、二层没有完成的纹样。因而根据第一、二层须弥台实测的线脚，可以初步复原第一、二层须弥台的装饰纹样。另外还有一个有趣的现象，各类线脚之间的交接关系似乎没有任何过渡而是生硬地相撞在一起。

---

[1] 详见本书第四章的相关内容。

Synopsis V Preliminary Study on the Reconstruction of Ta Keo Temple-mountain

1. Reconstruction of the Site Planning

It is presumed that Ta Keo temple-mountain is located in the center of the large-scale group within layout planning, and a causeway extends about 500 meters to the west embankment of Eastern Baray where have some ruins on the embankment. It is a rectangular area of about 1000 meters x 780 meters due to the Siem Reap River take shape of a natural barrier along the west direction.

King Indravarman I, King Yasovarman I and other Kings of the early Angkorian period had successively built Bakong, Preah Ko and other huge temple-mountains in the center of their capital like Hariharalaya, Yasodharapura since the middle of theninth century AD. During the next four hundred years, such as Phimeanakas, Pre Rup, Ta Keo, Baphoun, Angor Wat, Bayon, the magnificent temple-mountains had constructed the extremely significant symbols of Angkorian architectural style. For ancient Khmers, temple-mountains had creatively sought to develop their own architectural expressions and art characteristics, which are based on the prototype of Indianization.

The Angkorian temple-mountain style is provided with such invariable aspects as the compositions: foundations or platforms namely pyramid, prasats, galleries around the pyramid, affiliated buildings namely libraries or long-halls, causeway, ponds and moats, and these compositions can construct a special architectural complex with sacred symbolization. As a core, there lies a central prasat on the pyramid commanding the whole complex. Both the pyramid and the prasats are symbolizing the center of the universe as well as the home ofGod, Meru Mountain. And the galleries are symbolizing sacred mountains as God's territory as well as the water in the moat is the symbolization of ocean in the universe. Either planning or constructions processing of temple-mountains have intensively embodied the belief system and the sovereign politics during the Angkorian period. As national cult, the real sovereign power is endowed to the King due to Devaraja is showing a special symbolization of the God and the King to prove the legality of the sovereign.

The ancient Khmers had also created the layout planning which is mainly including prasat and sanctuary as the basic architectural element, as well as the various factors, which are contained in architectural arts, materials and techniques. Most prasats are rectangular and faced to the east direction, and its masonry structure is constructed layer by layer. It is mainly the sacred space to worship Linga of Shiva or Vishnu inside of the prasrat. Consequently, it is mostly highlighted the symbolization of the temple-mountain with the quincunx-shaped layout of five-prasat on the pyramid as well as the steep steps between the foundations and the different terraces.

Due to the centre of temple-mountains, the prasats location can be divided into two categories: one is on the top platform of the pyramid, and another is on the each level of the pyramid or around it. Similarly, the layout of the central prasat also can be divided into two categories: solo -prasat separately and quincunx-shaped prasats. The converging center above-mentioned is presumed to have a close relationship with the spacial diagram of Indian Mandala. Owing to the profound impact from Indian culture, as a regulation of layout planning, in concretely, Mandala contribute abundant religion meaning to the temple-mountains.

The planning and the construction of Ta Keo temple-mountain experienced a lot of changes not only in the architectural style but also the layout in the Angkorian architectural history. Firstly, Gopura acquire a new style combined with the pyramid and contract layer by layer with the false storey. So it can be presented that the function is dedicated to the god as an important sacred space. Secondly, unlike the preliminary rectangular plan of prasats, it is cross-shaped and opened to four directions at Ta Keo temple-mountain. Thirdly, it can make the layout more regular and entirely due to the appearance of gallery at in Ta Keo temple-mountain. Therefore, it had been an important characteristic of the ancient Angkorian architecture owing to the practice for combining the gallery with the corner tower. Such temple-mountain as East Mebon and Pre Rup, which built earlier than Ta Keo temple-mountain, both have long hall around the pyramid and the enclosure outside, which constitute a vivid architectural characteristic in this period. However, owing to the reform of King Suryavarman I, it can be presumed that a new trend had been emergence in the evolution of Angkorian temple-mountain as follows: Firstly, the enclosure is substituted by gallery, and becoming a vital factor in the composition of façade. Secondly, from the viewpoint of function, the gallery also plays the same role as the enclosure and the long-hall except for the form. Moreover, for the evolution of architectural style, the long-hall style had faded away since Ta Keo temple-mountain.

For the layout of Angkorian temple-mountain, it had been mature no later than the end of 10th century or the beginning of theeleventh century, as well as the ornaments and the architectural style mainly including sandstone gallery enclose the pyramid or the central parast as the embodiment of temple-mountain.

Compared with Ta Keo temple-mountain group, based on the characteristics of the long-halls which located in the outer enclosure, as well as the long-halls and the gallery located in the inner enclosure, moreover, which is completely replaced by the stone structure and sandstone masonry prasats, it can be inferred by Ta Keo temple-mountain should be later than the East Mebon and Pre Rup. However, in Baphuon which built in the middle of the 11th century AD, the gallery lies around the foundations, and there are cross-shaped gallery connecting the central prasats. These features show that the style of Angkorian temple-mountain style, which cites Angkor Wat as a model has acquired a rudiment.

Therefore, it is supposed that the dating of Ta Keo temple-mountain is between Pre Ruptemple and Baphoun temple. It is also regarded as a transitional form because of the appearance of brick roof and stone corbelling gallery, as well as the disappearance of the long-hall. So the brick-roof and the gallery have become more and more precious with the passage of time, which proving the style of Ta Keo temple-mountain is a connecting between the preceding and the following. In generally, it takes a long time to endow the temple-mountain with the perfect style embodied in the Angkor Wat with the duration of nearly three centuries.

2. Reconstruction on Architecture Form

Prasat (Central Prasat and Northeast Prasat)

For the reconstruction of theprasat, on account of the top false-storey of prasat and unfinished status, Jacques Dumarçay suggested that amount of the corner parast's false-storey should be four, but he did not explain this conclusion and the reason in further. Concerning reconstruction prasat's false-storey, it should firstly consider its architectural form as well as the top part that is connected. Due to the ruined status is relatively imperfection, many sandstone blocks are scattered around prasats, which may be belong to the top components of prasats. As the prompting clue in rarely, but these sandstone components can provide the crucial reference to the reconstruction process of prasats. In addition, the amount of the false-storey can also be obtained to speculate the design scale of the uppermost false-storey. Identifying and measuring of the scattered sandstone component's style, it can be concluded that the scale of prasat top part is closely 0.8 or 0.9 square meter. Concerning the false-storey of the northeast corner parast, for the measuring data of the ruined three-layer false-storey, it can be acquired with three-dimensional laser scanner, which is 6.10m, 4.75m, and 3.85m from the bottom to the top. Obviously, for 3.85m, it is not quite matched with the top components of the corner prasat (0.8 - 0.9m). If the fourth false-storey is 3.0 meters high, it is pretty matched with the scattered components. Therefore, it can be inferred that the false-storey amount of the corner parast should also be four. For central parast, its false-storey amount should be presumed with four layers at least owing to the higher architectural form. On the other hand, referencing to the layout planning as well as its elevation, it also can be speculated that the prasats reconstruction should be have the four false-storeys.

For Angkor monuments, it is normally crown style as example as Banteay Srei temple at the highest part of prasats. Many scattered sandstone components around the southern area of the pyramid, which can be conjunct into a plate (its diameter is 0.92m as well as itsthickness is 0.3m), which is supposed to be the crown's base. However, for the most Angkor monuments, due to the crowns are usually in relatively decayed condition as well as Ta Keo temple-mountain is even unfinished status, therefore, it seems difficult to reconstruct its crown form owing to the searching or matching works for scattered components is so far from completed. Fortunately, the crown of Banteay Srei temple and Baphoun temple are all in relatively preserved status, which the epochal characteristics and art style are both similar to Ta Keo temple-mountain. Studies on the scattered components around the prasats of Ta Keo temple-mountain, it can be proposed that its crown reconstruction should be in accordance with the Baphoun style.

According to choose the northeast corner parast as an example, its pediments is also relatively not well condition to maintenance. It can be conclusion that the front pediment consists of three-layer sandstone blocks from bottom to top. The sandstone blocks of the uppermost layer are triangular form. Between the layers, the horizontal parts are all with a bulging (1-2cm high) in order to enhance the structural stabilities. Compared with the front pediment, the pediment of false storey is considerably different with its two layers. And the lower layer consists of three stone blocks which are much larger than usual. The upper layer is a solo-rectangular stone block which is sticking into the inside of the false-storey in order to enhance the connection stabilities with the false-storey. Such construction skill is relatively similar to the pediment of false-storey in Banteay Srei temple. So it proposed that the prasat's pediment reconstruction of Ta Keo temple-mountain should be in accordance to the Banteay Srei temple. For the front pediment of porch, there should be four layers above the lintel of the central parast, however, its actual situation has just ruined two layers. Furthermore, the carvings can show that it maybe one more layer of pediment that is added. Moreover, the pediment of false-storey of the central parast is similar to the style of the corner parast, which style should be similar to the pediment in Banteay Srei temple.

Gopura

For the reconstruction of thegopuras of Ta Keo temple-mountain, based on the field investigation and surveying, many fragments of brick and tile were found in the filled earth, which is located in the moats and the ponds around Ta Keo temple-mountain. In fact, EFEO had ever carried out a series of archaeological clearances and excavations during 1920s, after the historical documents, which can show many broken bricks scattered in the soils at the outer enclosure and the inner enclosure of Ta Keo temple-mountain. Moreover, such as the western outer gopura, the eastern inner gopura, the southern inner gopura, the western inner gopura, the northern inner gopura, it even showed that the scattered sandstone components maybe match with the scale of bricks. It can be inferred that the roof form of Gopuras reconstruction should be analogous with the size and style of the bricks.

Many carved end-tileshad ever been scattered outside of the gopuras, which should be matching with the cornice of the eastern inner gopura as well as the top structure of the main hall. For the outside of foundation of the northern outer gopura, some scattered component with special L-shaped section and its scale can be matched with the end-tile's location in the cornice. Consequently, the continuous grooves are carved in the edge of top cornice of the gopuras. The pediment of gopuras can be assembled in primarily which carvings are relatively completed. According to its under part is little longer than the span of the lintel and the bilateral pilasters in the eastern porch entrance of the eastern outer gopura, so the pediment is exactly the outermost layer in the eastern porch of the eastern outer gopura. Lintel of the cornice is with false storey in the porch in the front and back of the eastern outer gopura. The roof of porch is covered with pediment outside the false storey, but between the two pediments can make a gap as 30cm in width. It also can be found that example is similar with this style in the false storey pediment in Banteay Srei temple. Furthermore, there still remain some bricks on the end of the western porch false storey in the eastern outer gopura, so it can be inferred that the gap should be filled with bricks. The gap between the pediments and the main-hall is the normal characteristic of the other gopuras in Ta Keo temple-mountain.

Affiliated Buildings

The affiliated buildings of Ta Keo temple-mountain is including oflong-hall, library and gallery as well as its turrets. For most affiliated buildings, their roofs had been entirely missing and in terms of roof structure it may be the first step in the reconstruction program. For the inner and outer long-halls, there are many mortises and grooves on surface of the pediment, which is probably relative to the wooden structure, however, there still remain some ruined brick-pieces under the pediment where the northern part of the western outer gopura meets the gallery. It is presumed that the roof of the gallery should be covered by brick-layers. And based on some scattered components at the northwest corner of the inner enclosure which look like part of the corner turret's roof, its roof form can be acquired. The corner turret's false-storey had been missing all and the amount of layers is just acquired by analyzing the amount of layers of central prasat and its ratio. In term of the each corner turret, based on the reconstruction of prasats, the regular scale of height can be proposed by the way of width contraction.

There aretwo kinds of the ruined end-tile in long-hall, its several models can be proposed for choosing by means of style and scale. So the end-tile form can be confirmed. In addition, the end-tile on the cornice of the gallery can be simultaneously concluded. Nevertheless for the library, it cannot be seen like the ruined end-tile as well as the other scattered components to acquire the details about the scale and the form. In the case, it mainly can be referenced to the library at Banteay Srei temple and other library examples at the earlier temples in the reconstruction processing.

Asabove-mentioned as the analysis on the measuring data, the similar depth-width ratio of pediment can be seen at Ta Keo temple-mountain. The eastern outer gopuras can be chosen as the model for the pediment reconstruction, due to it is relatively easy to measure as well as its good preserved condition with completed carvings. In terms of the pediments of affiliated buildings at Ta Keo site, it can be concluded that the depth-width ratio of the pediment form is relatively constant after measurement of the considerable scattered components. The reconstruction of pediments is adopted from the model of the eastern outer gopura, which had already been assembling together at the site.

In case of thewooden structure of the long-hall, perhaps due to the tropical weather, it is very few ancient wooden structures still remain at Angkor monuments. However, in the CA report during the clearance processing, Jacques Dumarçay had taken some details about the scattered tiles and the size of the purlins. The wooden structure of the long-hall is exactly based on his studies for the reconstruction. Consequently, the section scale of the wooden structure is not acquired from the mortise size of the ruined pediment in the northern inner longhall but the reconstruction of the wooden structure is the referenced to Jacques Dumarçay's relevant studies on Banteay Srei temple.

For reconstruction of its end-tile, the various hypotheses of form can be proposed and chosen by means of the measuring data.

Architecture Decoration

Concerning reconstruction of the architectural decorations, there are several aspects as follows: the ornament is relatively clear and complete; and the ornament is unfinished status, in addition to there is no any clue of ornaments to understand whether it is just an idea in the blueprint stage. For the classification of the architecture decoration, it can be divided into several steps to undertake its reconstruction as follows: first, make a summary of the clear, complete parts and then do the reconstructing. All these works can provide important reference for those unfinished ornaments or even the part which had never commenced. That is to say, it can be inferred that the whole architectural decoration style from the finished status. Besides, the possibility discussion of the unfinished ornament and whether there is ornament or not on those which had never begun is based on the collected data from the field investigation and relative analysis. To make a general survey of the various architectural ornaments, each part has its own historic religion and a process of the style evolution. Next, it is to analyze how to reconstruct the important architectural parts, like colonnette, pediment, false-storey architrave, pyramid architrave and so on. For the carving and the dermatoglyphic pattern of the colonnette, the decorative colonnettes belong to the part whose ornament is relatively complete in Ta Keo temple-mountain. Colonnette relics appear mainly in the entrance of the gopuras, inner and outer long-halls, and libraries. The entrance of prasats is furnished with colonnettes, but they are just carved roughly without ornament. The ratio and the size of the colonnettes mentioned above is suitable for exquisite decorative carving and because of the special location and the distinct change of the style, they are regarded as one of the important basis for analyzing the artistic style and its building age. From the status survey about the colonnettes of the gopuras, the long halls and the libraries, although most ornament of colonnettes are unfinished, it can be inferred that an insight into the initial design ideas based on the dermatoglyphic pattern and the symmetrical characteristic. And during the field investigation and the measurement, we find that the decorative subjects are quite similar of the colonnettes in Gopuras, long halls and libraries. And they all consist of chapter, column bottom, column base, and decorative repeating unit in the column body and so on. But they have different size and ratio, nothing more. So we reconstruct the gopura colonnette which is more topical. It is worth noting that the colonnettes of gopuras (eastern outer gopura, western outer gopura and western inner gopura) furnished with side door in the side room can be divided into two categories. The first type is on both sides of the entrance into the main hall, and the other type is a small colonnette, which is subject to door pillar on both sides of the entrance into the outer side room. Besides, there is not any clue of carving in the column bottom, that is to say, it is white space. We guess that the white space is not their initial design ideas. So what kind of the type do they want at first? Looking back the research on the evolution of the Angkorian architectural colonnette by J. Boisselier, a French Khmer art historian, it is found that the difference between the column bottom and the column body is not quite distinct in the early colonnettes and the column bottom has the same repeating architrave as the column body. And this is always symmetrical with the chapter to a large extent. But obviously Ta Keo temple-mountain is not the case. Firstly, which is to carve in the top of the colonnette? So maybe the top is not symmetric with the bottom. Secondly, if the bottom has the same architrave as the body, the white space mentioned above cannot hold water. And why don't make all parts carved at first? Based on the colonnettes in Northern Kelang, it can be inferred that the carving content in the white space of the colonnettes in Ta Keo temple-mountain should be different from the decorative dermatoglyphic pattern of the architrave to a large extent or the statue? Since the work is limited by the colonnette's location, the colonnette will be carved with rough architrave after simple, crude dispose. And then carry on exquisite dermatoglyphic pattern carving on the surfaces between the colonnette and the wall or the pilaster. Perhaps the architect with richer experience finishes the more complicated dermatoglyphic pattern in the white space mentioned above uniformly.

For the pediment carving and the dermatoglyphic pattern, compared with colonnette, pediment lies where best view lines get together. So pediment becomes the most important part of decoration in angkorian architecture. The pediments are mainly located in the gopuras, inner and outer long halls, libraries and central prasat, so are the turrets of the gallery.

Owing to the fact that the pediment of the central prasat are mostly sandstone with coarse surface and the turrets of the gallery are all gone, we classify and survey the scattering components of the gopuras, inner and outer long-halls, libraries in Ta Keo temple-mountain.

Among the scattered components, which constitute pediment, the pediment in long hall is in better conservation, we adopt the three-dimensional laser scanning to do the reconstructing with the help of the photographic surveying. The pediment of the gopura is a little bit complicated. It includes not only the pediment on the main hall's lintel and the outer side room lintel supported by the pilaster, but also small pediment supported by the pilaster in the decorative false storey. According to the change of the plan layout and the tectonic way, the design of the pediment in gopura is considerably flexible. And the change of the connection between the pediment and the stone will result in pediments of different types, too. Moreover, diverse pediment size will lead to a change of the decorative dermatoglyphic pattern subjects. As the pediments in Ta Keo temple-mountain have fallen mostly, it is very necessary for us to survey, identify and piece the scattering components together. At first, find the way about the layout of the various pediments, then classify and identify them orderly.

For architrave and carving, Jacques Dumarçay of EFEO once made detail narration about the architrave and the carving in the reports, which become important reference data for the reconstruction study of the architectural ornament in Ta Keo temple-mountain. The elevation carving architrave of the first foundation and the second foundation is basically complete and with no dermatoglyphic pattern, while the architrave and the decorative dermatoglyphic pattern of the pyramid is legible. So, firstly, analyze the congruent relationship between the architrave and the decorative dermatoglyphic pattern of the pyramid, and make a general survey of the relatively regular decorative dermatoglyphic pattern motifs. Secondly, adjust the work based on the importance of the location and the ratio scale accordingly. During the process, we correct several mistakes in the Ta Keo temple-mountain surveys drawn by EFEO. Besides, according to the congruent relationship, it also can be inferred that the first, the second terrace of the pyramid have no complete dermatoglyphic pattern. Therefore it can be reconstructed for the dermatoglyphic pattern of the first, the second terrace of the pyramid based on the measured architrave of the pyramid. Another interesting phenomenon is that the all connections between the various architraves seem like hard collisions, without any transitions.

第六章　茶胶寺庙山的营造尺度与设计方法探微

一、概述

如前文所述，茶胶寺庙山是诸多吴哥遗迹中最为雄伟且具有鲜明特色的庙山建筑之一。作为早期吴哥庙山建筑的重要遗构，古代高棉宗教与文化赋予茶胶寺庙山独特的建筑风格和艺术魅力。茶胶寺庙山依然保留着千年前创建之初尚未完工的状态，以其完全以砂岩砌石构筑庙山中央五塔的做法，或因其十字型平面且四面开敞皆出抱厦的塔殿以及环绕须弥台回廊平面格局的出现，开创了吴哥时代风气之先的庙山建筑形制。

在茶胶寺庙山建筑形制复原研究的过程中，一方面通过对公元十世纪末以降吴哥时代庙山建筑源流与变迁的梳理，可以大致了解古代高棉建筑的形制特征、风格演变、宗教象征之所在。然而，在另一方面倘若继续在营造尺度与规划设计方法等方面进行探赜索隐，则对于深入理解古代高棉建筑设计意匠与建造方法，以及全方位地剖析古代高棉建筑所集中呈现出的复杂信仰体系及其象征性，都具有较为可观的学术意义；对于本书所涉及的茶胶寺庙山建筑形制与复原研究而言，此项工作更应是研究过程不可或缺的环节。

关于茶胶寺庙山的设计尺度及其规律的研究存在诸多缺环和空白，加之笔者对于古代高棉建筑尺度与设计方法的研究尚处于刚刚起步的初级阶段，因此本章内容是在参考法国及日本学者研究成果的基础上，选择茶胶寺庙山作为分析和研究的对象，主要针对庙山平面布局、立面设计、建筑单体布局、细部构造等方面所涉及的营造尺度与设计方法进行初步的剖析和讨论，借此或可管窥古代高棉建筑整体面貌及其设计意匠之一斑。

根据二十世纪二十年代法国学者的研究，吴哥时代的建筑营造尺度可能是以源自古代印度的哈斯塔（hasta）或毗耶玛（vyama）作为标准度量单位的，并且推测1个哈斯塔的距离约为450毫米（1hasta=450mm）。作为一种非常古老的长度计量单位，哈斯塔亦可理解为"肘尺"，系指人体手臂自肘部关节突出处至手掌中指指端之间的长度；较之哈斯塔，毗耶玛则是一种尺度相对较大的长度计量单位，系指人体正常站立时伸展双臂至与肩部平行，测量双手中指指端之间的距离为1个毗耶玛，而哈斯塔与毗耶玛之间的换算关系是1vyama=4hasta。在公元十九世纪末沦为法国保护国之前的柬埔寨，作为通用的长度计量单位，哈斯塔和毗耶玛曾被广泛地应用于社会生活的诸多方面，尤其是在土地丈量、寺庙工程、器具制作、雕刻绘画等方面。在建造房屋的时候，甚至还利用建造房屋的主人自己的哈斯塔尺作为基准尺度进行房屋的设计和建造。[1] 古代高棉哲匠们专注于此道，薪火相传，积累了丰富的实践经验，因而能以敏锐而准确的尺度感和娴熟的艺术技巧，灵活而妥善地运用各种建筑体量，

---

[1] 参见 George Cœdès, *Etudes Cambodgiennes*, In: Bulletin de l'Ecole française d'Extrême-Orient (BEFEO), Tome24, 1924, pp. 345–358.

进行各种规模、尤其是大规模建筑组群的空间组织处理并达到很高的造诣。[1]

另外，在诸多遗存至今的吴哥时代碑铭之中，涉及当时建筑营造尺度的记载非常罕见。其中，以在达梅瑞寺（Prasat Damrei）、安东寺（Prasat Andong）、巴塞增空寺（Baksei Chamkrong）等三座小型寺庙遗存中发现的梵文碑铭最为重要。以上碑铭的主要内容虽亦多为敬献或赞颂神灵和国王的祝词，但其中最为弥足珍贵的部分却是碑铭内容提及了贡开地区（Koh Ker）一座规制很高的寺庙所供奉林伽的尺度，并明确提及了当时的尺度单位哈斯塔（肘尺）的存在及其使用情况，赛代斯的法文翻译将上述碑铭中涉及林伽尺度的部分摘录如兹：

其一，达梅瑞寺（Prasat Damrei）碑铭："……充满诚意的心让人喜悦，为去除傲慢自大之心，他（王）在9×9肘尺（hasta）之上修建了极其重要的乌尔加（Ugra）的林伽。"

其二，安东寺（Prasat Andong）碑铭："……虽然先王们没有获得成功，但是他（王）却很轻松地将贡布胡（Cambhu）的林伽及其雕像一起矗立到了9×9肘尺（hasta）的高度。"

其三，巴塞增空寺（Baksei Chamkrong）碑铭："……与卡楚姆卡（Caturmukha）和其他的雕像一起被安置的是9×9哈斯塔（hasta）的湿婆林伽，他们矗立在他（王）所选择的林伽补罗（Lingapura）的土地之上。"

赛代斯比较研究了上述三段梵文碑铭的内容，将原文中"navadha....navahastanistham（或者navahastantam）"初步解释为"安置在9哈斯塔的高度"，又通过分析考证推测"navadha"可能为"9次"的意思，认为林伽的尺度应表述为9hasta的9倍，即81hasta的高度。与上述碑铭的记述相符合的实例，或许即是位于贡开地区中央大塔（Prasat Thom）被称为普兰（Prang）的寺庙遗存。赛代斯当时还认为，1hasta的实际长度约为0.45米，因此他判断81hasta相当于36.45米。[2]

由此可见，吴哥时代关于建筑尺度的明确记载或可追溯至公元十世纪中叶。

近年来，以沟口明则为代表的日本学者通过对吴哥及贡开（Koh Ker）地区的多座重要寺庙建筑遗迹的营造尺度进行分析，并佐之以皇宫广场十二塔（Prasat Suor Prat）中N1塔在解体修复工程中所采集到的数据资料的分析，推测复原出吴哥时代的营造尺度，1哈斯塔的实际长度可能约为412毫米（1hasta=412mm）。关于寺庙的规模与建筑布局的尺度设计，皆是以1哈斯塔作为营造尺度的基本单位。另外，通过采用10进制为标准组成简明扼要的整数值进行尺度控制与设计，这也已经通过对诸多实例的

---

[1] 较之中国古代尺度的源流，亦有《大戴礼记·主言》转述孔子云："布指知寸，布手知尺，舒肘知寻，十寻而索"论述；而《说文解字》的解释则更为具体，如《寸部》云："寸，十分也。人手却十分动脉，谓之寸口。""寻……度人之两臂为寻，八尺也。"《尸部》言："尺，十寸也。人手却十分动脉为寸口，十寸为尺。尺，所以指尺规矩事也……周制，寸、尺、咫、寻、常、仞诸度量，皆以人之体为法。"《十部》又说："丈，十尺也。"《夫部》又称："周制八寸为尺，十尺为丈，人长八尺，故曰丈夫。"王充《论衡·气寿》则据此强调："人形一丈，正形也；名男子为丈夫，尊公姬为丈人。"如此等等。而以人体为基础构成建筑规划设计的模数方法进一步扩展，还形成了"形体之法"，即《周礼·地官·遂人》所谓"以土地之图经田野，造县鄙形体之法"。郑玄注："经、形体，皆telle制分界也。"形体就是土地和都邑规划设计，其涵义，则如《国语·楚语》明确指出："且夫制邑若体性焉，有首领股肱，至于手拇毛脉。"引自王其亨：《风水形势说与古代中国建筑外部空间设计理论探析》，载于王其亨主编：《风水理论研究》，天津：天津大学出版社，1992年版，第117页。

[2] H. 帕蒙蒂埃测绘普兰寺现存七层基台上端的高度为31.2米，若加上遗存顶部残留部分的高度，七最上部分的高度为35米。因此，他认为赛代斯的碑铭解释，如果将约尼基座的高度计算在内的话，那么可以将林伽的总高度理解为36米，并与赛代斯的主张一致。但由于林伽及约尼基座皆堙圮无存，林伽及约尼的尺度与碑铭记载一致的可能性已经无法确定。参见沟口明则：《设计技术复原考察》，载于沟口明则、中川武等编著：《贡开与崩密列：高棉帝国东部地区的两大遗迹组群（*KOH KER and BENG MEALEA: Two Large Mounments at the Eastern Portion of the Khmer Empire*）》，名城大学、早稻田大学、APSARA National Authority、JASA，2011年版，第76-113页。

分析逐渐得以确认。建筑组群布局中的主要单体建筑位置，可能都是沿着建筑组群轴线位置且依据简明严谨的整数比例关系进行定位的，而其他附属建筑的尺度规律或比例关系却似乎不甚清晰统一。[1]

在 2007 – 2010 年期间，中国文化遗产研究院及中国政府援助吴哥古迹保护工作队与天津大学合作，利用三维激光扫描技术对茶胶寺庙山进行了近年来最为详细的测绘记录工作。本章的主要内容即以上述前辈学者的研究成果作为依据或线索，利用最新发表的测绘数据资料，初步尝试对茶胶寺庙山建筑遗存的营造尺度与设计规律进行考察，借此或为深入剖析吴哥时代庙山建筑的尺度规律及其设计意匠提供参考。

---

[1] AKAZAWA Yasushi, NAKAGAWA Takeshi, MIZOGUCHI Akinori, *THE TECHNIQUES AND COMPOSITION OF PRASAT SUOR PRAT TOWER N1：Study of architectural techniques of Prasat Suor Prat tower in Angkor（Part 2）*, Journal of architecture and planning 73（628），2008 – 06 – 30, pp. 1327 – 1333; MIZOGUCHI Akinori, AKAZAWA Yasushi, NAKAGAWA Takeshi, CHUBACHI Yoshihiko: *ON THE DIMENSIONAL PLAN IN PRASAT SUOR PRAT TOWER N1：Study on the dimentionnal plan and the planning method of Kmer architecture No. 2*, Journal of architecture and planning（616），2007 – 06 – 30, pp. 175 – 181,; MIZOGUCHI Akinori, NAKAGAWA Takeshi, ASANO Takashi, SAITO Naoya: *ON THE DIMENSIONAL PLAN OF THE COMPLEX IN THOMMANON AND BANTEAY SAMRE：Study on the dimensional plan and the planning method of Kmer architecture No. 1*, Journal of architecture and planning（612），2007 – 02 – 28, pp. 131 – 138.

二、茶胶寺庙山平面布局的尺度分析

关于茶胶寺庙山的历史源流，在相当长的时间里始终知之甚少。自二十世纪二十年代以来，围绕茶胶寺庙山建筑源流及其变迁的探讨一直在进行。诸多法国学者尝试从平面布局、建筑装饰以及碑铭内容等诸方面入手，针对茶胶寺庙山的建筑形制及其始建年代开展了甚为详缜的梳理与考证，以期追溯茶胶寺庙山的历史源流与变迁历程。如前文所述，以诸如斯特恩、雷慕沙、格罗布维、赛代斯等为代表的法国学者分别从茶胶寺庙山的建筑装饰、平面布局、碑铭等方面探讨其建造年代，基本确定了茶胶寺庙山创建于公元十世纪末叶阇耶跋摩五世（Jayavarman V，公元968 – 1001年）在位期间，庙山的建造工程直至公元十一世纪初苏利耶跋摩一世（Suryavarman I，公元1002 – 1050年）统治时期可能仍处于兴建之中，但却由于某种尚未确定的原因导致庙山工程未能最终完工。

如本书第四章所述，茶胶寺庙山可能位于一座规划严整且规模宏大的建筑组群的中心，庙山东侧延伸大约500米直至东池西侧的堤岸之上，而堤岸之上仅存角砾岩砌石基址的建筑遗址及其与庙山相连的神道残迹，皆应是构成茶胶寺庙山整体格局的重要实物遗存。在占地面积约为1000米×780米的矩形区域之内，惟因茶胶寺庙山西侧由于暹粒河所形成天然屏障的阻隔，庙山建筑主体部分略偏向此区域的西侧。另外，茶胶寺庙山的整体布局还应包括：坐落于神道南北两侧的两座水池以及环绕庙山的壕沟。而茶胶寺庙山建筑主体部分自下而上主要包括：第一层基台（亦称庙山外院）、第二层基台（亦称庙山内院）、须弥台以及坐落于须弥台之上的中央五塔。其中，围墙和回廊分别环绕的第一层基台和第二层基台四周，在其正交轴线位置分别辟为八座塔门；另外，第一层基台的东侧南北对称布置有外长厅两座，第二层基台的东侧南北对称布置内长厅及藏经阁各两座。[1]

整体而论，茶胶寺庙山的平面布局简洁规整，建筑形制清晰，轴线分布明确，因而为分析其营造尺度与设计方法提供了比较有利的条件。

1. 茶胶寺庙山东西方向轴线的偏移

茶胶寺庙山主体部分平面布局的轮廓呈矩形且东西方向的距离大于南北方向的距离。在上述2007 – 2010年间利用三维激光扫描技术获取点云而绘制完成的CAD测绘图中，量取沿庙山东侧神道外端至西侧壕沟外皮之间轴线的长度为243.08米（东西两侧壕沟外皮之间轴线尺寸为229.77米），而沿南北两侧壕沟外皮之间轴线的长度为193.59米[2]。根据日本学者对古代高棉建筑营造尺度的研究成果，采用假定1hasta = 0.412米进行分析整理，如果能够以此为依据分析并推测出合乎逻辑的设计规律与方法，反之也可佐证营造尺度的设定是否妥当。由此，茶胶寺庙山建筑主体部分东西方向的轴线尺寸可折算为590hasta（东西两侧壕沟外皮轴线尺寸折算为558hasta），南北方向轴线尺寸则折算为470hasta，其长宽比非常接近整数比5∶4。由此或可推测茶胶寺庙山的平面布局及其尺度设计应将神道及壕沟的范围包括在内。

---
[1] 详见本书第四章《茶胶寺庙山建筑形制简述》的相关内容。（著者注）
[2] 量取尺寸的标准皆以建筑基座的基底尺寸为准。（著者注）

茶胶寺庙山第一层基台基底尺寸为135.96米×111.24米（东外塔门东侧踏道外缘至西外塔门西侧踏道外缘），其东西方向距离折算为330hasta，南北方向距离折算为270hasta，但是第一层基台南北两侧外皮（外塔门外侧踏道外缘）至中央主塔东西轴线距离分别为55.93米（136hasta）与55.31米（134hasta），二者差值为2hasta；第二层基台基底尺寸为94.7米×88.7米（230hasta×215hasta），其南北两侧外皮至中央主塔东西轴线距离分别为44.77米（109hasta）与44米（107hasta），二者差值为2hasta；须弥台的基底尺寸为60.7米×59.1米（147hasta×143hasta），其南北两侧外皮至中央主塔东西轴线距离分别为29.89米（73hasta）与29.36米（71hasta），二者差值亦为2hasta；由此可见，茶胶寺庙山建筑主体部分并非沿东西方向轴线对称，而是其南侧比北侧增出了约2hasta的距离。根据近年来日本学者的研究，在吴哥时代的建筑遗迹中，沿东西方向轴线两侧的建筑布局大多并非完全对称，此轴线往往略微向南或北侧偏移1hasta的长度，从而在轴线两侧产生2hasta的距离差值。由于偏移量较小，加之建筑体量巨大，因而1-2hasta的长度差值多不引人注意。然而，通过对诸多吴哥时代建筑遗迹实测值的分析确认，上述轴线的偏移现象十分普遍，反倒是完全轴线对称的平面布局实例却非常罕见。[1] 因此，在茶胶寺庙山平面布局中出现的东西方向轴线偏移现象，多与其他吴哥时代建筑遗迹所呈现出的规律颇为契合。[2]

2. 茶胶寺庙山平面设计尺度与构成

由于茶胶寺庙山所具有的形制格局及其体量关系所限，主要的建造与设计流程理应是由外及内依次展开的。作为统领庙山建筑组群的核心及主旨所在，中央主塔也理应一以贯之地处于庙山建造与设计过程最为关键的环节之中。参照1hasta = 412毫米的营造尺度，在茶胶寺庙山总平面图上覆以4120毫米（10hasta）为模数单位的平格网，通过使用这种平格网的分析，或可对茶胶寺庙山的营造尺度及其规律进行初步的归纳与解读，进而可以对茶胶寺庙山平面布局最初的设计方法进行推测或复原。

首先，古代高棉匠师大概利用他们所掌握的天文学和测量学知识，并可能经过极为复杂的仪轨或

---

[1] 在吴哥时代的建筑遗址中，并非沿东西方向完全对称的平面布局方式多种多样：若等分寺庙南北宽度的中心线，则在向北侧偏移1-2hasta设置东西方向轴线。其中，有些寺庙具有南北方向等分的中心线，则要在寺庙北侧去掉1-2hasta宽度的带状面积；而有些寺院因其在中央设置南北对称的中心建筑，则就要在设置建筑物的一侧减掉几个hasta以改变建筑中轴线的位置。另外，东西轴线位置的变化都与其寺庙规模相呼应，多数仅为1hasta到1.5hasta，即使是吴哥窟规模的寺庙，其平面布局的南北尺寸差也不过是3.5hasta而已。根据在吴哥地区的调查，发现的唯一例外是托玛侬寺（Thomamnon），其南北全长100hasta，在南北等分的位置设置东西走向的中心线，然后从规划用地的北端去掉一条1.5hasta的宽度。结果寺庙的用地变为南北宽98.5hasta，围墙之内的面积变小了。因此，这种技法虽然没有直接触及最初设定的中心线，但结果是使其相对位置发生了偏移，其中心线变为了向北偏向的中轴线，在此轴线上布置东西的塔门、拜殿及中央祠堂。参见沟口明则、中川武、佐藤桂：*The ancient Khmer's dimensional planning at the Prasat Thom in Koh Ker: study on the dimensional plan and the planning method of Khmer architecture*（no.5）Journal of architecture and planning 75（653），1751-1759，2010-07；*The ancient Khmer's dimensional planning at the Prasat Pram in Koh Ker: study on the dimensional plan and the planning method of Khmer architecture*（no.4），Journal of architecture and planning 75（651），1273-1278，2010-05；或参见沟口明则：《设计技术复原考察》，载于沟口明则、中川武等编著：《贡开与崩密列：高棉帝国东部地区的两大遗迹组群（*KOH KER and BENG MEALEA: Two Large Mounments at the Eastern Portion of the Khmer Empire*）》，名城大学、早稻田大学、APSARA National Authority、JASA，2011年版，第76-113页。

[2] 这种将主要的建筑轴线进行偏移的做法，至今在柬埔寨的建筑工程做法之中仍多有遗存。无论是寺庙殿堂，还是普通的民居住宅，承担施工的匠师都会有意将建筑物主要轴线向某一侧偏移30-50mm。因为在柬埔寨的民间传说中，建筑物的轴线位置是设置神位之所在，因此人之居所理应避开具有神性的轴线避免与神位发生冲突，以祈求平安。感谢柬埔寨吴哥古迹保护与发展管理局（APSARA Authority）考古学家SoChheng先生提供的相关信息。（著者注）

仪式，将东池之水与茶胶寺庙山的建造过程密切地关联起来，进而茶胶寺庙山平面布局的东西方向轴线也随之得以确定下来。可能依据事先拟定的形制与规模（或暂假定以30hasta为基本模数单位H），以此轴线为基准分别向其南北两侧偏移135hasta（4.5H）的距离从而形成第一层基台外缘尺寸。接下来似乎是利用的惯用设计方法，将此首先确定的轴线向北侧偏移1hasta而形成新的轴线（或可将其称之为基准轴线），轴线的南侧较之北侧即增出2hasta的距离，而沿着这条东西方向的基准轴线可以设置四座塔门及中央主塔的位置。

关于吴哥时代庙山建筑运用中轴线偏移进行平面布局的方法，迄今为止，已发现并分析了包括贡开地区寺庙在内的多个实例，然而对茶胶寺庙山轴线偏移的平面布局方法此前却知之甚少。在上述茶胶寺庙山的东西方向轴线确定之后，遂以其南北方向的总长度270hasta（9H）作为基准，确定270hasta×270hasta（9H×9H）的正方形平面的中心位置，并以此作为庙山平面设计与构成的中心点或基准点。或许事先已经考虑到在茶胶寺庙山的东侧设置藏经阁、长厅等附属建筑，因而将此正方形的东侧边界沿东西方向轴线向东扩展60hasta（2H或15vyama）的长度，进而增至330hasta（11H）的位置即是东外塔门的踏道外缘，若以此位置再沿上述轴线向东延伸120hasta（4H或30vyama）的长度，可以直抵庙山东侧壕沟的内侧边缘，最终再以此向东延伸70hasta的长度终至神道外端的起点部分，实际上这其中包括至神道起点的60hasta（2H或15vyama）及神道亭10hasta（1/3H或2.5vyama）的尺寸之和。包括茶胶寺庙山在内的诸多吴哥时代的庙山建筑皆有以东为尊的传统，庙山主入口与主要单体建筑皆朝向东方，想必对于庙山东侧部分而言，其设计与建造的过程应是工程最为重要的阶段或环节。因此，首先通过茶胶寺庙山东西方向轴线的确定，庙山东侧的设计尺度与形制规模得以基本明确。

对于茶胶寺庙山南北两侧的尺度及规模，以270hasta×270hasta（9H×9H）的正方形的南北两边界分别向南北两侧延伸100hasta（3H+1/3H或25vyama）的长度可抵壕沟的外缘，而以上述正方形西边界向西延伸70hasta（2H+1/3H或17.5vyama）至壕沟外缘。因此，庙山本体东西方向总长度590hasta（19H+2/3H或147.5vyama），南北方向总长度570hasta（19H或142.5vyama），至此，茶胶寺庙山建筑的总体规模得以基本确定，抑或包括壕沟范围在内茶胶寺庙山建筑的总体布局是东西方向比南北方向增出20hasta（2/3H或5vyama）的矩形平面。

继续以270hasta×270hasta（9H×9H）的正方形平面作为基准平面，将其南北两侧边界分别向内收进50hasta（12.5vyama），共计100hasta（25vyama），而在东西两侧边界中，东侧边界收进60hasta（15vyama），西侧边界收进40hasta（10vyama），亦为100hasta（25vyama），遂将170hasta×170hasta（5H+2/3H或42.5vyama，须弥台的设计尺寸包含了踏道尺寸）须弥台的位置确定。南北踏道外缘尺寸也为170hasta，恰为庙山建筑南北方向整体长度的2/3，东西方向整体长度的1/3。而以须弥台东侧踏道外缘位置为基点，即可确定藏经阁抱厦中心轴线，若向东侧偏移20hasta则可以确定内长厅位置，再向东偏移3hasta又可确定回廊的位置。但值得注意的是，回廊的尺度规律与整体平面布局的尺寸比例关系不甚明显，显得颇有些牵强，倒是内长厅与第二层基台外缘距离为10hasta。[1]

---

[1] 一般认为长度相当于4倍哈斯塔（hasta）的毗耶玛（vyama），在古代高棉碑铭中多有出现。vyama一般是在土地丈量及测量围墙长度时所使用，在庙山建筑总体尺度设计中以其作为尺寸单位是完全可能的。但是茶胶寺庙山的建筑平面及立面设计尺寸可能是以hasta为单位且原则上遵循某种简洁的整数比的设计方法。作为土地丈量单位的vyama是如何从总体建筑布局及其规模设计过渡到建筑设计尺寸单位hasta，其中的设计程序及方法，则是需要进行继续研究的课题。（著者注）

第六章 茶胶寺庙山的营造尺度与设计方法探微

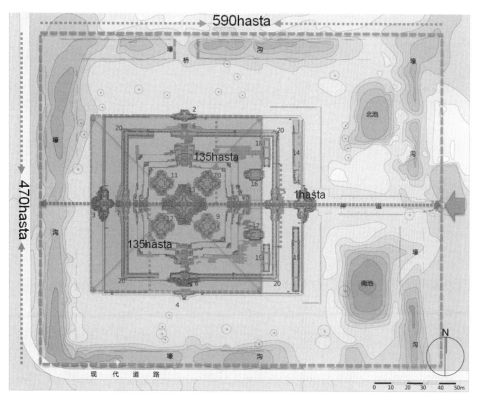

图 6-1 茶胶寺庙山平面基准尺度的推测

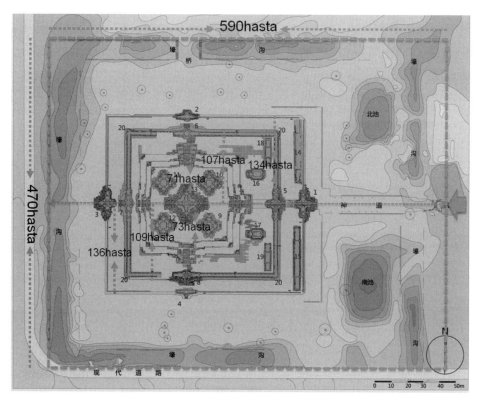

图 6-2 茶胶寺庙山东西方向轴线的偏移

茶胶寺庙山建筑研究

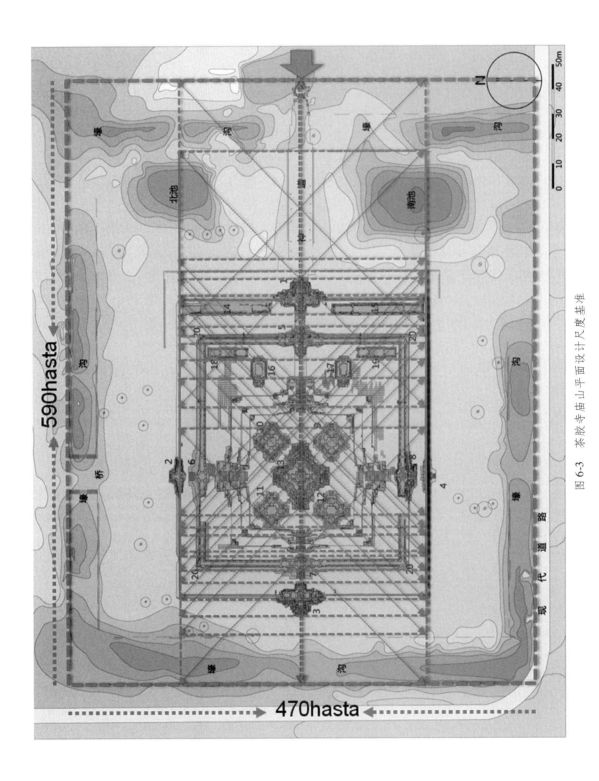

图 6-3 茶胶寺庙山平面设计尺度基准

由上述简略的尺度分析或可进行大胆的推断，环绕茶胶寺庙山第二层基台四周的回廊或许未包括在最初设计意图之中，最初的设计可能依然延续着东梅奔寺、比粒寺时代的庙山平面布局形制，以各类长厅型建筑环绕须弥台四周设置，今之第一、第二层基台上诸多类似建筑基址的遗迹现象估计可能也与此相关。从平面布局的设计尺度分析来看，回廊的位置多与茶胶寺庙山整体平面布局的尺度规律有所抵牾而颇显异类，由此推测在苏利耶跋摩一世继续茶胶寺庙山建造工程之际，很有可能对庙山最初的设计方案进行过修改或调整，通过拆除多数长厅型建筑并以四周环绕须弥山的回廊样式取而代之，进而由此引发了苏利耶跋摩一世以降庙山建筑形制的一系列革故鼎新。

三、茶胶寺庙山建筑立面设计的尺度分析

参照1hasta＝412毫米的营造尺度，在茶胶寺庙山总立面图上覆以4120毫米（10hasta）为模数单位的平格网，再次通过使用这种平格网的分析，或可对茶胶寺庙山的立面设计尺度及其规律进行初步的分析，进而对茶胶寺庙山最初的立面设计或竖向设计意图进行推测或复原。

根据茶胶寺庙山平面布局的分析可知，在庙山的东西方向轴线确定之后，遂以其南北方向的总长度270hasta（9H）作为基准，确定270hasta×270hasta（9H×9H）的正方形平面的中心位置，并以此作为庙山平面设计与构成的中心点或基准点。那么，在茶胶寺庙山的立面设计或竖向尺度中，以此为基准是否依然有效呢？可以尝试进行如下的分析：[1]

以270hasta×270hasta（9H×9H）的正方形边长的1/2（135hasta）作为竖向设计的标高位置，此位置要远高出今之中央主塔遗存的顶部标高将近33hasta，似乎与五塔假层的形制与构造逻辑不符；而若以正方形边长的1/3（90hasta）作为竖向设计的标高位置，恰好是中央五塔之角塔顶部的标高位置（各座角塔顶部结构基本完整，惟塔刹皆不存）；再以正方形边长的1/4（67.5 hasta）作为竖向设计的标高位置，此位置恰好是一个重要位置的标高，即中央主塔基座的顶面；再若以正方形边长的1/5（54hasta）作为竖向设计的标高位置，此位置恰好为另一个重要位置的标高，即须弥台第三层的顶面；采用同样的方式，若以正方形边长的1/6（45hasta）作为竖向设计的标高位置，此位置又恰为须弥台第二层的顶面；而若以正方形边长的1/8（33.75hasta）作为竖向设计的标高位置，此位置则恰为须弥台第一层的顶面。

若再以170hasta×170hasta的须弥台基底平面尺寸进行分析，遂以其1/2（85hasta）作为竖向设计

---

[1] 根据近年来日本学者的系列研究，吴哥时代庙山建筑的竖向设计尺度或与其相应的平面尺度有所联系。例如，通过对吴哥皇宫广场十二塔（Prasat Suor Prat）中N1塔在解体修复工程中所采集到的数据资料的分析，塔高为50hasta，基台高为4hasta（实测值接近3.5hasta），塔身第一层高度为20hasta（实测值接近20.5hata），第二层高度为10hasta，第三层高度为8hasta，最上屋顶高度8hasta）。由该塔设计尺度特征可以看出，当时可能采用比较简明的整数尺寸进行设计的，而塔高尺度的确定或以塔之平面尺度的1/2作为计算方法。参见沟口明则：《プラサート・スープラN1塔の寸法計画：クメール建築の造営尺度と設計技術に関する研究》（ON THE DIMENSIONAL PLAN IN PRASAT SUOR PRAT TOWER N1：Study on the dimensional plan and the planning method of Khmer architecture），载于Architectural Institute of Japan, Journal of architecture and planning (616)，2007-06-30. pp. 175-181. 另外，对于贡开地区的普兰寺（Prang）的中央大塔（Prasat Thom）而言，其基本设计是按照平面的缩小减去高度，减去2hasta，全高就变为78hasta。此时不是重新调整各层的高度，而是很简单的从高度最大的初层减去2hasta。这个结果所实现的各层的高度就是基座1.0hasta，初层12.0hasta，第2、3层13.5hasta，第4、5层13.0hasta，下部基坛12.0hasta。然而，各层平面的递减是从基座的轮廓，各面减去1hasta，以边长为154hasta的正方形作为初层平面，第2层平面的递减采用该层的高度13.5hasta。结果第2层的边长变为127hasta。第3、4层从该数值中顺序各减1hasta，分别递减12.5hasta和11.5hasta。第3、4层的平面边长分别变为102hasta和79haasta。第5层，从实测数值的整理来看，应该是递减10hasta，变为边长为59hasta的正方形。但是，设计意图若根据第3、第4层的设计，应该是10.5hasta比较自然，此时的平面边长应该是58hasta。从下部基坛开始递减发生了较大变化，如果是7hasta（实测值是7.5hasta，但如果把第5层看作10.5hasta时，就是7hasta），那么平面就变为边长44hasta的正方形，高度较低的上部基坛，在突出四方的短阶梯的外面是40hasta的四方形，连接四角的正方形的边长可以看作是33hasta。虽然每个数值都是让人意外的复杂，但是递减的设想是可以理解其意图的设计方法，可以认为第2层13.5hasta的高度，是平面递减设计的出发点。参见沟口明则：《设计技术复原考察》，载于沟口明则、中川武等编著：《贡开与崩密列：高棉帝国东部地区的两大遗迹组群（KOH KER and BENG MEALEA：Two Large Mounments at the Eastern Portion of the Khmer Empire)》，名城大学、早稻田大学、APSARA National Authority、JASA，2011年版，第76-113页。

的标高位置，恰与今之中央主塔遗存的顶层标高位置相符；若以须弥台顶部平面尺寸120hasta的1/2（60hasta，中央主塔塔基的平面尺寸亦为60hasta）作为竖向设计的标高位置，却似与完整状态中央主塔顶层的标高位置（包括推测复原的第四级假层与塔刹高度）[1]大致相符。

由上述分析可见，在茶胶寺庙山的立面设计过程中，平面尺寸为270hasta×270hasta（9H×9H）的正方形似乎仍然具有与其在平面设计中同样重要的尺度控制作用，其立面尺度的控制与调整应与上述重要的基准尺度密切相关且可能按照相应的比例进行设计。例如，作为茶胶寺庙山最重要的组成部分之一，至少对于须弥台的竖向设计而言，通过对其平面尺寸按照诸如1/3、1/4、1/5、1/6、1/8等简洁的比例关系进行划分而获取竖向设计尺寸的方法，其表现甚为清晰。而作为同样重要的中央主塔，其立面高度的确定可能与须弥台的平面尺度息息相关，今之中央主塔的现状标高位置，即与须弥台基底平面尺度的1/2相匹配，而推测复原的中央主塔完整状态的标高位置，却与须弥台顶面平面尺度的1/2相吻合；另外，由于今之茶胶寺庙山中央主塔内的林伽皆已不存，几乎无法了解或复原中央主塔内所供奉林伽的尺度及其与中央主塔的比例关系。然而，倘若参照前述碑铭中所提及的林伽的尺度81hasta进行推测，却可以隐约透出贡开大塔（Prasat Thom）与茶胶寺庙山之间的某种关联。如图所示，自茶胶寺庙山地坪至其立面高度为81hasta的标高位置，恰为接近庙山中央主塔完整状态立面高度的1/2，此位置非常可能就是中央主塔内所供奉林伽的顶端标高。而自抱厦入口门槛石顶面视线起点（假定高度165毫米）与林伽顶端的标高位置之间的夹角值非常接近27°，而这个角度正是观瞻建筑物立面形象的最佳视角[2]。上述推测过程及结果似乎并非巧合，至于其中的原因及规律，尚有待继续作进一步的探究。

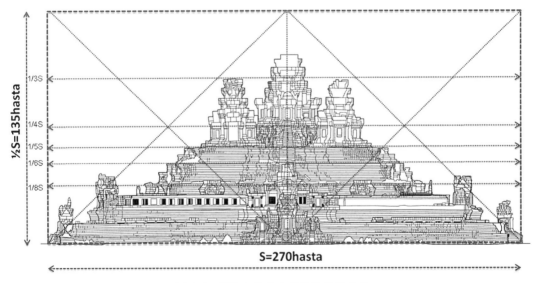

图6-4 茶胶寺庙山立面设计尺度分析

---

[1] 参见本书第五章关于中央主塔复原的相关内容。（著者注）
[2] 不同视距（D）和建筑高度（H）的关系所决定的不同空间感：如在D/H=1，仰角为45°时，空间围合感很强，人倾向于观看建筑立面的局部或细节；当D/H=2，仰角为27°时，空间围合感适中，倾向于观看整幢建筑的立面构图及其细部；当D/H=3，仰角为18°时，围合感下降，人倾向于观看单幢建筑与周围景物的关系，或观看一群建筑；而到了D/H=4，仰角为14°时，空间围合的容积性特征趋于消失，倾向于把建筑看成是突出于整个背景的轮廓线。参见王其亨：《风水形势说与古代中国建筑外部空间设计理论探析》，载于王其亨主编：《风水理论研究》，天津：天津大学出版社，1992年版，第117-137页。

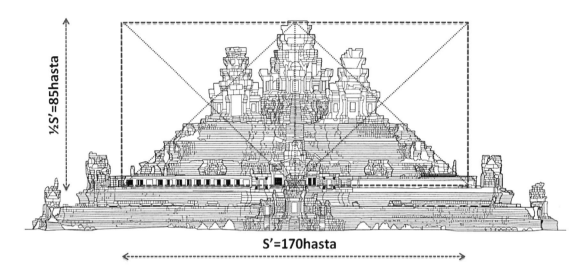

图 6-5 茶胶寺庙山立面设计尺度分析——须弥台底面

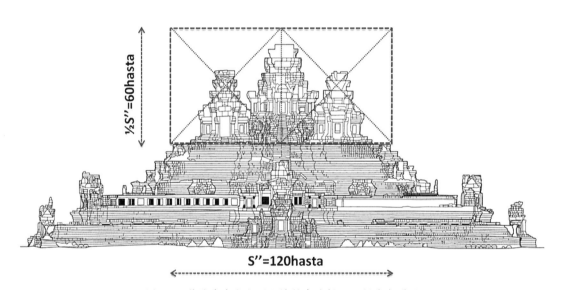

图 6-6 茶胶寺庙山立面设计尺度分析——须弥台顶面

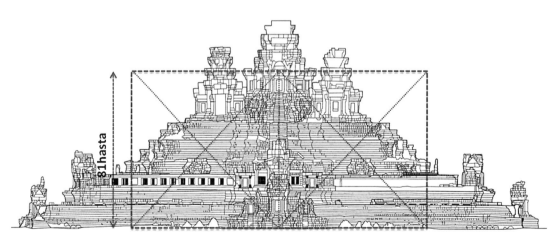

图 6-7 茶胶寺庙山立面设计尺度分析——林伽高度

综上所述，与吴哥时代其他建筑遗迹的情况多有类似，茶胶寺庙山的立面设计亦与其相应的平面尺寸密切关联。其中，庙山建筑核心部分的须弥台和中央主塔尤为关键：首先，须弥台的整体及各层的标高想必应是依据第一层基台的平面尺寸，并且按照非常简洁的比例关系经过细致详缜的划分得以确定的，而其中比例关系及其设计方法有可能为当时的匠师所熟稔。在另一方面，无论须弥台基底抑或顶面的平面尺寸，就其对于统领整座庙山建筑组群的中央主塔的立面设计而论，其发挥的主导作用是决定性的。另外，通过诸如上述平面尺寸、立面比例、林伽高度、观瞻视角等方面尺度规律的推测或复原，隐约闪现其间的某种关联，或许现在虽尚不能进行清晰而完整的解释，但对于深入探赜茶胶寺庙山设计方法与建造流程而言，以上这些线索毫无疑问皆是极好的提示。

四、茶胶寺庙山建筑单体设计尺度分析

1. 中央主塔的平面设计

茶胶寺庙山的五塔坐落于须弥台顶面之上，其构造完全以砂岩石块砌筑而成。在五塔的整体布局之中，中央主塔踞于中心位置，而在规模尺度方面略逊于中央主塔的四座角塔则分别踞于须弥台之四隅。五塔的平面形式皆为四面出抱厦的十字型平面，较之四座角塔，中央主塔在其抱厦和主厅之间增出过厅一间。五塔的建筑形制大致相近，惟因四座角塔的规模较小，因而中央主塔在平面、塔基、山花的构造方面显得更为繁复。作为茶胶寺庙山建筑组群的核心，中央主塔理应一以贯之地处于庙山建造与设计过程最为关键的环节之中。

茶胶寺庙山中央主塔东西方向实测尺寸为 26.91 米（65hasta），南北方向实测尺寸为 24.02 米（58hasta）。参照上述 1hasta＝412 毫米的营造尺度，以同样的方法在茶胶寺庙山总平面图上覆以 4120 毫米（10hasta）为模数单位的平格网，可对茶胶寺庙山中央主塔的营造尺度及其规律进行初步的分析，进而可以推测复原茶胶寺庙山中央主塔的平面设计方法。

由图示可以看出，基底平面尺寸为 170hasta×170hasta 的须弥台，其顶面平面尺寸为 120hasta×120hasta，系由须弥台的基底平面尺寸每边向内收进 25hasta 而来。对于中央主塔的平面设计而言，在茶胶寺庙山总体平面设计阶段所确定下来的东西方向的轴线，亦可称为基准轴线。通过这条轴线不仅可以将中央主塔在须弥台顶面的位置进行定位，并将古代高棉匠师细致入微的意匠通过中央主塔的尺度设计展示出来。现将中央主塔平面设计方法推测复原如下：

首先，将平面尺寸为 120hasta×120hasta 须弥台顶面平分形成两条正交轴线，将东西方向轴线向北侧偏移 1hasta 的距离确定中央主塔的主轴线，另将南北方向轴线向西侧偏移 0.5hasta 的距离而形成次轴线。以主轴线为基准，分别向其南侧偏移 30hasta 以及向其北侧偏移 28hasta 的距离皆可至中央主塔塔基南北侧踏道的外缘。由此可见，中央主塔南北方向尺度的确定，依然恪守着将主要轴线进行偏移的高棉建筑传统。而在中央主塔的南北方向次轴线的偏移方面，以其次轴线为基准向西侧偏移 30hasta 之后似乎并未达及设计位置，在此基础上再向西侧偏移 2.5hasta 才至中央主塔塔基西侧踏道的外缘；再以次轴线为基准向东侧偏移 30hasta 似乎仍然未达及设计位置，亦再向东侧偏移 2.5hasta 才可至中央主塔塔基东侧踏道的外缘。因此，中央主塔的平面尺寸为 65hasta×58hasta，但其平面布局并非对称设置，其西侧较之东侧增出 0.5hasta 的距离，其南侧较之北侧增出 1hasta 的距离。

中央主塔室内平面设计，其进深实测尺寸沿东西方向轴线依次为：东抱厦 2267mm（5.5hasta）、东过厅 2068mm（5hasta）、主室 5813mm（14hasta）、西过厅 2078mm（5hasta）、西抱厦 2285mm（5.5hasta）；沿南北方向轴线依次为北抱厦 2114mm（5hasta）、北过厅 1855mm（4.5hasta）、主室 5837mm（14hasta）、南过厅 1885mm（4.5hasta）、南抱厦 2156mm（5hasta）。通过实测尺寸似乎看不出其中的规律，而将其以 hasta 为单位折算再进行比较，则可出较为明显的变化规律。可能由于中央主塔东西方向尺度略大于南北尺寸且并非对称布局，因而中央主塔的室内空间尺度也随之进行了相应的调整，在南北轴线方向的抱厦及过厅的进深尺寸皆缩减了 0.5hasta。

通过茶胶寺庙山中央主塔各部分平面尺寸的分析可知,其设计原则可能是按照整数尺寸进行平面布局控制的,惟有细部设计采用了 0.5hasta 的尺度标准进行调整。由于这种空间尺度的变化非常细微,几乎很难让人感觉到它的存在,很显然这绝非偶然为之,也并非由于空间功能的调整所致。个中缘由,颇为耐人寻味。[1]

2. 藏经阁的平面设计及其立面尺度

如本书第四章所述,藏经阁位于茶胶寺庙山第二层基台(庙山内院)的东侧,沿庙山东西向轴线南北两侧对称设置,按其位置分别称其为南藏经阁与北藏经阁。从整体风格来看,以上两座对称布局的藏经阁形制相同且保存状况类似,其空间布局、使用功能及其创建年代理应完全相同。

首先以北藏经阁为例,其东西方向实测尺寸为 12370 毫米(30hasta),南北方向实测尺寸为 7035 毫米(17hasta)。参照上述 1hasta=412 毫米的营造尺度,以与中央主塔同样的分析方法在茶胶寺庙山北藏经阁平面图上覆以 412 毫米(1hasta)为模数单位的平格网,推测复原茶胶寺庙山藏经阁的平面设计方法如兹。

北藏经阁的平面布局设计尺度为 30hasta×17hasta,系自西侧抱厦踏步基底外缘至东侧门头踏步基底外缘之间的距离,并且以此作为藏经阁规模的依据。以西侧抱厦踏步基底外缘为基准,向东偏移 5hasta 至抱厦主入口抱框石的外皮(向东侧偏移 4hasta 至抱厦花柱外缘),向东偏移 10hasta 可至主室入口抱框石的外皮,向东偏移 30hasta 则至东侧门头踏步基底外缘。若以东侧门头踏步基底外缘为基准,向西偏移 4hasta 至东侧门头外缘(向西侧偏移 3hasta 至东门头花柱外缘),向西偏移 6hasta 可至主室砌石壁龛内皮,主室两侧砌石墙厚为 2hasta,主室东侧壁龛深为 0.5hasta,而主室西侧抱框石内侧与砌石墙体转折处厚度亦为 0.5hasta。因此,在北藏经阁东西方向轴线上,西抱厦平面尺寸为 6hasta,主室平面尺寸 15hasat(净尺寸为 11hasta),东门头平面尺寸为 1hasta。而其东西两侧基座分别皆为 4hasta。也可以进行如下归纳:西侧抱厦平面尺寸为 10hasta、主室平面尺寸为 15hasta、东侧门头平面尺寸为 5hasta,其中 2∶3∶1 的比例关系似乎可以一目了然。

总之,可能与中央主塔平面设计原则与流程一致,茶胶寺庙山北藏经阁平面设计可能也是按照整

---

[1] 在贡开地区的普兰寺(Prang)也发现类似的细节:"基台上的三座塔殿都供奉湿婆和祖先像。三座塔殿形制相同,惟南北两侧塔殿较之中央塔殿尺度略小。在这两种塔殿的平面图,覆以 412mm 为单位的平格网来分析各部分的尺寸:周长较小的侧塔平面,其附属基座的东西边长为 6484mm(15.74hasta),南北宽 6654mm(16.15hasta),大致是 16 hasta 的正方形。但是基座细部形制略有不同,东西长度短 1/4 hasta。连接台座四角的长度,东西与南北基本没有差别,都是 5687mm(13.80 hasta),14 hasta 的四方形。塔殿室内净尺寸,东西长 2805mm(6.81 hasta),南北长 2783mm(6.75 hasta),可以看作是 7 hasta 的正方形。最大壁厚 1455mm,约为 3.5 hasta。入口的净宽为 961mm(hasta),可以看作是 2.5 hasta 或 2 又 1/3 hasta。很难想象类似采用砂岩加工精致的入口抱框构件,初看应为尺寸相同的标准构件,但也会发现很多意想不到的尺寸微差。而尺度较大的中央塔殿,其附属基座东西长 7601mm(18.5hasta 哈斯塔),南北长 7799mm(19hasta),虽然通常认为是正方形平面,但东西长度的差值约为 0.5 hasta。中央塔殿与两侧塔殿形制相同,唯有东西与南北基座的形状略有不同。连接基座四角的矩形,却没有方向的尺寸差,边长为 6667mm(16hasta 哈斯塔),从而形成边长为 16 hasta 的正方形。加上入口构造的墙体最大长度,东西为 6773mm(16.44 hasta),南北为 6824mm(16.56 hasta),可以认为它是按照 16.5 hasta 设计的。墙体四角形成的矩形,边长 6028mm(14.63 hasta),四边基本相称。可以看作这是从 16.5 的最大值中去掉附柱等入口构造的结果,亦可看作是 14 又 2/3 hasta 比较妥当。塔殿室内的净尺寸在东西与南北方面也没有尺寸差,为 3715mm(9.017 hasta)。即是边长为 9 hasta 的正方形平面,壁厚最大值为 1553mm(3.77 hasta),即 3 又 3/4 hasta,入口的净尺寸为 1316mm(3.19 hasta),可以看作是 3 hasta。参见沟口明则、中川武编著:《贡开与崩密列:高棉帝国东部地区的两大遗迹组群(KOH KER and BENG MEALEA: Two Large Mounments at the Eastern Portion of the Khmer Empire)》,名城大学、早稻田大学、APSARA National Authority、JASA,2011 年版,第 76-113 页。

数关系进行尺度控制的，唯有在细部设计方面采用了0.5hasta的尺度标准进行细微的调整。

南北两座藏经阁的定位，在南北方向上两座藏经阁基座外缘之间的距离为69hasta，若参照茶胶寺庙山东西方向主轴线，其南侧为35hasta，北侧为34hasta，很显然这两座藏经阁的总体平面布局也与庙山主轴线向北侧偏移1-2hasta的规律相符，因而推测茶胶寺庙山附属建筑的设计与建造应是与中央主塔、须弥台等庙山建筑的核心部分一并进行考虑的。另外，根据图示南北两座藏经阁平面布局的东侧基底定位，可能是依据茶胶寺庙山第二层基台顶面外缘为基准向西偏移20hasta得以确定的。然而，两者之间却有1hasta的尺寸差值，即至北藏经阁东侧基底的距离为20hasta，而南藏经阁的情况却是19hasta。这是由于两座藏经阁西侧定位基准一致，东西方向尺度北藏经阁比南藏经阁增加1hasta的缘故。由此可见，两座藏经阁的平面布局在正交轴线方向上皆不对称，分别有1hasta的微小差距。

关于藏经阁的立面设计，可能也是遵循与之相应平面尺度的1/2进行确定的。由于两座藏经阁的屋顶部分及山花皆损毁无存，所以无法窥其旧时全貌。若以北藏经阁东西方向平面尺寸30hasta的1/2计，其立面设计可接近至主室墙体顶面标高位置（16hasta），但尚有0.5hasta的差距。根据现场散落的山花构件进行的建筑形制复原，北藏经阁主室外侧山花顶端的标高为33hasta（比东西方向平面尺寸30hasta增出约3hasta），内侧山花顶端的标高为19hasta（比南北平面尺寸17hasta增出约2hasta），而抱厦山花顶端标高为16hasta（与主室砌石墙体顶面标高差值为0.5hasta），上述竖向尺度的调整与变化或许也与平面尺度设计存在某种联系。

3. 长厅的平面设计（以南内长厅为例）

茶胶寺庙山的四座长厅分别位于第一层基台（庙山外院）与第二层基台（庙山内院）的东北角及东南角。第一层基台上两座长厅依其位置分别称为南外长厅和北外长厅，第二层基台上的两座长厅则分别称为南内长厅和北内长厅。基台上的南北长厅皆沿庙山东西轴线对称布局，其形制描述详见本书第四章的相关内容。

首先以南内长厅为例，其南北方向实测尺寸为16680毫米（40.5hasta），东西方向实测尺寸为4740毫米（11.5hasta）。参照上述1hasta=412毫米的营造尺度，以与中央主塔同样的分析方法在茶胶寺庙山北藏经阁平面图上覆以412毫米（1hasta）为模数单位的平格网，推测复原茶胶寺庙山内长厅的平面设计方法如兹。

南内长厅平面布局设计尺度为40.5hasta×11.5hasta，其南北方向系自北侧抱厦踏步基底外缘至南侧砌石墙体基底外缘之间的距离，东西方向则以主室墙体两侧基底外缘为基准，并且以此作为南内长厅规模的依据。自南藏经阁南侧基底外缘偏移4hasta即为南内长厅北侧抱厦踏步基外，以此为基准位置，向南偏移2hasta至抱厦主入口抱框石的外皮（向南侧偏移1.5hasta至抱厦花柱外缘，主要建筑平面尺寸已皆为整数），向南偏移11hasta可至主室入口抱框石的外皮，向南偏移14hasta则至后室入口抱框石的外皮，再向南偏移12hasta终至南侧砌石墙体基底外缘。可以进行如下归纳：南内长厅抱厦平面净尺寸为10hasta、主室平面净尺寸为15hasta、后室平面净尺寸为10hasta，其中2∶3∶2的比例关系似乎也是一目了然[1]。再分析整理南内长厅的东西方向尺寸，如图示所见，南内长厅也并非对称布局，自其西侧砌石墙体基底外缘至中心基准线尺寸为5.5hasta，而自其东侧砌石墙体基底外缘至中心

---

[1] 运用同样的分析方法可归纳出南外长厅抱厦、主室、后室三者之间平面尺度比例约为1∶3∶1。

基准线尺寸为6hasta。这显然表明，内长厅平面布局也存在以其自身中心轴线偏移1hasta的特征，再次佐证了古代高棉建筑平面布局以其中心轴线偏移1-2hasta的基本规律。

与上述进行尺寸控制的规律类似，茶胶寺庙山内长厅的平面设计依然表现出简洁清晰的整数关系，仅在细部设计方面同样采用了0.5hasta的尺度标准进行略微的调整。

南北两座内长厅的定位，在南北方向上两座内长厅基座外缘之间的距离为45.05米（109hasta），若参照茶胶寺庙山东西方向主轴线，其南侧为22.67米（55hasta），北侧为22.25米（54hasta），这两座内长厅的总体平面布局也与庙山主轴线向北侧偏移1-2hasta的规律相符。另外，根据图示南北两座内长厅平面布局的东侧基底定位，可能是依据茶胶寺庙山第二层基台顶面外缘为基准向西偏移10hasta得以确定的。

4. 塔门的平面设计（以南内塔门为例）

如前所述，在茶胶寺的第一层基台（庙山外院）围墙的四面中央沿庙山轴线分别辟为外塔门四座（东外塔门、南外塔门、西外塔门、北外塔门），第二层基台（庙山内院）回廊四面中央沿庙山轴线分别辟为内塔门四座（东内塔门、南内塔门、西内塔门、北内塔门）。对于塔门的形制描述详见本书第四章。

如图所示，南内塔门由主厅、内侧室及外侧室构成。主厅与南北内侧室相连，再两侧则为外侧室各一。其南北方向实测尺寸为9566毫米（23hasta），东西方向实测尺寸为16918毫米（41hasta）。参照上述1hasta＝412毫米的营造尺度，以与南内长厅同样的分析方法在茶胶寺庙山南内塔门平面图上覆以412毫米（41hasta）为模数单位的平格网，推测复原茶胶寺庙山南内塔门的平面设计方法如兹。

南内塔门平面布局设计尺度为23hasta×41hasta，其南北方向系自基座北侧踏道外缘至基座南侧踏道外缘之间的距离，东西方向则以东西外侧室外的基座外缘为基准，并且以此作为南内塔门规模的依据。因须弥台的南侧踏道破损比较严重，所以从中央主塔的南侧踏道开始测量尺寸。自中央主塔的南侧踏道最南侧外缘向南偏移65hasta即为南内塔门北侧踏道的最北侧外缘。以这个位置为基准，向南偏移3hasta即为南内塔门内侧室基座的北侧外边缘，再向南偏移0.5hasta即为外侧室基座的北侧外缘。回到这个基准，向南偏移5hasta即为主厅的北侧内边缘，再偏移1hasta即为南内塔门内侧室和外侧室的北侧内边缘，再偏移4hasta即为南内塔门内侧室和外侧室的南侧内边缘，再偏移1hasta即为主厅的南侧内边缘，最后偏移12hasta即为南内塔门基座南侧外边缘。

接下来再分析南内塔门东西方向的尺寸。若以西外侧室墙体的西侧外缘为基准向东移1hasta，恰为墙厚的尺度；再向东侧移动13hasta，即为主厅左侧门洞西侧外缘，再向东移10hasta即为主厅右侧门洞东侧外缘，再向东移动13hasta，即为东外侧室东侧墙体西侧外缘，再东移1hasta的墙厚，即为墙体东侧外缘。量取西侧外侧室和内侧室的尺寸分别为8hasta和5hasta，量取东侧，数据与西侧一致，主厅的尺寸为7 hasta。判断南内塔门本身是否关于中心对称时，因其南北两侧形制不一致，所以只分析东西方向。量取主厅东西两侧内墙边缘到中心基准线的尺寸分别为3hasta，4hasta。这表明内塔门平面布局也存在以其自身中心轴线偏移1hasta的特征，再次佐证了古代高棉建筑平面布局以其中心轴线偏移1-2hasta的基本规律。

另外，南北两座内塔门的定位亦非以茶胶寺庙山的东西主轴线对称布置，其中，北内塔门基座最南侧外缘距主轴线的距离是38.15米（93hasta），而南内塔门基座最北侧外缘距主轴线的距离为38.94米（95hasta），所以这两座内塔门的总体平面布局也与庙山主轴线向北侧偏移1-2hasta的规律相符。

五、关于茶胶寺庙山营造尺度与设计方法的初步结论

通过茶胶寺庙山的建筑平面布局及立面设计尺寸分析结果，可以取得如下引人注目的特点。

首先，以 270hasta×270hasta 尤其是南北方向长度 270hasta 作为茶胶寺庙山建筑设计的基本尺度规模，通过轴线的偏移而在庙山南侧增出 1hasta 的方法，确定庙山平面布局的基准轴线，并沿此轴线设计定位四座塔门及中央主塔的位置。

其二，茶胶寺庙山的建筑平面布局设计多采用非常简洁的整数比，总体布局中较大的尺度可能以 10hasta 为模数单位并按照整数比例关系进行设计；各座单体建筑平面的尺寸权衡，同样采用简洁的整数比而一目了然，细部尺寸则以 1/2hasta 或 1/3hasta 为模数单位进行调整。总之，对于茶胶寺庙山的平面尺度设计而言，无论从庙山总体规模抑或直至建筑单体的细部，皆是以 hasta 作为基本的尺度单位，并采用简明扼要的整数比进行规模及布局设计与调整的。

其三，茶胶寺庙山的立面设计亦与其相应的平面尺寸密切关联。其中，庙山建筑核心部分的须弥台和中央主塔尤为关键。一方面，须弥台的整体及各层的标高应是依据第一层基台的平面尺寸，并且按照非常简洁的比例关系经过细致详缜的划分得以确定的，而其中比例关系及其设计方法有可能为当时的匠师所熟稔。在另一方面，无论须弥台基底抑或顶面的平面尺寸，就其对于统领整座庙山建筑组群的中央主塔的立面设计而论，皆发挥着决定性的主导作用。

其四，通过诸如上述平面尺寸、立面比例、林伽高度、观瞻视角等方面尺度规律的推测或复原，隐约闪现其间的某种关联，对于深入剖析茶胶寺庙山设计方法与建造流程皆是极好的线索与提示。

Synopsis VI The Preliminary Study on the Scale Regulation and Design Method of Ta Keo Temple-mountain

1. Introduction

As previously mentioned, Ta Keo is the most majestic and distinctive feature of temple-mountain at Angkor site. Takeo temple-mountain is assigned to unique architectural style and artistic charm by the ancient Khmer religion and culture. Ta Keo temple-mountain has still kept the unfinished state since a thousand years, and its sandstone masonry, cruciform plan, opening porch of quincunx prasats, in addition to enclosed gallery, are the leading architectural style of temple-mountain during the Angkorian period.

In the process of the reconstruction studies on Ta Keo temple-mountain, it firstly can generally understand the ancient Khmer architectural characteristics, evolution of style, religious symbols which should be accordance with the evolution of temple-mountain during the Angkorian period. However, it absolutely has more important academic significance if we are deeply exploring the scale regulation and design methods for understanding the architectural design concept and construction method which get along with comprehensive analysis of the complex system of belief and symbolic of the ancient Khmer architecture. In the matter of the architectural reconstruction studied on Ta Keo temple-mountain in this book, it can be an indispensable link in the program.

For the studies on the scale regulation and design method of Takeo temple-mountain, there are many missing links and gaps, which combined with the preliminary stage of author's research. In this chapter, based on the reference to the research results of both French and Japanese scholars, it is mainly focused on layout plan, façade and details involving the scale analysis and design methods of Ta Keo temple-mountain, to take a glimpse of the ancient Khmer architecture and its design conception of the evident.

According to the studies of French scholars, the construction scale of the Angkorian period may be originated in ancient India Hasta or Vyama as a standard unit of measure, and speculated approximately 1hasta = 450mm. As a kind of very old length measuring unit, hasta can be understood as the length refers to the distance from the elbow joint to the protruding fingertip. Compared to Hasta, and Vyama is a relatively large scale of length measuring unit, refers to the normal human standing with arms outstretched to shoulder in parallel, measured both finger tips of hands as the distance as 1vyama = 4hasta. Both hasta and vyama are widely used in many aspects of social life of Cambodia as length measuring unit, especially in the land surveying, temple construction, sculpture, and fresco painting in the end of the nineteenth century AD as well as before the French colonization, even then the utilization of owner's hasta length is still as the benchmark for housing design and construction.

In addition, it is very rare that referring to construction scale in the inscriptions of Angkor era, which in Prasat Damrei, Prasat Andong, Baksei Chamkrong, it is most important of the Sanskrit inscriptions. The main content of these inscriptions are the homage presentations, but it is the most precious issue that scale of Linga is mentioned in Koh Ker site within the clear reference to the scale unit of hastain addition to its utilization.

According to the comparative study on the three Sanskrit inscriptions, G. Cœdès proposed that the inscription is directing to the Prasat Thom of Prang temple at Koh Ker site. He was also considered that actual length of 1hasta maybe about 0.45 meters, so he suggested that the height of Linga can be 81 hasta or equivalent to 36.45 meters.

It is therefore, the records about architectural scale of Angkor era can be traced back to the middle of the tenth century. And in recent years, as the representatives of the Japanese scholars, Prof. AKAZAWA Yasushi and Prof. NAKAGAWA Takeshi had attempted to restore the construction scale of Angkor era in presumably, and 1 hasta actual length may be about 412 mm (1hasta = 412mm) which involving the series of data analysis on the temples at Angkor and Koh Ker site, especially in the dismantling process of N1 tower of Prasat Suor Prat. It is gradually distinct that 1 hasta is the basic unit of control for the layout planning and construction process of the Angkor temples. In term of the standard of components, it can be concise and to the point of scale control and design through the decimal use of integer values, and it also gradually has the many examples to be confirmed.

During the period of 2007-2010, CACH&CSA had been conducted the documentation and survey campaign of the whole complex of Ta Keo temple-mountain in cooperation with the Tianjin University by using 3D laser scanning technology. The very detailed data and results had been acquired in recent years. Based on the predecessors' results or clues, this chapter mainly aims to the initial studies on construction scale and design method of Takeo temple-mountain by using of the latest or extensive documentation data. With these backgrounds of in mind, the author hopes to provide a very personal perspective and probably enlighten some useful horizon for the in-depth analysis of scale regulation and design conception of Ta Keo temple-mountainover and above the other Angkor monuments.

2. Scale Analysis on Layout Planning of Ta Keo Temple-mountain

In the matter of the site of Ta Keo temple-mountain, it is the rectangular layout that the distance of east-west axis is greater than that of north-south axis. In the CAD drawing that obtained by 3D laser scanning point cloud, the length of east-west axis (from the outer side of causeway to the outer side of western moat) is 243.08 meters, over and above the length is 229.77 meters between the outer side of eastern moat and the outer side of western moat. And the length of south-north axis is exactly 193.59 meters between the outer side of southern moat and the outer side of northern moat. According to Japanese scholar's studies on the construction scale and design method of ancient Khmer architecture, using the assumption that the 1hasta = 0.412 meters and analyzed, if it can be inferred to the logical regulations and methods and it also conversely can prove the unit scale whether right or not. Thus, the length of east-west axis of Ta Keo temple-mountain can be converted into 590 hasta (the length can be converted into 558 hasta between the outer side of eastern moat and the outer side of western moat), the length of north-south axis is converted into 470 hasta. It may be inferred that the aspect ratio is very close to an integer ratio 5 : 4 as well as the layout planning and scale design should be the range of moat and causeway included at Ta Keo temple-mountain site.

The outer enclosure with dimensions of 135.96 meters x 111.24 meters (from east edge of the first step of eastern outer Gopura to west edge of the first step of Outer Western Gopura), it can be converted into 330 hasta in the east-west axis as well as 270 hasta in the north-south axis. It is 55.93 meters (136 hasta) with dimension between southern edge of the first terrace and the east-west axis of central prasat, and however it is 55.31 meters (134 hasta) between northern edge of the first terrace and the east-west axis of central prasat.

The inner enclosure with dimensions of 94.7 meters x 88.7 meters (230 hasta × 215 hasta), similarly as the outer enclosure, it is 44.77 meters (109 hasta) with dimension between southern edge of the second terrace and the east-west axis of central prasat, and however it is 44 meters (107 hasta) between northern edge of the first terrace and the east-west axis of central prasat. And then, the pyramid is with dimensions of 60.7 meters x 59.1 meters (147 hasta × 143 hasta), it is 29.89 meters (73 hasta) with dimension between southern edge of the second terrace and the east-west axis of central prasat, and however it is 29.36 meters (71 hasta) between northern edge of the first terrace and the east-west axis of central prasat.

Therefore, in the case of above-mentioned, it is not symmetrical in accordance with the east-west axis of Ta Keo temple-mountain, and the dimension of southern area is increasing 2 hasta than that of northern area.

In terms of numerous Angkor monuments, along the direction of axis on both sides of the architectural layout are not completely symmetrical according to the recent studies of Japanese scholars. The axis is often slightly offset in length of 1 hasta between southern side and northern side, resulting in 2 hasta length difference at two sides of the east-west axis. Due to the offset smaller, together with huge range or volume of building, and 1 or 2-hasta lengths difference is more obscure. However, through to the scale analysis of many Angkor monuments, it approximately can be confirmed that the offset phenomenon of architectural axis is very common, and on the contrary, the examples of completely axis-symmetric layout are very rare. Thus it can be seen, for Ta Keo temple-mountain's layout, the offset phenomenon of east-west axis is exact matching with the scale regulation of other Angkor monuments.

Due to the limitation of general layout characteristics, for the process of construction or design, it is supposed to be from outside to inside of Ta Keo temple-mountain. As core and essence of the temple-mountain group, the central prasat is also supposed to be crucial point of the construction and design process. According to 1 hasta = 412 mm of the design scale, if the general layout of Ta Keo temple-mountain is covered with the grid network as module unit size 4120 mm (10 hasta), it is summarized that original design method can be reconstructed more accurately owing to the analysis of scale regulation.

Firstly, ancient Khmer architects utilized the astronomy and surveying knowledge, also probablyover and above extremely complex rituals or ceremonies, they had closely linked with Eastern Baray and Ta Keo temple-mountain. Thereby the east-west axis of temple-mountain had been identified. According to the presetting of the ranges and dimensions (as a basic module unit, it is temporarily assumed to 1H = 30 hasta), it is probably referenced that the offset of the east-west axis is respectively to the north and south sides of length in 135 hasta

(4.5H) to designate outer dimensions of the first terrace or enclosure. Subsequently, it seems to be the use of conventional design methods, and make the determined axis 1 hasta offset to setup a new axis (or can be called a reference axis) in the north side, consequently, the length of the south side is increasing 2 hasta than that of the north side. Along the east-west direction of the reference axis, it can be assigned for the east and west gopuras, as well as the location of central prasat.

For the utilization of axis-offset method of layout planning in the Angkor monuments, up to now, have been found and analyzed several examples including Koh Ker temples, however, it is extremely little knowledge of axis-offset method of layout planning of Ta Keo temple-mountain. As above-mentioned, the east-west axis of Ta Keo is firstly determined, it is the total length of 270 hasta (9H) in south-north direction as a reference, to determine the center position of a square in the scale of the 270 hasta × 270 hasta (9H × 9H), and it can be central-point or reference-point of planning scale of the entire temple-mountain.

Perhaps, it can be prior to consideration of the assignments of long-halls and libraries on the east side of Ta Keo temple-mountain. As a result, it can reach the east-step's edge of Eastern Outer gopura if it extend the east side of the 270 hasta × 270 hasta square in length of 60 hasta (2H or 15vyama) along the reference axis, and then to 330 hasta (11H) location. if it is from this position along the axis of the eastward extension of the 120hasta (4H or 30vyama) length, it can reach the inner boundary of eastern moat of temple-mountain, and then eventually it also can reach into the beginning of causeway with eastward extension length of 70 hasta, over and above the total is including of both to the causeway starting point 60hasta (2H or 15vyama) and to causeway pavilion 10hasta (1/3H or 2.5vyama). Including Ta Keo temple-mountain, the main entrance and main buildings are toward the orient due to the most of Angkor monuments is respecting toward the orient in the tradition, and presumably for the east side of the temple-mountain, it can be most significant aspects of the process of design and construction. Therefore, firstly according to the determination of east-west axis assignment of Ta Keo temple-mountain, it can be basically definite of the scale of the layout planning and design rules.

For the two side of southern and northern of Ta Keo temple-mountain, the scale and the dimension with the 270 hasta × 270 hasta (9H × 9H) square, if it is respectively extending 100 hasta (3H + 1/3H or 25 vyama), it can reach into the outer boundary of northern moats. And if it is extending 70 hasta (2H + 1/3H or 17.5vyama), it can reach into the outer boundary of western moats. Therefore, it is total length of 590 hasta (19H + 2/3H or 147.5vyama) in the east-west direction, as well as total length of 570 hasta (19H or 142.5vyama) in the north-south direction of Ta Keo temple-mountain. Consequently, the general scales or dimensions of Ta Keo temple-mountain can be virtually designated, or including the moats, for the overall layout of Ta Keo temple-mountain, it is the rectangular plane which length in the east-west direction is more 20 hasta (2/3H or 5vyama) than that in the north-south direction.

If it is continued to regard the 270 hasta × 270 hasta (9H × 9H) basic-square as a reference plane, and take northern and southern sides of the basic-square's boundary are respectively inwards into the 50 hasta (12.5vyama), totally 100hasta (25vyama), while as in the east and west sides of the basic-square's boundary, eastern border is inwards into 60 hasta (15vyama), and western border is inwards into 40 hasta (10vyama), also 100hasta (25vyama), then the pyramid terrace in scale of 170 hasta × 170 hasta (5H + 2/3H or

Synopsis Ⅵ The Preliminary Study on the Scale Regulation and Design Method of Ta Keo Temple-mountain

42.5vyama, dimensions of the steps is included) is determination. For peripheral dimensions of the steps in 170 hasta, it is just as 2/3 of overall length in the north-south direction of temple-mountain, as well as the 1/3 of overall length in the east-west direction. And then, if it is regard the outer edge of the pyramid terrace's east steps as a basic point, the central axis of libraries' porch can be assignment, and in case the axis is 20 hasta offset towards the east direction, it can determine the long-hall position, and then with continuing to eastward 3 hasta offset, it can determine the position of galleries. But notable is, for the galleries, either the scale or the proportion, the relationship can't be obviously observed with overall layout of temple-mountain instead of appearing quite far-fetched, merely it is 10 hasta in length from inner long-hall to the outer boundary of the second platform.

In accordance with the above scale analysis, it can be hypothetical inference that the galleries around the second platform maybe not included in the original design conception. So the initial design scheme of Ta Keo temple-mountain can be still continuing the plane layout like as East Mebon, Pre Rup with a series of long-hall around the pyramid, many doubtful ruins at the first or second platform in estimates may be related.

From the scale analysis of layout planning, the location of gallery is far greater divergence comparing with the scale regulation of Ta Keo temple-mountain's layout somewhat show quite heterogeneous. Presumably it can be some modifications or adjustments for the initial design scheme during the reign of King Suryavarman I, and probably it had been replaced with the galleries around pyramid by removing the majority of long-hall buildings, which caused a series of innovation of temple-mountain's architectural style after the reign of King Suryavarman I.

3. Scale Analysis on Façade of Ta Keo Temple-mountain

Referring to scale of 1 hasta = 412 mm to analysis, if the general elevation of Ta Keo temple-mountain is covered with the grid network as module unit size 4120 mm (10 hasta), it is summarized that original method of elevation planning can be reconstructed more accurately owing to the analysis of scale regulation.

As the layout analysis above-mentioned, while the east-west axis of temple-mountain is determined, and then it regards the overall length of 270 hasta (9H) as a reference to determine the center position of 270 hasta ×270 hasta (9H×9H) basic-square plane, and as the geometric central point of whole temple-mountain. On this basis, is it still valid in the facade design or elevation scale of Ta Keo temple-mountain? It can be analysis as following:

If with its 1/2 (135hasta) of 270 hasta ×270 hasta (9H×9H) basic-square scale as the elevation design, the point position is much higher closely 33 hasta than that on top elevation of the central prasat, appears the illogical rules to be associated with its false layer structure. However, if it regards its 1/3 (90hasta) length of the basic-square as elevation design, the position is affirmatively located in the top surface of ruin state of the corner prasat. And if it regards its 1/4 (67.5 hasta) length of the basic-square as elevation design, the position is exactly located in the top surface of the central prasat's foundation. And then if it regards its 1/5 (54hasta) length of the basic-squire as elevation design, the position is exactly located in another important position of the elevation, the top surface of the third layer of pyramid. And then if it regards its 1/6 (45hasta) length of the basic-square as elevation design, the position is exactly located in the top surface of the second layer of pyramid. And the last if it regards its 1/8 (33.75hasta) length of the basic-square as elevation design, the position is exactly located in the top surface of the first layer of pyramid.

On the basic analysis on the pyramid's base dimensions, 170hasta × 170hasta, and then if it regards its 1/2 (85hasta) as design elevation, the position is exactly located in the top surface of ruin state of the central prasat. And if it regards its 1/2 (85hasta) of the pyramid's top dimensions as design elevation, the position is roughly located in the top surface of the central prasat but like with complete state, which the fourth false layer and crown height is hypothetically included.

From the above-mentioned analysis, in the process of elevation design of Takeo temple-mountain, the basic-square with size of 270 hasta × 270 hasta (9H × 9H) still seems to play same scale control function as it in layout planning. It can be closely related with elevation scale control and adjustment as well as in accordance with proportion design. For example, as one of the most important parts of Ta Keo temple-mountain, at least for the elevation design of pyramid terrace, its performance is very clear based on its dimensions such as 1/3, 1/4, 1/5, 1/6, 1/8 and other simple proportions. Moreover, as equally important to the central prasat, its elevation height determination may be closely associated with the scale planning of pyramid terrace. It can be seen that the elevation height of central prasat (ruin state) is matched with 1/2 dimensions of the basal plane of pyramid terrace, and the hypothesized complete state of the central prasat is exactly matched with 1/2 dimensions of the top plane of pyramid terrace. In addition, due to the Linga has not been existed in central prasat of Ta Keo temple-mountain, almost unable to understand or to reconstruct the Linga scale, which is related with the proportion of the central prasat. However, it can be uncovered some indistinct link between Prasat Thom and Ta Keo temple-mountain if the referencing to the Linga's scale 81hasta is speculated in the above-mentioned inscriptions. As shown in the figure, it is close to 1/2 elevation height of central prasat (complete state) from the ground level to 81hasta elevation height. This position can be the hypothetical top-elevation of Linga in the central prasat. Moreover, between the threshold stone of porch entrance (assuming the view height of 165 mm) and the top elevation position of Linga, the view angle is very close to 27°, and this angle is exactly the best view angle of building's façade image. It is not seem to be a coincidence, as for the reasons and regulations, still needs to make further exploring.

主要参考文献

1. Aubin, N. , *Le Phimeanakas*: *Temple – Montagne à Angkor*. la thèse de l'école d'architecture de Nantes.
2. AYMONIER Étienne, (1900), *Le Cambodge*: *I. Le royaume actuel*, Vol. 1, E. Leroux, Paris, XXIII.
3. AYMONIER Étienne, (1904), *Le Cambodge*: *III. Le groupe d'Angkor et son histoire*, Vol. 3, E. Leroux, Paris, XXIII.
4. Boisselier, J. , (1966) . *Asie du Sud – est* (*t*. 1) , *Le Cambodge*. Paris: Picard.
5. Briggs, (1951) . *The Ancient Khmer Empire*. Philadelphia: American Philosophical Society.
6. Coedès, G. , (1952) . *Inscriptions du Cambodge IV*. Paris: E. de Boccard
7. Coral – Remusat, G. de, (1940), *L'art khmer. Les grandes étapes de son évolution*, Étude d'art et d'ethnologie asiatique, Paris,.
8. Coral – Remusat, G. de, Goloubew V. , and Coedès G. , 'La date de Takèo', BEFEO XXXIV, Hanoï 1935.
9. Cunin Olivier, (2000) *Le Bayon*, *contribution à l'histoire architecturale du temple*, Mémoire de travail personnel de fin d'étude en architecture (en deux volumes), École d'Architecture de Nancy.
10. Dagens B. , (1995) *Angkor*: *Heart of an Asian Empire*, London, Thames&Hudson.
11. Dagens, B. (1968), 'Étude iconographique de quelques fondations de l'époque de Suryavarman Ier', Arts asiatiques XVII, Paris.
12. Delaporte, L. , (1924) . *Les monuments du Cambodge*: *étude d'architecture khmère*. Paris: Delagrave.
13. DumarçayJacques & Royère P. , Smithies M. transl. and ed. , (2001) . *Cambodian Architecture*: *eighth to thirteenth centuries*. Leiden, Boston, Köln: Brill.
14. Dumarçay Jacques, (1967), *LE BAYON*, *histoire architecturale du temple*: *Atlas et notice des planches*, MAEFEO N° 3, Adrien – Maisonneuve, Paris.
15. Dumarçay Jacques, (1971) . *Ta Kèv*: *Etude architecturale du temple*. Paris: EFEO.
16. Dumarçay Jacques, (1973) . *Charpentes et tuiles khmères*. Paris: EFEO.
17. Dumarçay Jacques, GROSLIER Bernard – Philippe, (1973), *LE BAYON*, *histoire architecturale du temple* & *Inscriptions du BAYON*, EFEO, Adrien – Maisonneuve, Paris.
18. DumarçayJacques, GROSLIER Bernard – Philippe, (1973), *Charpentes et tuiles Khmères*, EFEO, Adrien – Maisonneuve, Paris.
19. Dumarçay, Jacques, (1998) . *L'architecture et ses modèles en Asie du sud – est*. Paris: Librairie Orients.
20. EFEO & Musée Cernuschi, (2010), *Archaeologists in Angkor*, Paris.
21. FINOT Louis, GOLOUBEW Victor, (1923), *Le symbolisme du Neak Pean*, BEFEO XXIII.

22. FINOT Louis, GOLOUBEW Victor, (1925) *Temple de Mangalârtha à Angkor Thom*, BEFEO XXV.
23. Freeman, M. and Jacques, C., (2003). Ancient Angkor (revised), Bangkok: River Books.
24. Glaize, M., (1948). *Les monuments du groupe d'Angkor*. Saigon: Portail.
25. Groslier, B. P. (1974), 'Agriculture et religion dans l'empire angkorien', Etudes Rurales,
26. Groslier, B. P. (1979), 'La cité hydraulique angkorienne: exploitation ou surexploitation du sol', BEFEO LXVI/1.
27. Groslier, B. P., (1961). *Indochine carrefour des arts*. Paris: Albin Michel.
28. Groslier, B. P., (1968). *Angkor, Hommes et Pierres*. Paris: A. Arthaud.
29. Groslier, B. P., (1997). *Mélanges sur l'archéologie du Cambodge*. Paris: EFEO.
30. Groslier, G., (1921). Recherches sur les Cambodgiens. Paris: Augustin Challamel.
31. Groslier, G., (1921-23). *Arts & Archeologie Khmers* (t. 1). Paris: Augustin Challamel.
32. Groslier, G., (1924-26). *Arts & Archeologie Khmers* (t. 2). Paris: Augustin Challamel.
33. *L'évolution de la couverture de tuiles à Angkor, du IXe au XVIe siècle*, (2000) Udaya, APSARA, Phnom Penh.
34. Parmentier, H. (1935), 'La construction dans l'architecture khmere classique', BEFEO XXXV, Hanoï.
35. Parmentier, H., (1939). *L'art khmer classique. Monuments du quadrant Nord-Est*. Paris: EFEO.
36. Stern, P., (1934), 'Le temple-montagne khmèr, le culte du linga et le Devarâja', BEFEO XXXIV (1).
37. （元）周达观原著，夏鼐校注：《真腊风土记校注》，北京：中华书局，1981年版。
38. D. G. E. 霍尔著，中山大学东南亚历史研究所译：《东南亚史》（D. G. E. Hall, *A History of Southeast Asia*），北京：商务印书馆，1982年版。
39. G. 赛代斯著，蔡华等译：《东南亚的印度化国家》（G. Cœdès, *LES ÉTATS HINDOUISÉS D'INDOCHINE ET D'INDONÉSIE*），北京：商务印书馆，2008年版。
40. 沟口明则、中川武等编著：《贡开与崩密列：高棉帝国东部地区的两大遗迹组群（*KOH KER and BENG MEALEA: Two Large Mounments at the Eastern Portion of the Khmer Empire*）》，名城大学/早稻田大学/APSARA National Authority/JASA，2011年版。
41. 陆峻岭、周绍泉编注：《中国古籍中有关柬埔寨资料汇编》，北京：中华书局，1986年版。
42. 马克思·韦伯著、康乐等译：《印度的宗教：印度教与佛教》，桂林：广西师范大学出版社，2005年版。
43. 中国文物研究所著：《世界遗产·柬埔寨吴哥古迹——周萨神庙》，北京：文物出版社，2007版。
44. 梁英明著：《东南亚史》，北京：人民出版社，2010年版。
45. 尼古拉斯·塔林主编，贺圣达等译：《剑桥东南亚史》，昆明：云南人民出版社，2003年版。
46. 黄南津、周洁编著：《东南亚古国资料校勘及研究》，北京：中国社会科学出版社，2011年版。

附录 I

柬埔寨吴哥古迹保护处
茶胶寺庙山清理发掘月报摘录
Conservation des monuments d'Angkor RAPPORT
(1920 – 1923)

（温玉清、张春彦编译）

茶胶寺 1920 年 5 月

本月，此前在达布隆寺的工作队开始进驻茶胶寺进行清理，由于茶胶寺第一层基台西侧与道路之间并没有通路，现在则要挖土垫出一条通往茶胶寺西外塔门的道路。在挖土过程中，发现土层中混有很多残砖。

茶胶寺 1920 年 6 月

本月，工作队派出 25 个工人继续清理庙山的西侧区域，第一层基台的塔门彻底清理完毕，清理了第一层基台塔门与上层踏道之间的狭窄空间，在塔门内设置了一些混凝土的支护框架。

茶胶寺 1920 年 7 月

本月，工作队 26 个工人开始清理和搬移上部基台的积土和散落构件，清理挖掘出的积土先被运到第一层台面上，稍后再行运走将其最终堆放在庙山的北侧和南侧底部，本月茶胶寺的西侧立面已经完成清理。

茶胶寺 1920 年 8 月

本月，工作队 25 个工人完成了须弥台顶部的清理，工作包括挖掘积土、搬运塌落石块、清理植被等，建筑结构的破坏是由于植被生长形成裂缝进而引发垮塌的。目前正在清理北侧，积土和构件残片被直接运到下面两个基台，并将被逐层运出堆放至丛林之中。Corporal Huot 是这个项目的监工，此人极其张狂，现命其 8 月 17 日前必须返回工作队。他的位置已经被一位工人取代，此人从前当过兵，能说一点儿法语，每天给他 0.65 美元的报酬使他非常满意。至于加固工作，本月完成对茶胶寺塔殿几处加固。

茶胶寺 1920 年 9 月

从本月 13 日开始，原本 25 个工人的工作队增加至 40 个工人，新的下士充分展示出了他的重要性和管理现场的才能。庙山南侧和东侧覆盖的植被包括树干和根系已经被清理完毕。北侧正在实施过程中，南内塔门与东内塔门室内也被清理，回廊也沿着从南内塔门向西、东两侧清理。回廊的砂岩铺地保存状况非常好，根据回廊内的积土混有碎砖，推测其屋面可能是砖砌的，还发现散落的砂岩挡头瓦也是来自屋面部分。临近东南回廊的藏经阁和南内长厅正在清理过程中，与在回廊中的发现类似，其屋顶也可能是砖砌的。在塔门与两座藏经阁之间的院子内，我们利用挖掘探槽调查土层情况。十分规则且铺装均匀的砂岩铺地保存状况相当好，踏道附近的积土比院内更多。通过第二层台搭设在垮塌回廊空隙的右侧的陡直木架，工人们利用垂直升降推车直接将积土清运出去。在东内塔门内的碑铭被发现和标识（详见9月23日及10月3日的电报）。

茶胶寺　1920年10月

本月，工作队的工人数量增加至50人是相当重要的因素，因此项目进展非常顺利。包括两座藏经阁、内长厅以及靠近回廊的南内塔门和东内塔门在内的庙山内院清理工作进展顺利。除了南藏经阁西北侧的大树在清理积土之前仍然起到一定的临时支撑作用之外，院内的大型植被皆已砍伐清理。

除了树冠即将被清理之外，南内塔门的状况如前所述类同。在环绕内院的四周，地面积土的清理工作一直到铺石地面。院内东北角以及北内长厅的清理和挖掘工作也一直在持续。从一开始起，茶胶寺各层基台上挖出的积土都是逐层向下清运的，其具体方法在九月份的报告中业已说明。架子搭设在北侧塌毁回廊所形成的空隙之间，质量不太好的竹板已经被替换，新换的架子足以承担一个工人将三辆推车连续使用。此外，最新发现的碑铭在10月9日86号电报中已经报告。

1. 新发现的一座砂岩雕像，立像，1.42米高，包括0.32米高的基座与基柱，雕像戴冠冕的头部与躯干分离，且前臂缺失，另一只前臂及手与胳膊分离。

2. 新发现的一座砂岩雕像，女性立像，1.25米高，包括0.30米的基座和基柱，戴冠冕的头部也与躯干分离，前臂全部缺失，这座雕像的部分碎片被埋在南内长厅附近的墙角部位。

3. 新发现的一座砂岩神牛雕像，牛背上有明显的凸起，除了牛角和基座之外，雕像保存基本完整。雕像基座没有榫，其具体尺寸如兹：0.6米（高）×1.1米（长）×0.5米（宽），其中包括基座0.05米，这座雕像发现于须弥台根部靠近北内长厅的地方。

茶胶寺　1920年11月

本月，工作队的50个工人继续清理环绕茶胶寺主台基的平台，围合成南内塔东南侧区域南院的部分已完全清理完成。靠近东院东北侧区域北回廊缺口处设立了第二个运输通道。在最后这座院子里只剩东北侧区域及其上建筑要清理。

茶胶寺　1920年12月

本月，工作队的50个工人继续清理茶胶寺东院东北侧区域，以及南院南内塔门对面接近西侧部分。

茶胶寺　1921年1月

本月，50个工人的工作队队伍扩大了，在一周的时间内，没有工作的（10人）"混凝土"工作队,完成了东院东北角及角上建筑的清理,继续北院的清理工作,已经超过了北内塔门的位置。回廊内部的清理工作仍然在继续。第三条运输通道在靠近北院西北角的缺口处搭设起来。在北院,清理已经超过了西南角的位置。

茶胶寺 1921年2月

本月,工作队的50个工人完成了第二层基台四座院子的清理工作,即围绕茶胶寺主要基台的院落和平台清理工作全部完成。积土和灌木植被所覆盖的回廊和各座建筑单体也被逐一清理发掘出来,工人们遂开始将基台上的积土和灌木植被清除并运送出去。

茶胶寺 1921年3月

本月,工作队的50个工人一方面开始东侧第二层基台的院落和平台的清理工作,另一方面开始清除前几个月清理出来的积土和废墟。我们设置的运输路线有两条:其一,沿着庙山轴线一直延伸至庙山东侧;另外一条则是沿着通向达布隆寺道路的方向,这样的安排便于利用些积土作为填充材料铺设进入大吴哥城东侧主要寺庙的道路,此项工作使用事先铺装轨道 Decauville 推车完成的。

茶胶寺 1921年4月

本月,工作队的50个工人被专门安排在寺庙的东侧继续清除积土和散落石块,为了打开东塔门外部与其垂直的东向宽通道,此项工作暂停了两周。工作是为了使埋在下面2米的东塔门能够通畅。并能够在下个月从主入口进入建筑,甚至天气干燥的话能够开车直达台阶底部。

在须弥台踏道底部与内塔门之间的积土已经清理完毕(参见3月份的报告)的情况下,探槽A与探槽B的深度为3米,探槽C与探槽D的深度为2米,至今还剩下30-40厘米未完成。通过探槽C与探槽D,可以清晰地看到角砾岩基台的状况,在距离东外塔门大约50米之处,挖掘出土了石狮残件及大量的散落石块。与此同时,安排另一支工作队的20个工人计划从5月2日(星期一)开始工作,主要任务是外部的土方工程,以便使第一支工作队全部来完成第二层基台上的积土运送工作。

茶胶寺 1921年5月

本月,安排两支工作队同时开展工作。

第一支工作队的50个工人,继续清理寺庙第二层基台的东侧区域(在第二层基台内,沿着须弥台东南角至东北角一线的地面系由非常规则的角砾岩石块铺设),两座建筑(藏经阁?)的墙体部分也进行了清理,完成了五分之四。塔门的东面和西面的前面部分以及三个厅也被清理完毕。另外,这支工作队承担的寺庙东侧两座塔门之间探槽的挖掘已经接近完成。随后由第二支20人的工作队来执行这些工作。

开挖了一条通道可以延伸至通往达布隆寺的道路,两者的地坪一致,道路布满标杆。在这支队伍完成任务后,工人们将替换前述的已经解散的老队伍,参与大吴哥城内的清理工作。

茶胶寺 1921年6-7月

在两个月内,安排两支工作队同时开展工作。

第一支工作队的 30 个工人（平均），全部完成了寺庙第二层基台东侧区域的积土清理和南北对称布局的两座单体建筑（藏经阁）的清理工作，并对散落的石块进了堆放整理。此后，他们继续在寺庙第一层基台的东南角大约 20 米的范围内实施清理工作。

第二支工作队的 17 个工人（平均），全部完成了新通道以及排水护坡的施工任务。在完成这项任务后，他们开始沿着第一支工作队相反的方向进行清理工作，即寺庙第一层基台的北侧区域。为迎接总督前往达布隆寺参观，其间的清理工作被中断了十天，工人们被借调去参与清理加固达布隆寺内某些塌落的部位。

茶胶寺　1921 年 8 – 9 月

两个月内，安排两支工作队同时开展工作。

第一支工作队的 50 个工人，开始将寺庙第二层基台南侧区域的积土同时向西开始清理。工作进度已经超过了南外塔门，该塔门已经清理完毕，以及进入第一层基台院落的踏道。第二支工作队的 40 个工人在第一支工作队工作的对称位置，即寺庙第二层基台的北侧清理积土，但是由于人手不够进度较慢，现距离与第一层基台连接的北外塔门的踏道还有 10 米，如果 10 月份的雨水对工作影响不是很大，外部基台建筑的清理，我们认为应在 11 月 1 日之前全部完成。

寺庙内的积土通过南侧的环路运送出去，而寺庙北侧纵深 100 米的区域基本全部被森林覆盖。

茶胶寺　1921 年 10 月

本月内安排两支工作队同时开展工作。

第一支工作队的工人人数由上月的 50 人削减至 20 人，他们继续清理寺庙第二层基台南侧区域的积土，主要完成了南内塔门至第二层基台西南角区域的清理工作。第二支工作队工人人数由上月的 40 人也削减至 20 人，继续在第一支工作队工作的对称位置，即寺庙第二层基台的北侧清理积土，北塔门及连接第一层基台踏道清理完毕，开始在北塔门至西北角区域进行清理。

茶胶寺　1921 年 11 – 12 月

本月工作队的 30 个工人，接着完成第二层基台南侧清理工作并继续西侧的清理工作，几乎到了西轴线的塔门。清理积土和灌木沿寺庙西侧外的通道运出去。

茶胶寺　1922 年 1 月

本月，两个分别有 15 个工人的工作队，完成了寺庙第二层基台北侧和西侧的积土和散落石块的清理工作，并将积土和砍伐的灌木沿着寺庙西侧及北侧的道路运出去，以上工作由其中一支工作队继续在本月末之前完成；另一支工作队则去忙于使用混凝土加固寺庙第二层基台的回廊以及班蒂珂黛寺的松散结构。

茶胶寺　1923 年 1 月

本月，工作队派出 22 个工人，在完成寺庙南侧神道区域的清理任务之后，继续清理寺庙东南侧水池的角砾岩驳岸（参见照片 175、照片 176）。在 B 区发现有两座大型水池遗迹，虽然这绝对很难切分与寺庙整体的联系，但是我们还是从寺庙东侧区域清理出了大量积土，并揭示出一段角砾岩砌筑的墙基。这段墙基（报告中草图以红色绘制）看似隐藏在寺庙的东外塔门前面，而同时发现的一对砂岩狮子雕像残迹推测应也对称放置在 G 区。发掘清理覆盖在这段角砾岩墙基上积土的情况，可以参考如下原始报告中所绘制的墙基断面草图。一条清晰的界限可以分辨夯土的构成，在接近生土的土层中包含着大量的砖瓦碎片。通过照片 177 和照片 175，都可以清楚地显示出墙基断面的土层分布情况。另外，在此区域的发掘过程中，我们还发现了几片装饰直细齿状突起的陶瓦，并出土了砂岩雕像的一支手臂及一只手。

茶胶寺 1923 年 2 月

与清理寺庙神道南侧的任务相同，本月工作队的 20 个工人开始清除寺庙神道北侧的积土。也发现了一段时断时续的、与神道南侧情况相同的角砾岩砌筑墙基遗存，或者说与角砾岩相比，在神道端部似乎是更多地利用砂岩砌筑。在神道端部之南侧的 B 区，一座平台被巨大的树木所覆盖。这座砂岩砌筑的平台，其台阶向左右两侧扩展截断，其形制可以参见报告附图以及照片 187。在神道北侧的对称位置，角砾岩遗迹台阶是否与水池砂岩驳岸连接，这需要进一步的发掘，但是保存现状没有表现出适当的地坪平面。神道端部的角砾岩墙基（神道南北两侧皆可以铺设轨道）甚为简洁，其状况可以参考 C 区拍的照片 188（报告中的草图）：这张照片显示的是清理之前神道北侧的状况。本月末拍摄的照片显示了清理积土之后墙基的遗存状况。我们计划在下个月继续进行拍摄。

茶胶寺 1923 年 3 月

本月，工作队的 24 个工人完成了神道北侧积土、大块角砾岩与砂岩石块的清理工作。这些遗迹的形制几乎和神道南侧发现的情况对称设置（参见照片 203）。照片 204 则显示了角砾岩踏步上部边缘的情况，这些踏步一直转折并延续至寺庙东立面的北侧的 A 区。虽然我们曾经推测有环绕寺庙的台阶，但是实际情况看起来这些台阶好像只是靠近寺庙第一层基台，而且在转向北侧 C 区截止。为确保通过轨道运输积土，我们不能在此继续挖掘。另外，如同 C 区一样，B 区也有巨大的树木生长，很遗憾地阻挡了我们继续进行发掘的兴趣。然后，我们等待这次调查中关于角砾岩墙基的报告，通过他们可以了解 C 区确切的范围（参见照片 209）。而实际上，到处都是遗存重叠的情况，而且在地面之下也未能考虑这些台阶。这些遗迹仅能根据水池驳岸的位置进行大致的定位，另外根据土层中的砖瓦碎片判断可能建造年代较晚。对于寺庙的清理和发掘所出现的问题是：D 区缺口为何与寺庙第一层基台、神道没有形成任何联系，这让我们非常困惑。周围地形情况绘制在原始报告之中。

茶胶寺 1923 年 4 月

本月，工作队的 24 个工人完成了寺庙东侧神道建筑构件碎片的整理分类。我们在东部发现了更多的角砾岩石块（参见 2 月份的报告照片 188）。新的基台就像一个道路凹处入口台阶。在照片 213 及照片 213）都显示了台阶构造的总体面貌。由于本月末暴雨的侵袭，无法对神道平台两侧的角砾岩边缘部位进行发掘和推断。照片 214（即将尽快寄达）会让我们看到这些台阶的停顿状况。他们看起来不像与其他建筑有所联系，而是独立地与 A 区的砂岩边界毗邻，还可以看到一些地面上的冲刷痕迹，似乎与寺庙周围的空隙相关（详见原始报告的插图）。下一步调查工作计划：建议继续针对茶胶寺庙山的神道北侧角砾岩遗迹、神道东侧端部所发现的大型角砾岩石块进行调查。

茶胶寺　1923 年 5 月

本月，林业处的工作主要在我的指导和控制下进行，实施了帕蒙提埃先生要求的砍伐工作，以便自道路穿过的桥梁处能够看到整体的茶胶寺景象，同样道理，阇耶跋摩七世时代的医院自道路也就可以看见了，当人们从胜利门到达这里的时候，这一规划使茶胶寺进一步给人以美好的印象。

附录 Ⅱ

茶胶寺庙山历史照片

(法国远东学院提供)

茶胶寺庙山二层台回廊东南侧
Enceinte III, galerie sud – est
拍摄日期：1952 年
作者：佚名
典藏号：法国远东学院摄影档案馆
CAM01667ⓒEFEO

茶胶寺庙山二层台北内长厅解体维修情况
Enceinte II, salle longue nord-est, dépose des blocs de grés
拍摄日期：1924 年
作者：佚名
典藏号：法国远东学院摄影档案馆 CAM08693ⒸEFEO

附录 Ⅱ 茶胶寺庙山历史照片

茶胶寺庙山西北侧全景 Vue générale angle nord – ouest
拍摄日期：1950 年
作者：佚名
典藏号：法国远东学院摄影档案馆
CAM08698ⓒEFEO

茶胶寺庙山神道南侧之发掘情况
Chaussée IV/est, moitié sud
拍摄日期：1923 年
作者：佚名
典藏号：法国远东学院摄影档案馆 CAM08713ⓒEFEO

附录Ⅱ 茶胶寺庙山历史照片

茶胶寺庙山二层台
2e étage
拍摄日期：不详
作者：佚名
典藏号：法国远东学院摄影档案馆 CAM08723ⓒEFEO

茶胶寺庙山西侧两座塔门俯瞰
3e étage, pavillon d'entrée (gopura) vue prise du massif central
拍摄日期：不详
作者：佚名
典藏号：法国远东学院摄影档案馆 CAM08725ⓒEFEO

附录 Ⅱ　茶胶寺庙山历史照片

茶胶寺庙山东侧全景
Face est (vue d'ensemble)
拍摄日期：不详
作者：佚名
典藏号：法国远东学院摄影档案馆 CAM08757ⓒEFEO

茶胶寺庙山神道发掘情况
Chaussée IV/est, en cours de dégagement (vue vers l'ouest).
拍摄日期：1922 年
作者：佚名
典藏号：法国远东学院摄影档案馆 CAM08780ⓒEFEO

附录 Ⅱ 茶胶寺庙山历史照片

茶胶寺庙山西侧全景
Face ouest du temple
拍摄日期：1962 年
作者：佚名
典藏号：法国远东学院摄影档案馆 CAM08784ⓒEFEO

茶胶寺庙山西侧阇耶跋摩七世时期的医院遗址
Tour recouverte de végétation, située à l'ouest de la chapelle d'hôpital
拍摄日期：不详
作者：佚名
典藏号：法国远东学院摄影档案馆 CAM08761ⓒEFEO

附录III

吴哥时代国王世系及其主要建筑遗迹

	在位时间 Reign Period	君主 King	都城 Capital	风格 Style	主要建筑遗迹 Pricinpal monuments
800	802-850	阇耶跋摩二世 Jayavarman II	因陀罗补罗,摩诃因陀罗山, 阿摩罗因陀罗补罗,诃里诃罗洛耶 Indrapura, Mahendraparvata, Amarrendrapura, Hariharalaya	古伦山 Phnom Kulen	
	853-877	阇耶跋摩三世 Jayavarman III			巴空寺第一阶段 1st stage of Bakong
	877-889	因陀罗跋摩一世 Indravarman I	诃里诃罗洛耶 Hariharalaya	神牛寺 Preah Ko	神牛寺 Preah Ko 巴空寺第二阶段 2nd stage of Bakong 罗莱池 Indratataka
	889-900	耶输跋摩一世 Yasovarman I	耶输陀罗补罗 Yasodharapura		巴肯寺 Bakheng 克罗姆寺 Phnom Krom 柏克寺 Phnom Bok 东池 Yaçodharatataka
900	900-921	曷利沙跋摩一世 Harshavarman I		巴肯寺 Bakheng	巴塞增空寺 Baksei Chamkrong
	921-941	阇耶跋摩四世 Jayavarman IV	乔格吉厄 Chok Gargyar	贡开 Koh Ker	中央大塔 Prasat Thom
	944-968	罗贞陀罗跋摩二世 Rajendravarman II	耶输陀罗补罗 Yasodharapura	比粒寺 Pre Rup	东梅奔 Eastern Baray 比粒寺 Pre Rup
	968-1001	阇耶跋摩五世 Jayavarman V		女王宫 Banteay Srei	女王宫 Banteay Srei
1000	1001-1002	优陀耶迭多跋摩一世 Udayadityavarman I	耶输陀罗补罗 Yasodharapura		北仓 North Khleang 南仓 South Khleang 空中宫殿 Phimeanakas 茶胶寺 Ta Keo
	1002-1010	阇耶毗罗跋摩 Javaviravarman			
	1002-1050	苏利耶跋摩一世 Suryavarman I			
	1050-1066	优陀耶迭多跋摩二世 Udayadityavarman II		巴方寺 Bphoun	柏威夏 Preah Vihear 艾克寺 Vat Ek 巴塞寺 Vat Baset 巴方寺 Bphoun 西梅奔 Western Baray
	1066-1080	曷利沙跋摩三世 Harshavarman III			
	1080-1107	阇耶跋摩六世 Jayavarman VI			
1100	1107-1112	陀罗尼因陀罗跋摩一世 Dharanindravarman I	耶输陀罗补罗 Yasodharapura	吴哥窟 Angor Wat	瓦普寺 Vat Phu 菜山寺 Phnom Rung 披迈 Phimai 吴哥窟 Angkor Wat 周萨神庙 Chausay Tevoda 托玛侬神庙 Thommanon 崩密列寺 Beng Mealea 班蒂萨穆雷 Banteay Samre
	1113-1150	苏利耶跋摩二世 Suryavarman II			
	1150-?	陀罗尼因陀罗跋摩二世 Dharanindravarman II			达布隆寺 Ta Prohm 圣剑寺 Preah Khan 龙蟠水池 Neak Pean 班蒂克黛寺 Banteay Kedi 班蒂通寺 Banteay Thom 达内寺 Ta Nei
	-1160	耶输跋摩二世 Yasovarman II			
	1165-1177	特利布婆那帝跋摩 Tribhvanadityavarman			
	1181-1217	阇耶跋摩七世 Jayavarman VII	耶输陀罗补罗 Yasodharapura 吴哥通王城 Angkor Thom		
1200				巴戎寺 Bayon	班蒂布瑞寺 Banteay Prei 达逊寺 Ta Som 克鲁考寺 Krol Ko 布瑞寺 Prasat Prei 布瑞蒙蒂寺 Prasat Prei Monti 班蒂奇玛寺 Banteay Chhmar 巴南寺 Vat Banan 通王城城门 Gates of Angkor Thom 巴戎寺 Bayon 大象平台 Elephant Terrace 癞王台 Leppking Terrace
	1221-1243	因陀罗跋摩二世 Indrvarman II			
	1243-1295	阇耶跋摩八世 Jayavarman VIII			
	1295-1307	室利陀罗跋摩 Srindravarman		后巴戎寺 Post-Bayon	普拉比图寺 Preah Pitu 巴里莱寺 Preah Palilay
1300	1307-1327	室利陀罗阇耶跋摩 Srindraayavarman			
	1327-?	阇耶跋摩底波罗密首罗 Jayavarman Paramesvara			
	1353 吴哥被其他国家攻占				
	1357-?	苏利耶跋娑 Suryavarmsa			
	1393 吴哥被其他国家攻占				
1400	1431 吴哥被其他国家攻占				

实测图

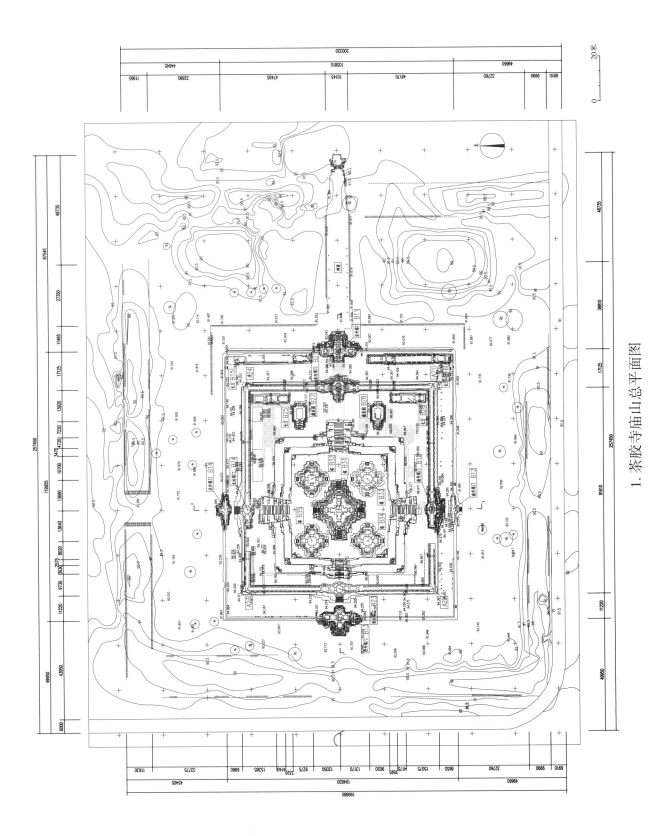

1. 茶胶寺庙山总平面图

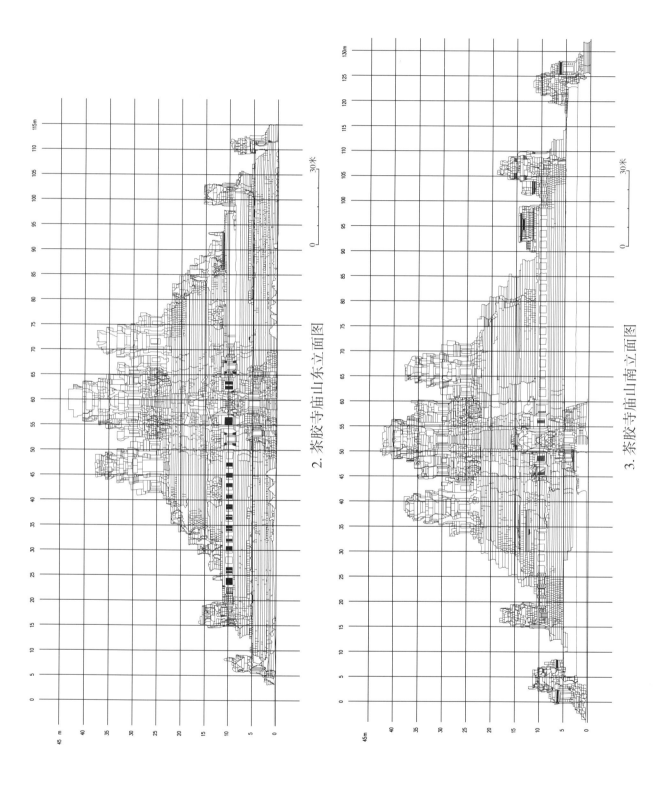

2. 茶胶寺庙山东立面图

3. 茶胶寺庙山南立面图

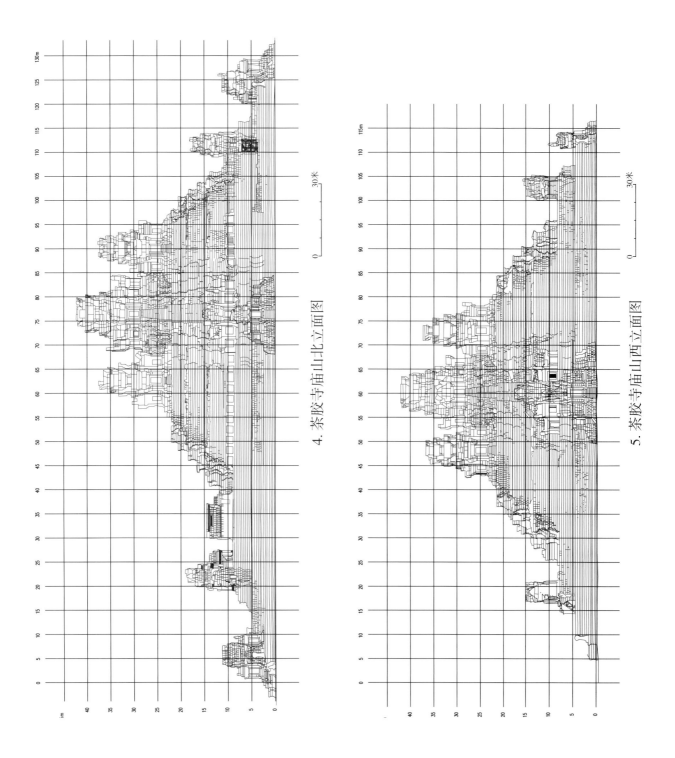

4. 茶胶寺庙山北立面图

5. 茶胶寺庙山西立面图

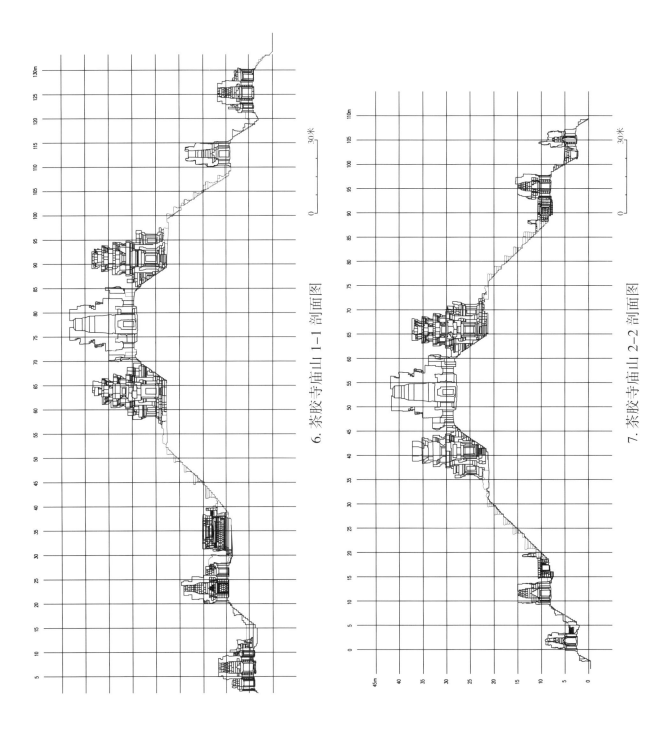

6. 茶胶寺庙山 1-1 剖面图

7. 茶胶寺庙山 2-2 剖面图

———— 实测图

8. 茶胶寺庙山散落构件分布示意图

9. 须弥台西南角平面图

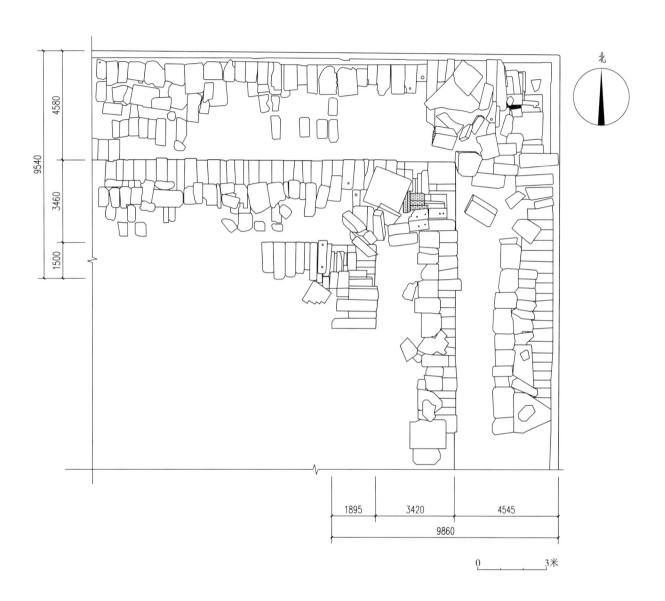

10. 须弥台东北角平面图

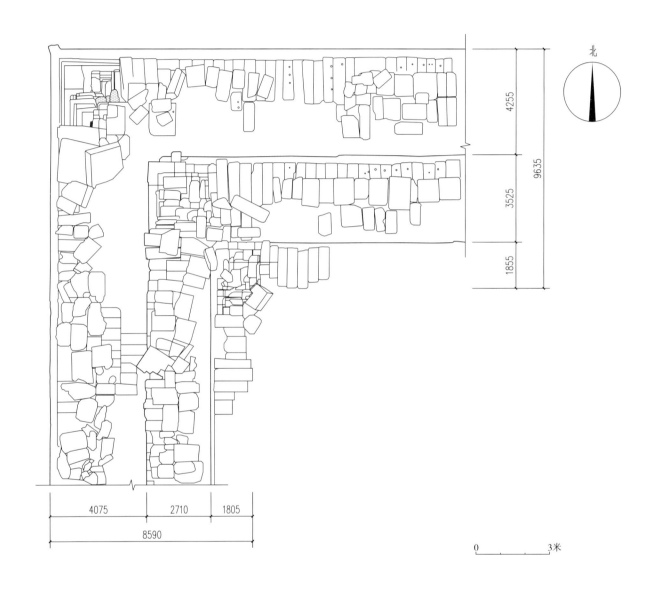

11. 须弥台西北角平面图

实测图

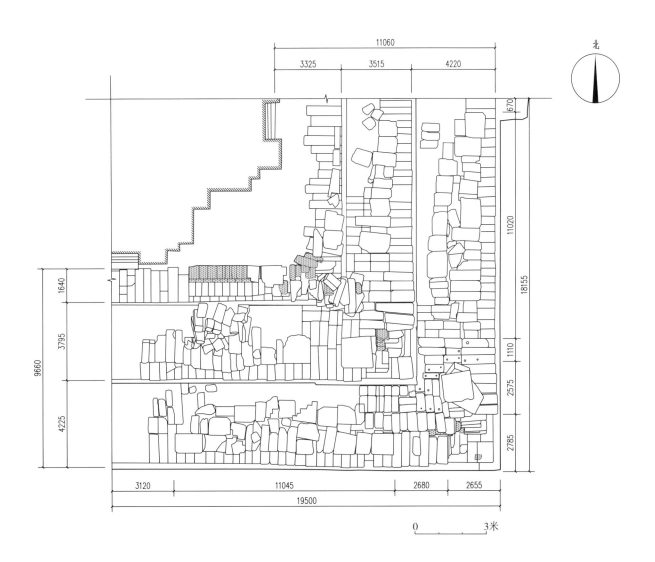

12. 须弥台东南角平面图

茶胶寺庙山建筑研究

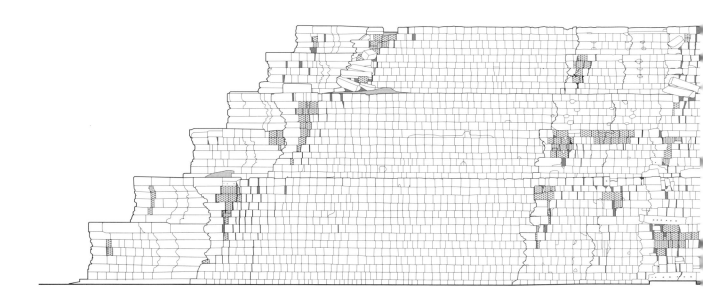

———— 实测图

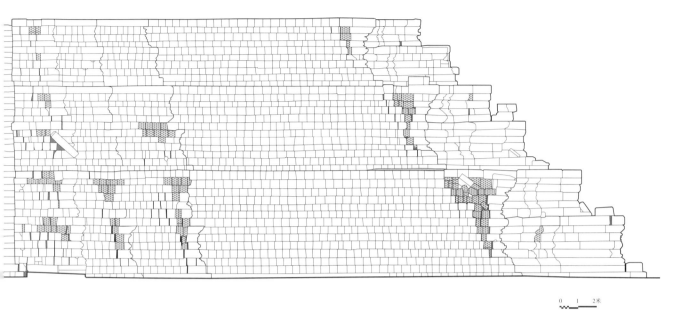

13. 须弥台东立面图

14. 须弥台东立面影像图

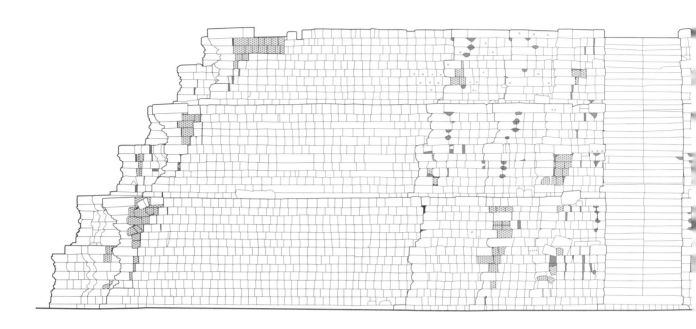

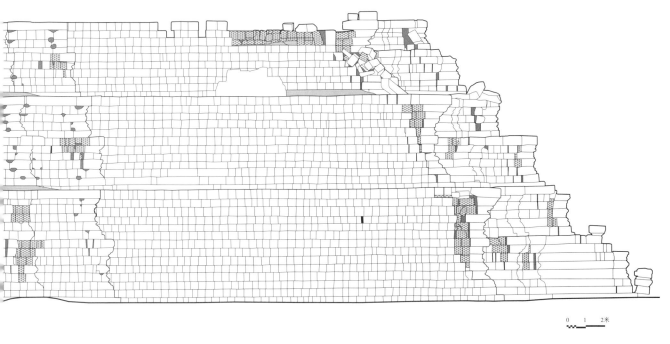

15. 须弥台南立面图

16. 须弥台南立面影像图

茶胶寺庙山建筑研究

———— 实测图

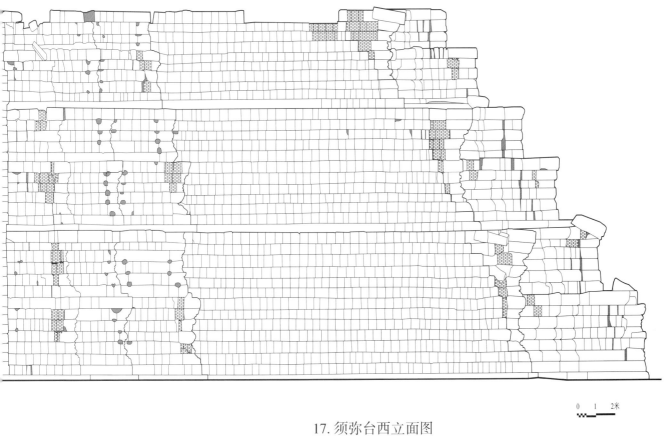

17. 须弥台西立面图

18. 须弥台西立面影像图

茶胶寺庙山建筑研究

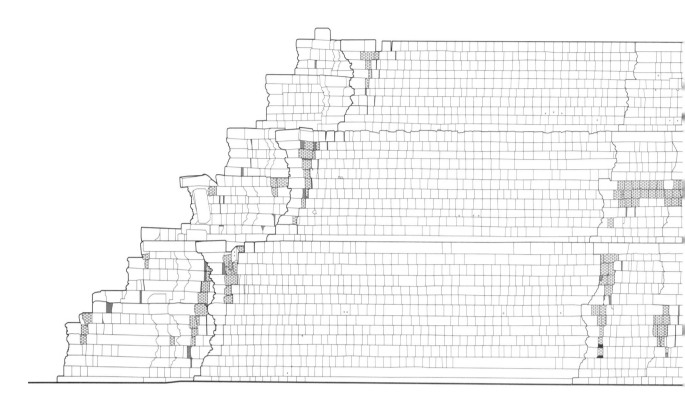

———— 实测图

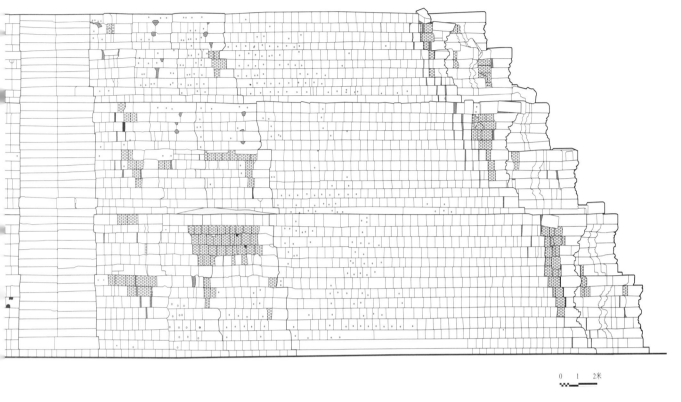

0 1 2米

19. 须弥台北立面图

20. 须弥台北立面影像图

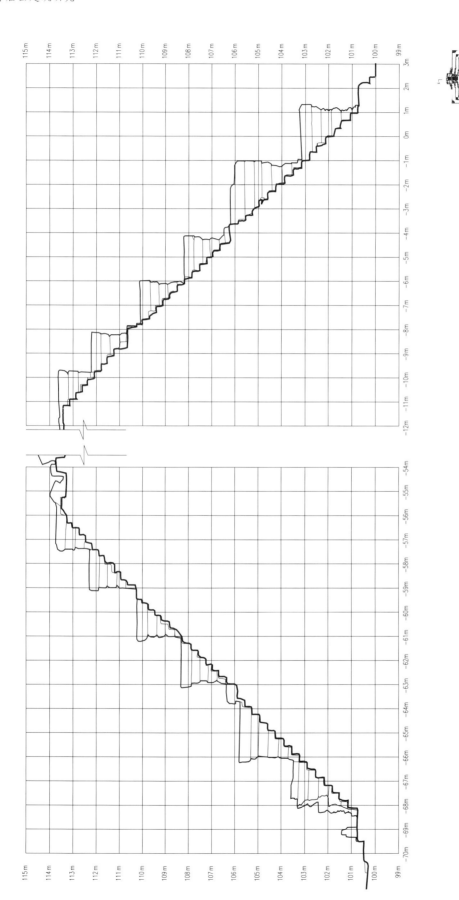

21. 须弥台 1-1 剖面图

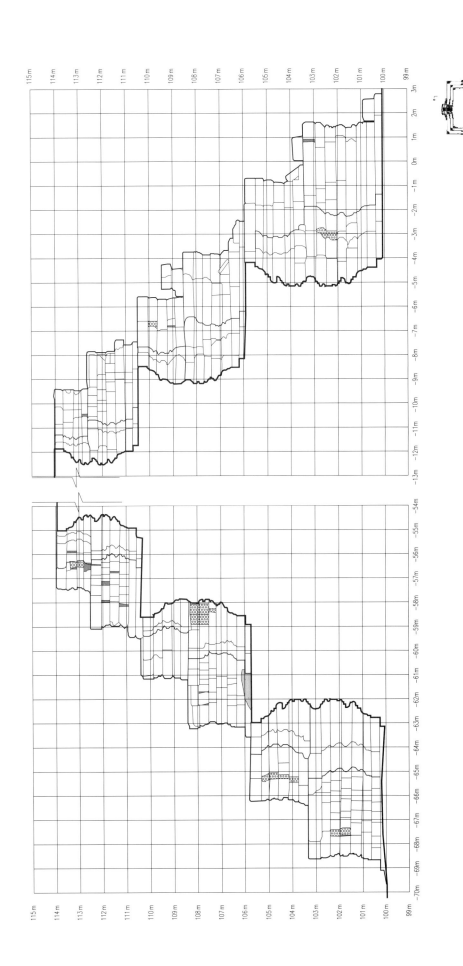

22. 须弥台 2-2 剖面图

茶胶寺庙山建筑研究

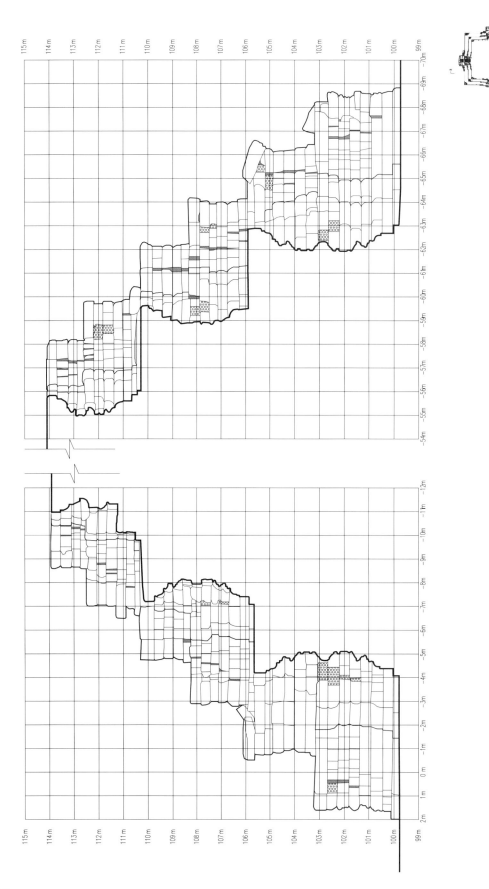

23. 须弥台 3-3 剖面图

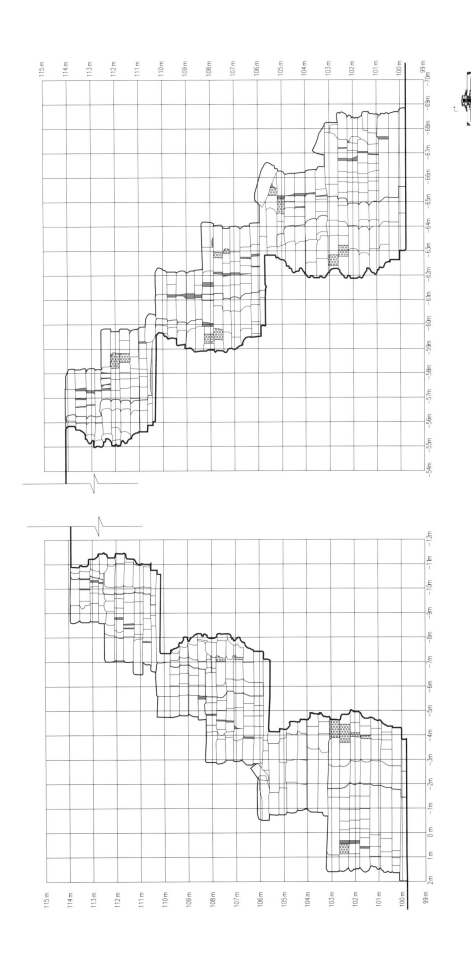

24. 须弥台 4-4 剖面图

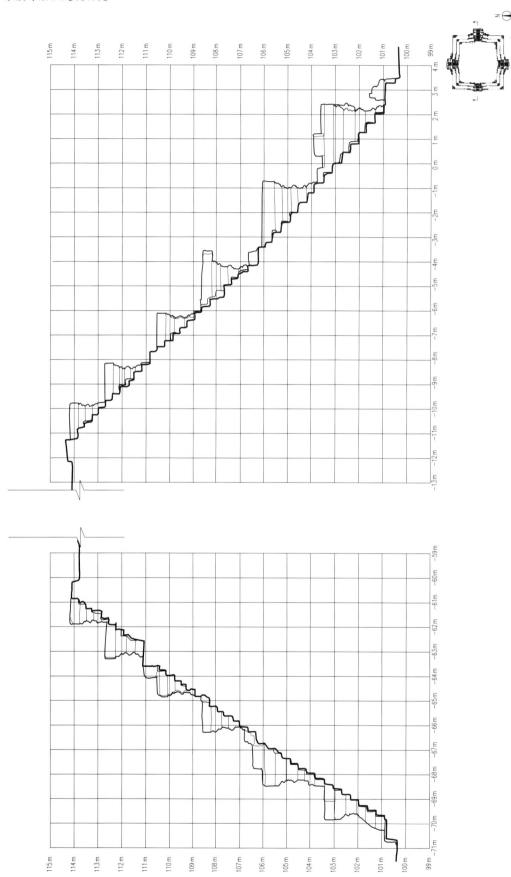

25. 须弥台 5-5 剖面图

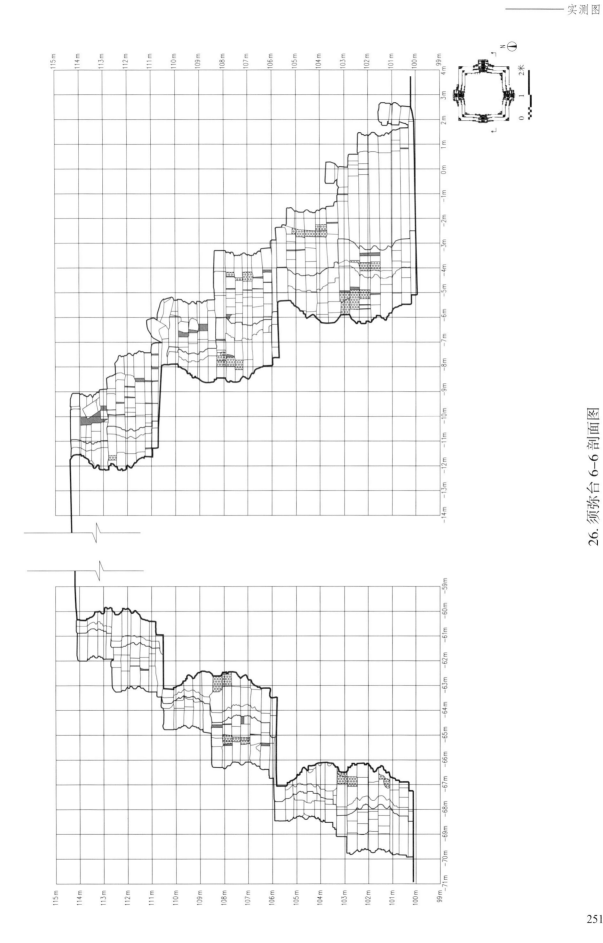

26. 须弥台 6-6 剖面图

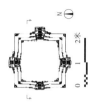
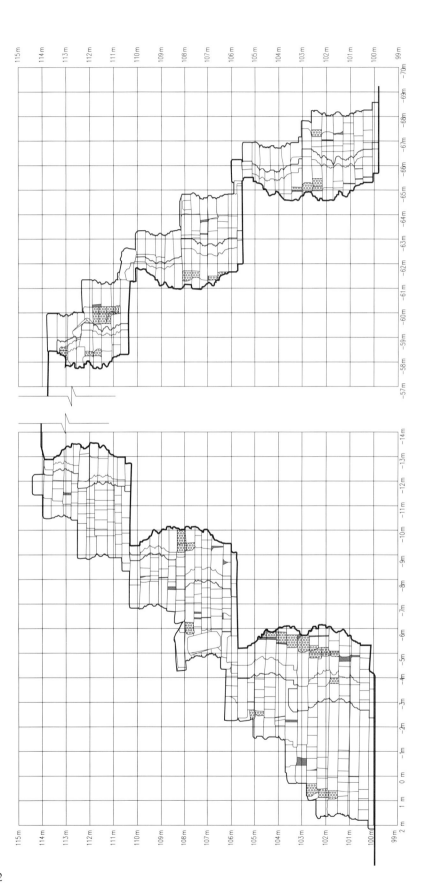

27. 须弥台 7-7 剖面图

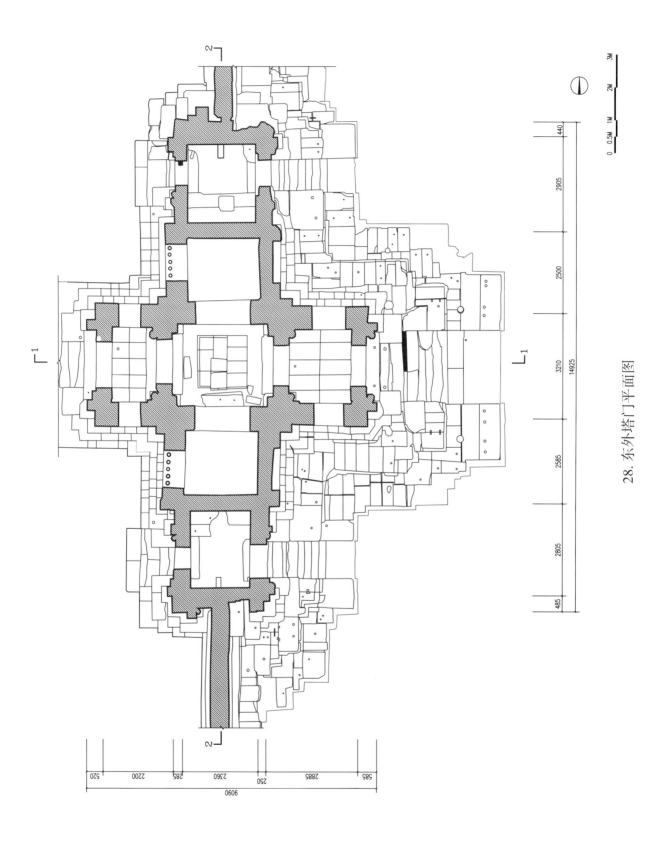

28. 东外塔门平面图

——— 实测图

茶胶寺庙山建筑研究

29. 东外塔门东立面图

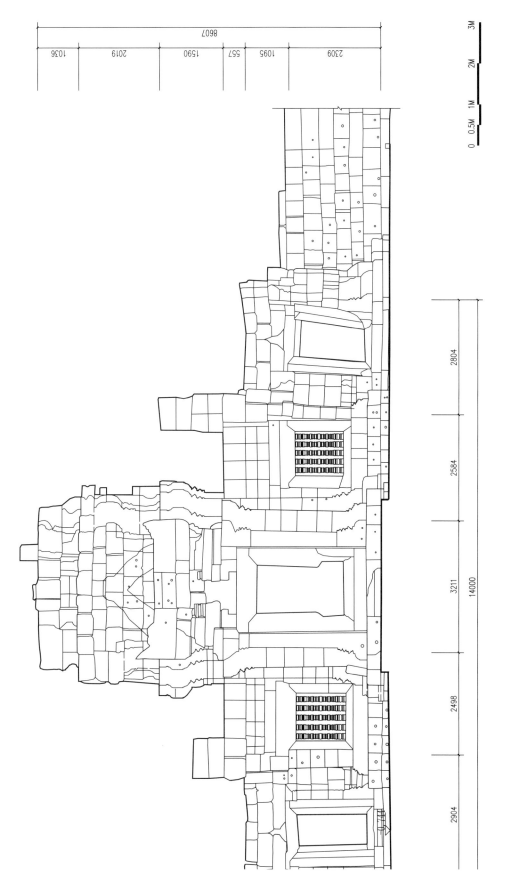

30. 东外塔门西立面图

31. 东外塔门南立面图

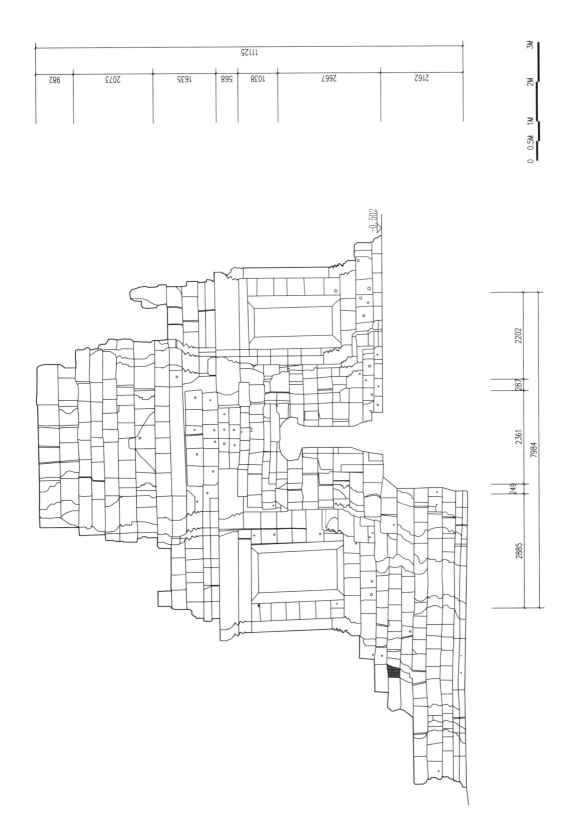

32. 东外塔门北立面图

——— 实测图

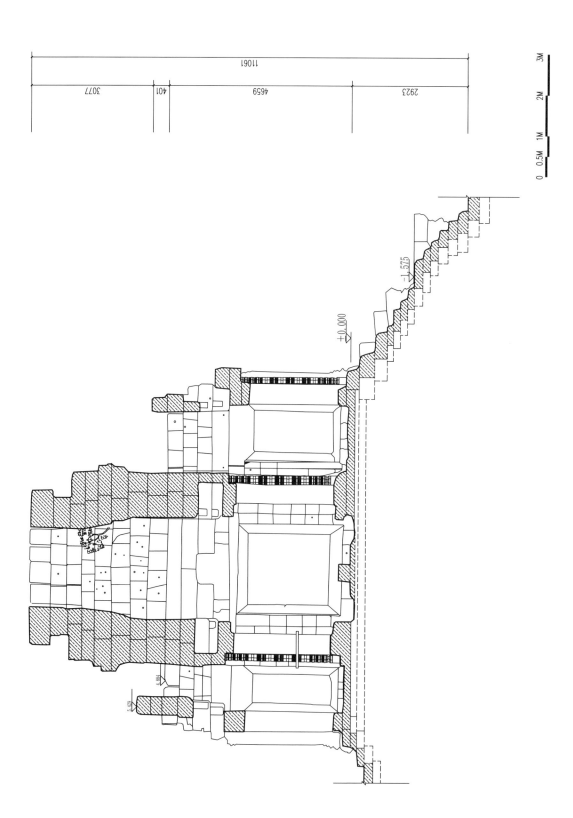

33. 东外塔门 1-1 剖面图

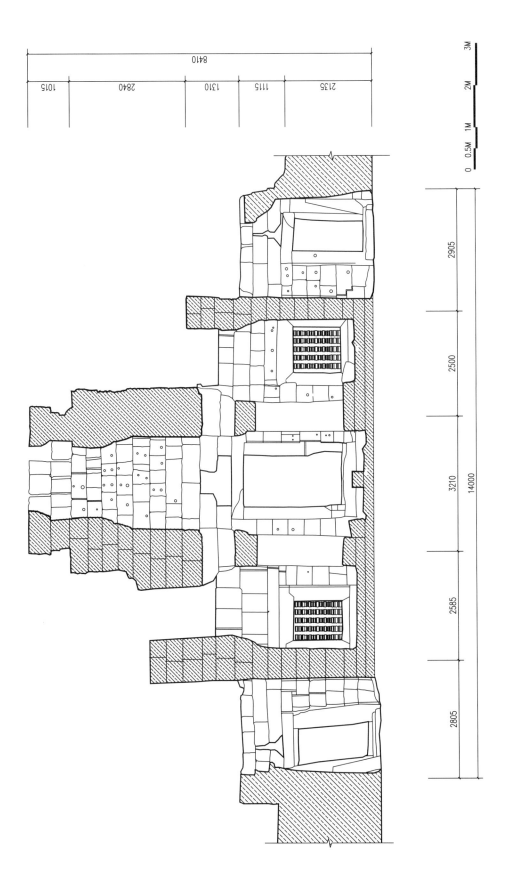

34. 东外塔门 2-2 剖面图

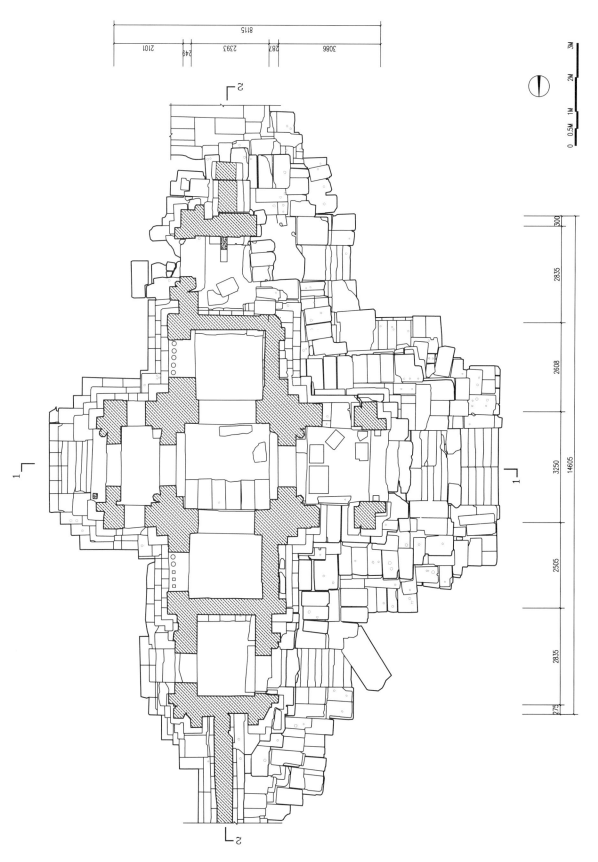

35. 西外塔门平面图

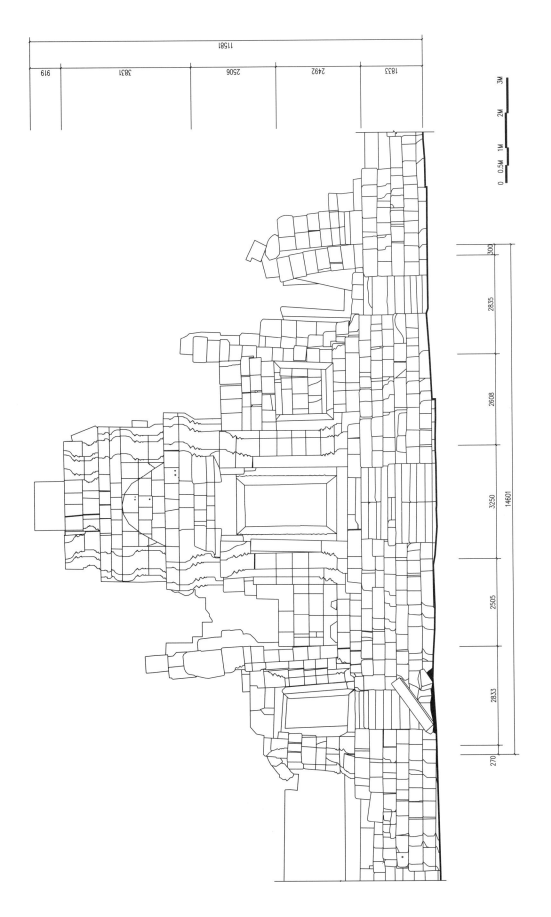

36. 西外塔门西立面图

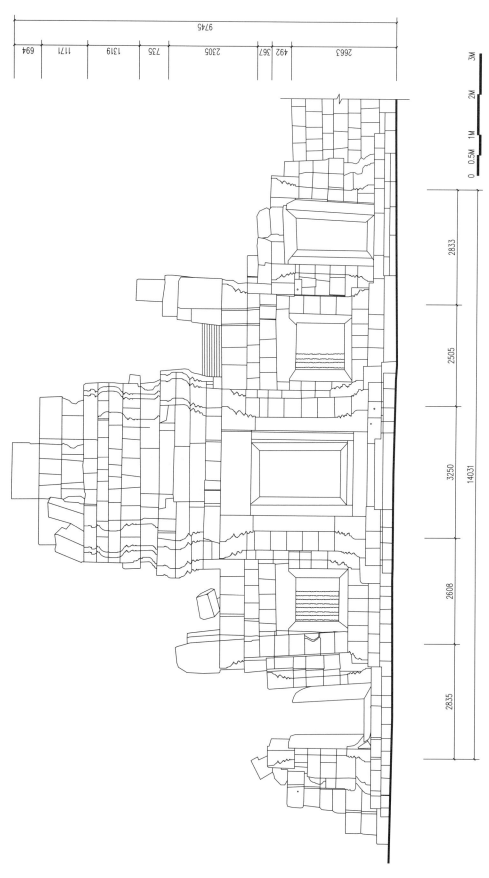

37. 西外塔门东立面图

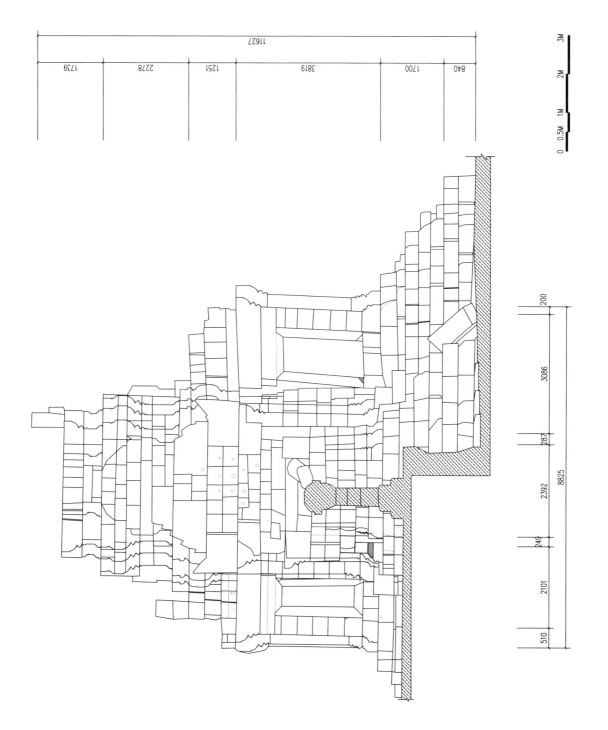

38. 西外塔门北立面图

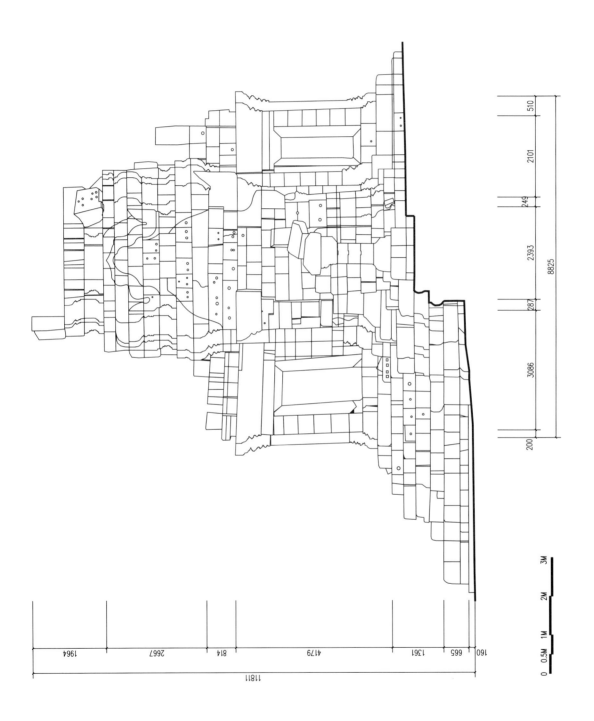

39. 西外塔门南立面图

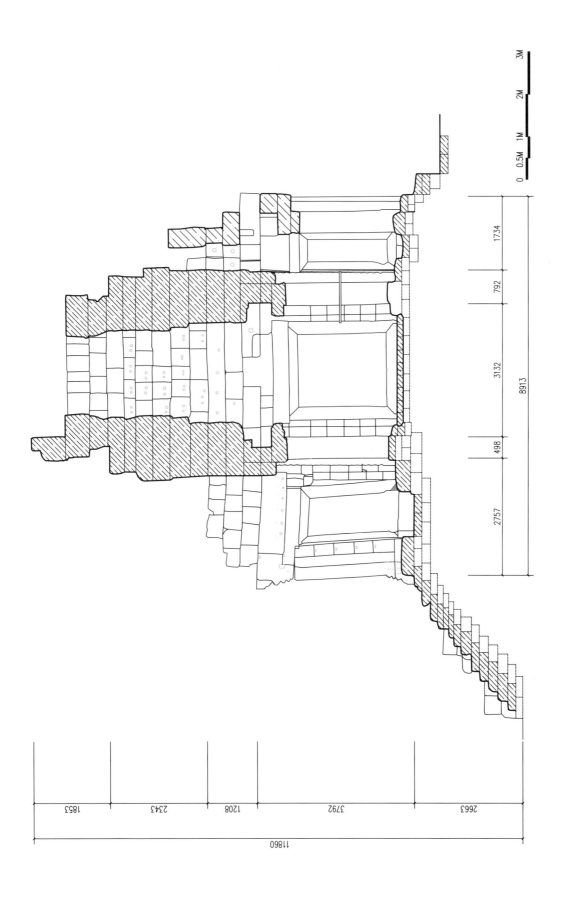

40. 西外塔门 1-1 剖面图

茶胶寺庙山建筑研究

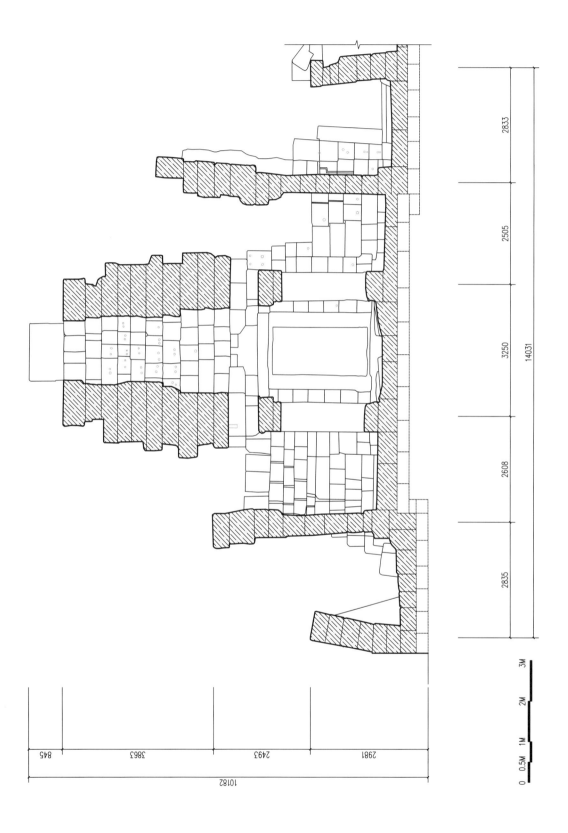

41. 西外塔门 2-2 剖面图

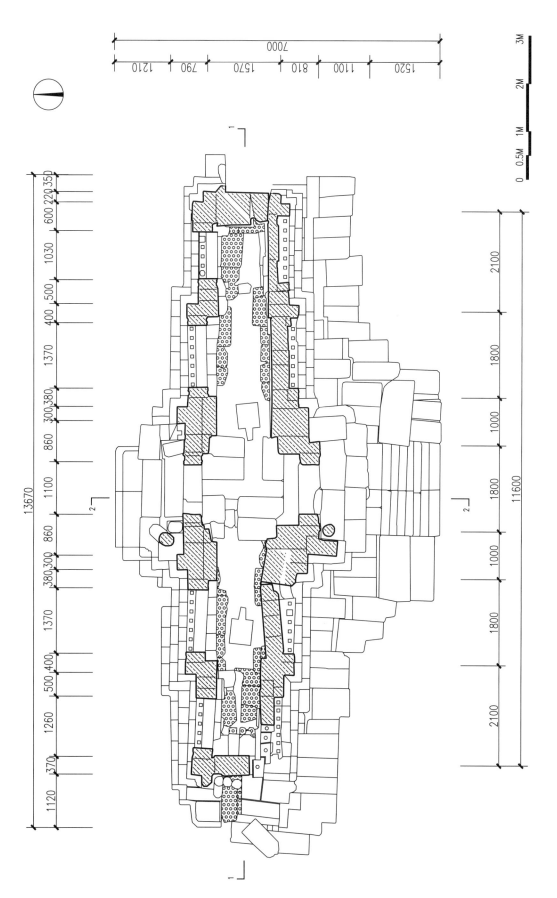

42. 南外塔门平面图

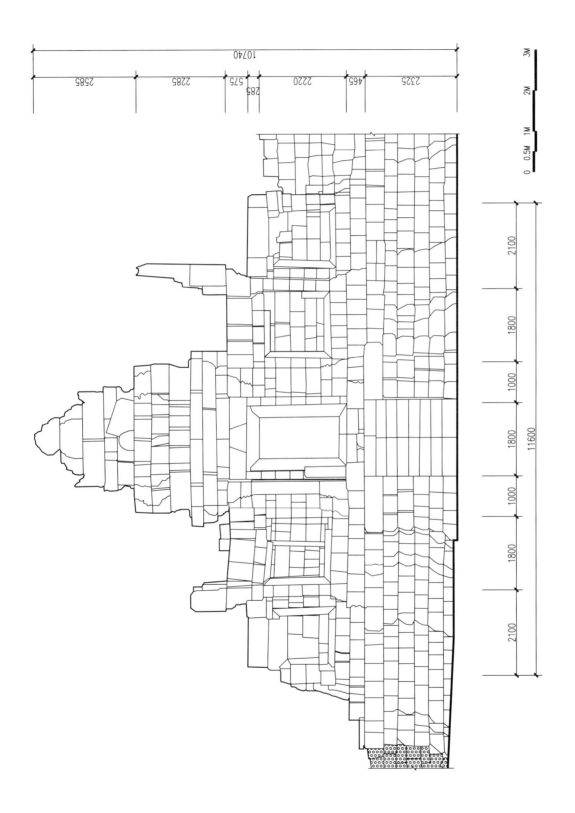

43. 南外塔门南立面图

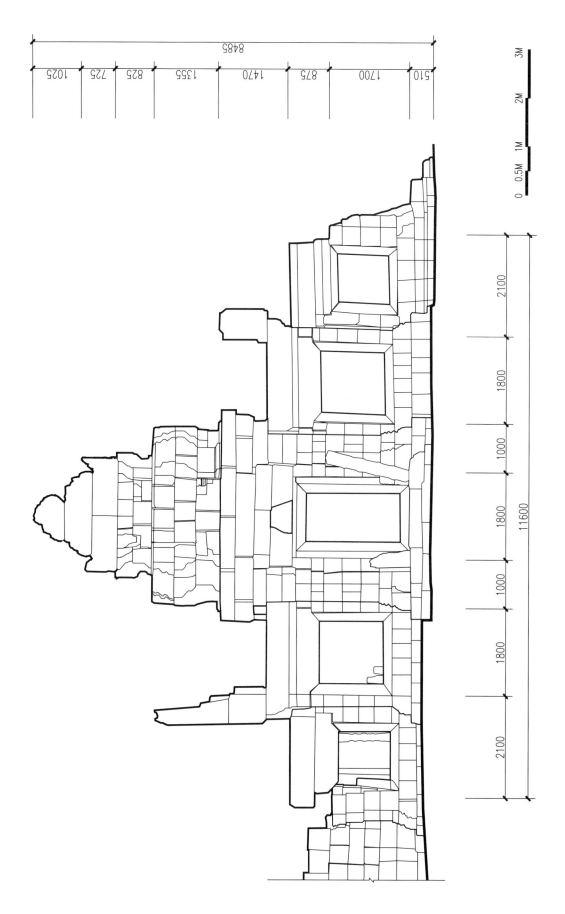

44. 南外塔门北立面图

——— 实测图

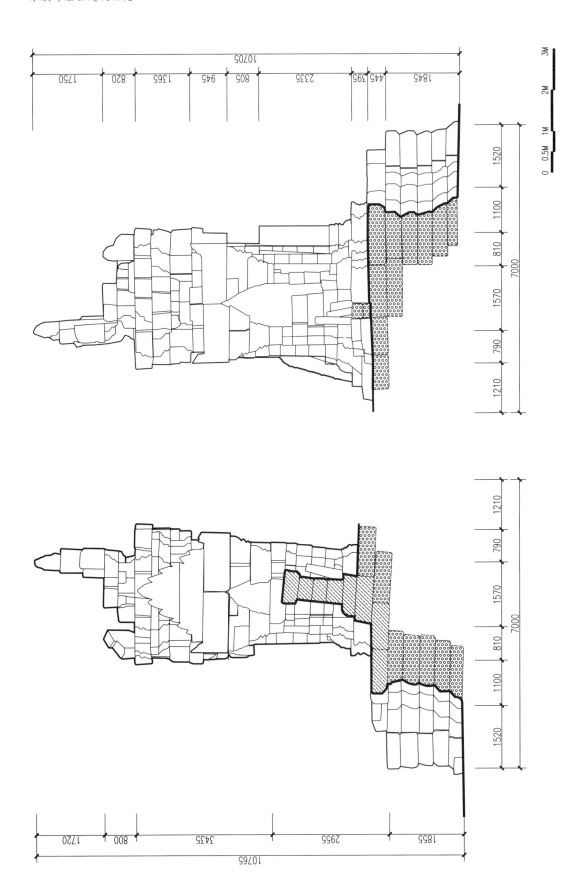

45. 南外塔门东立面图

46. 南外塔门西立面图

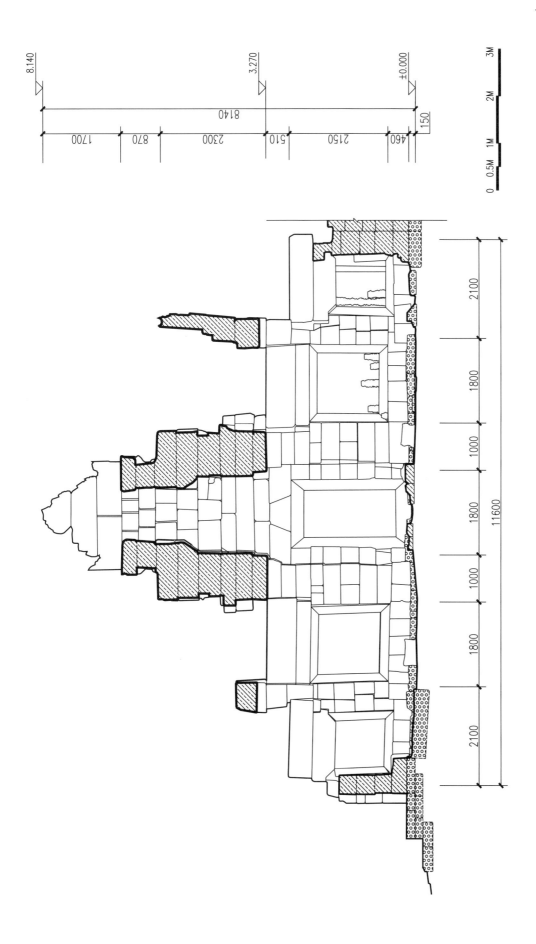

47. 南外塔门 1-1 剖面图

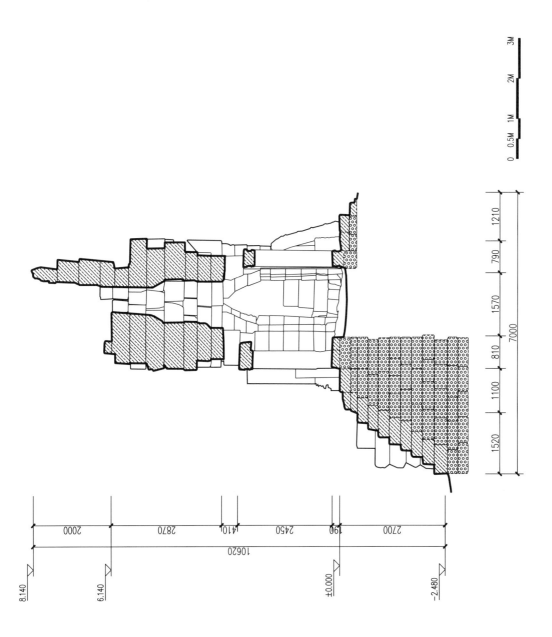

48. 南外塔门 2-2 剖面图

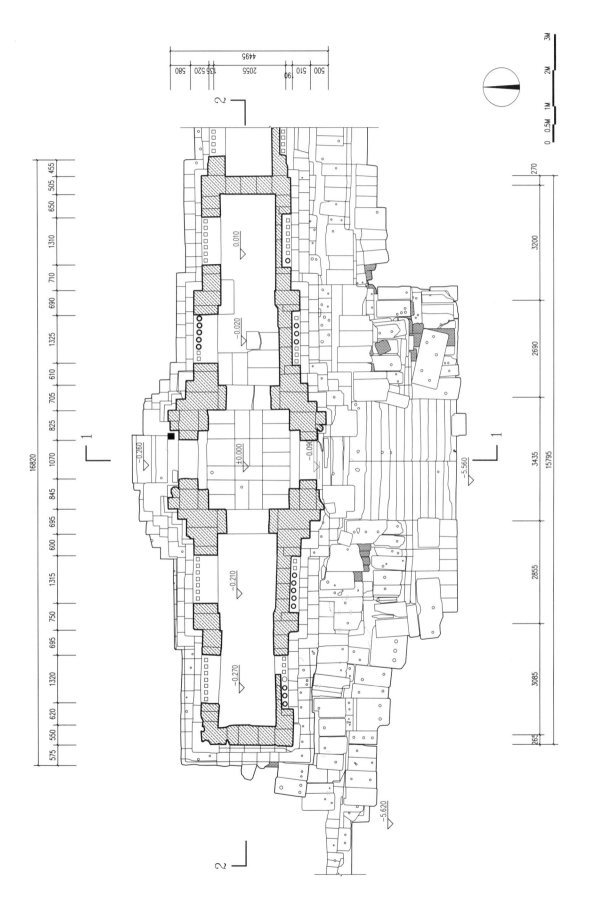

49. 南内塔门平面图

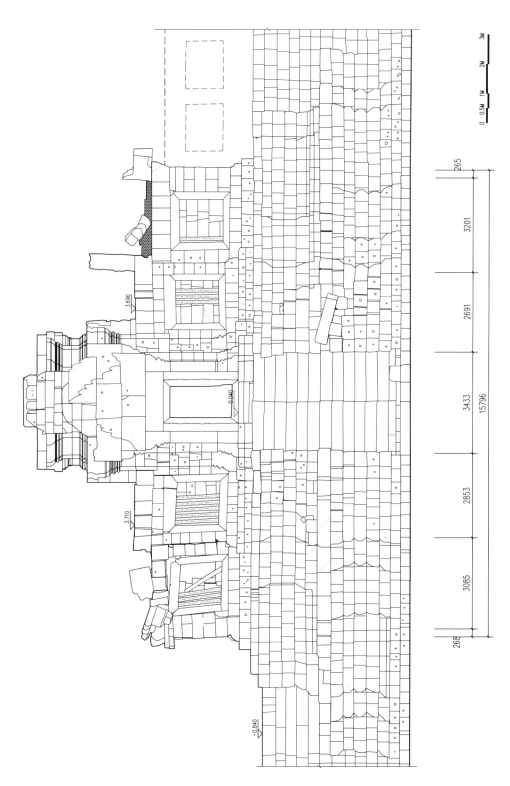

50. 南内塔门南立面图

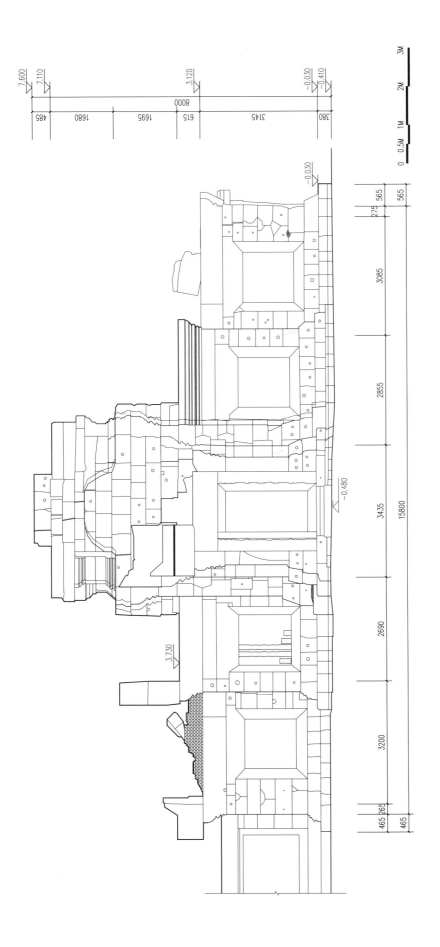

51. 南内塔门北立面图

275

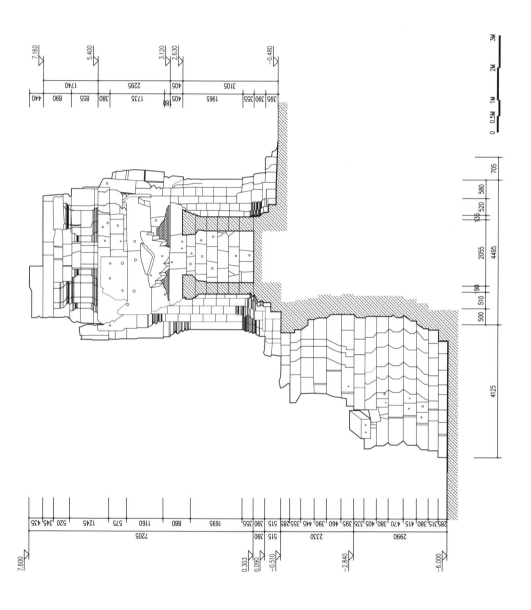

52. 南内塔门东立面图

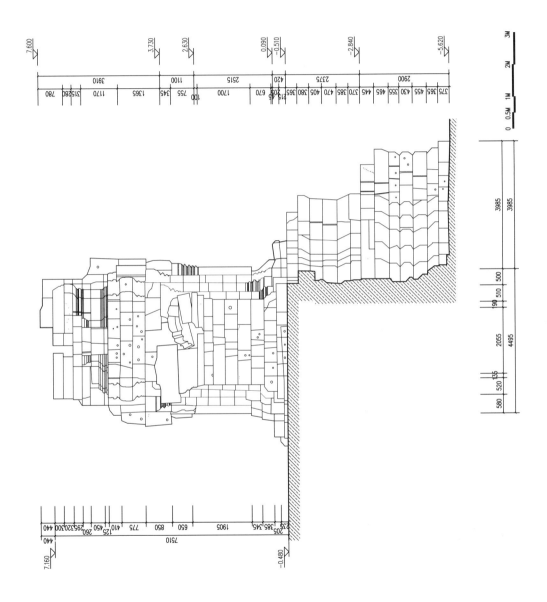

53. 南内塔门西立面图

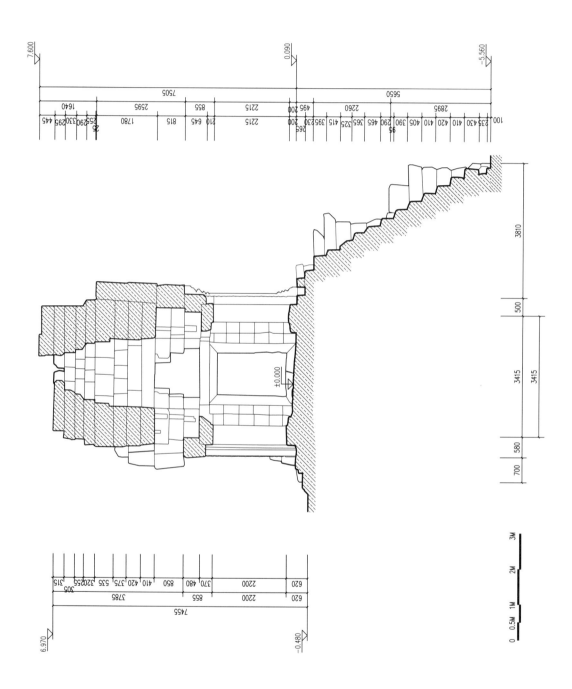

54. 南内塔门 1-1 剖面图

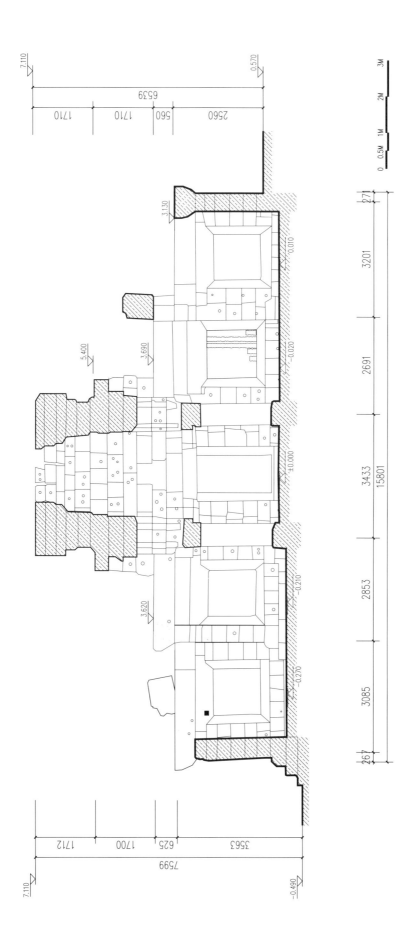

55. 南内塔门 2-2 剖面图

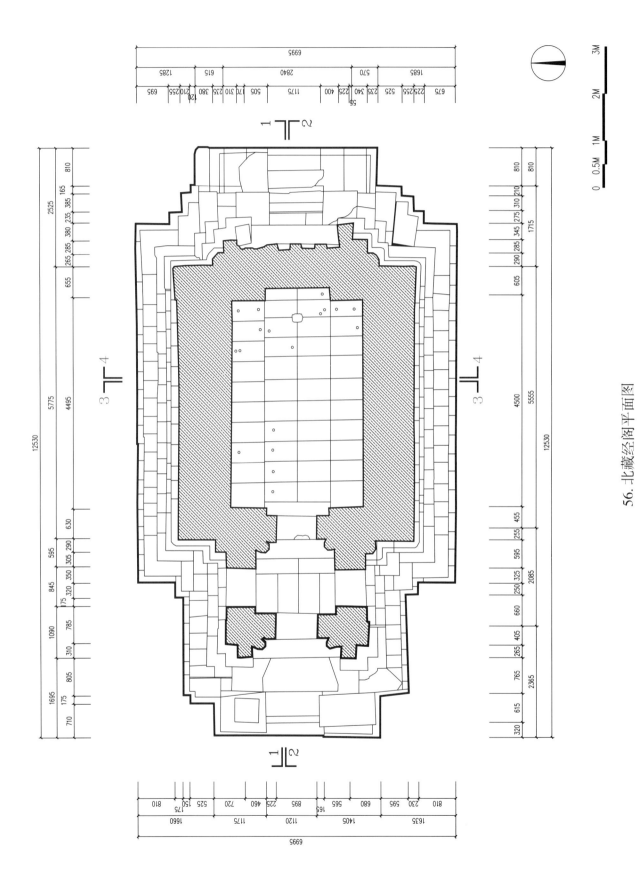

56. 北藏经阁平面图

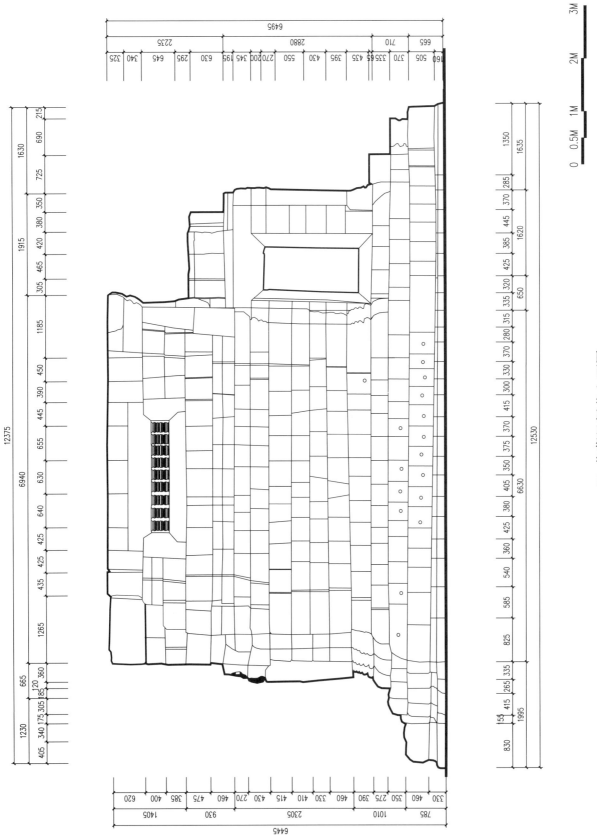

57. 北藏经阁北立面图

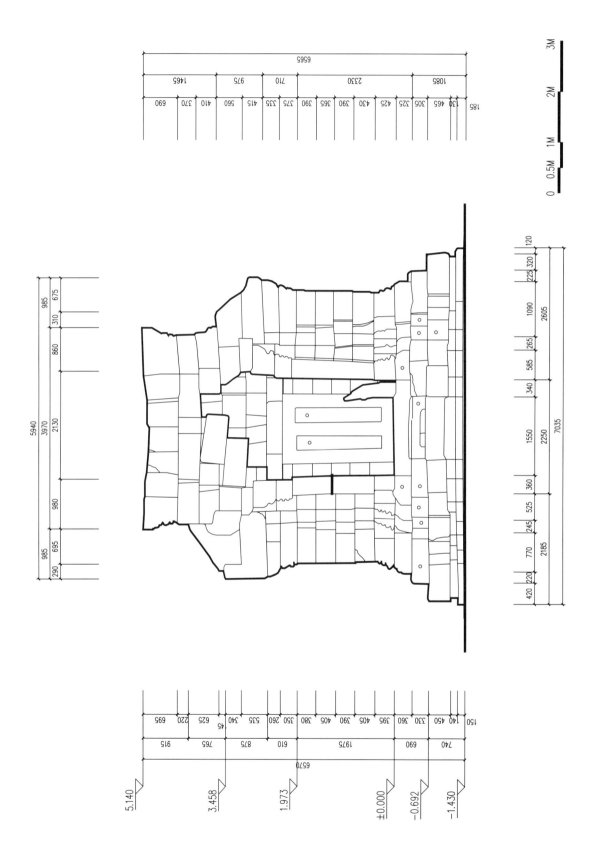

58. 北藏经阁东立面图

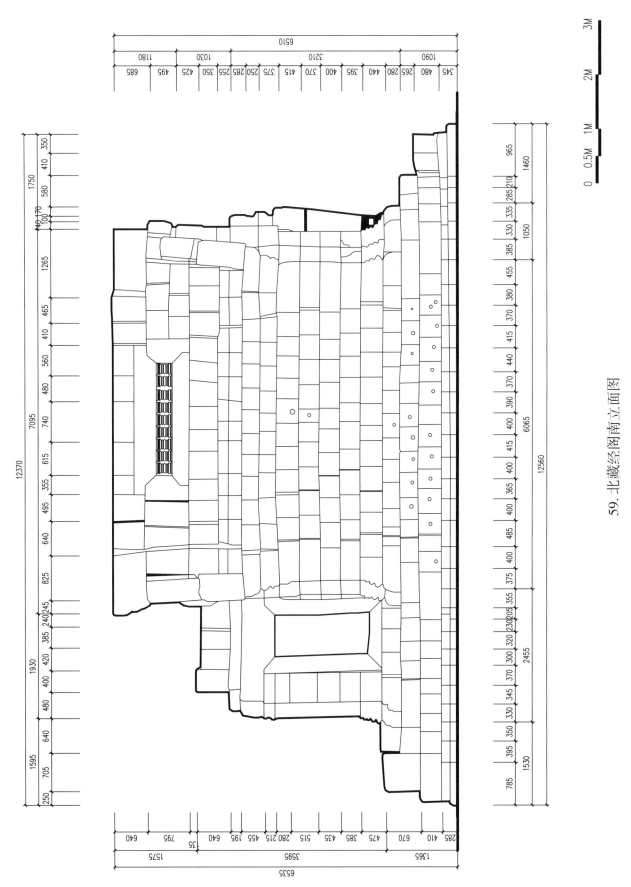

59. 北藏经阁南立面图

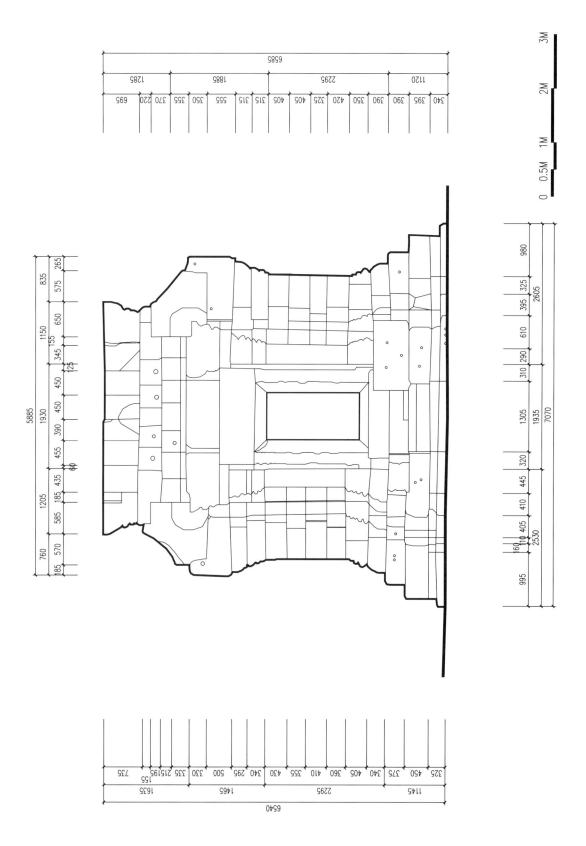

60. 北藏经阁西立面图

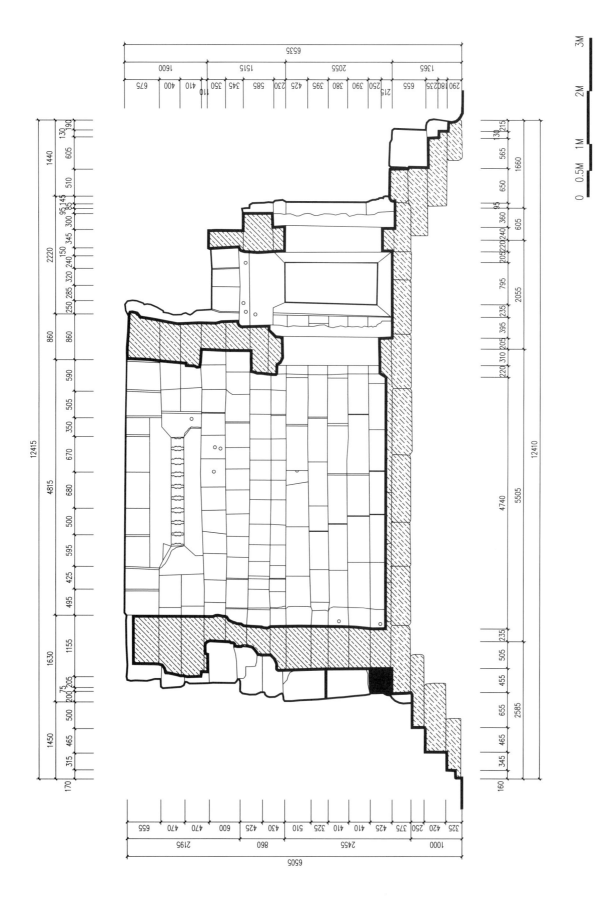

61. 北藏经阁 1-1 剖面图

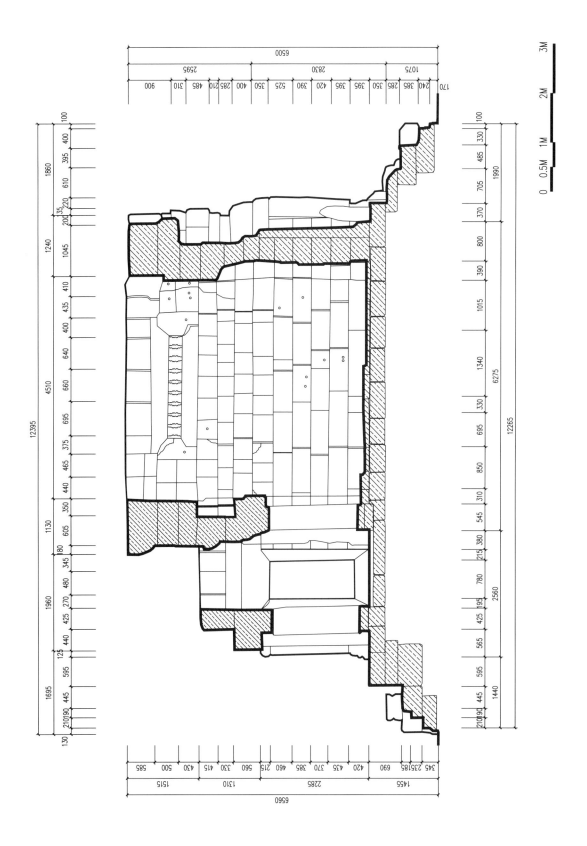

62. 北藏经阁 2-2 剖面图

———— 实测图

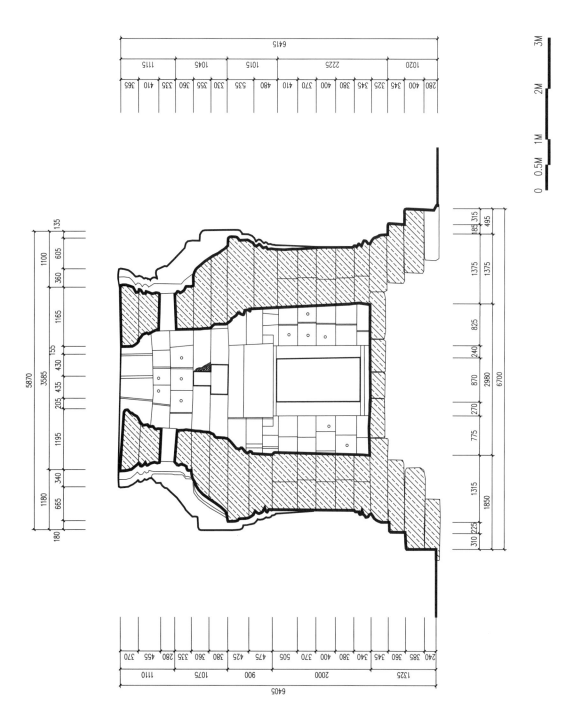

63. 北藏经阁 3-3 剖面图

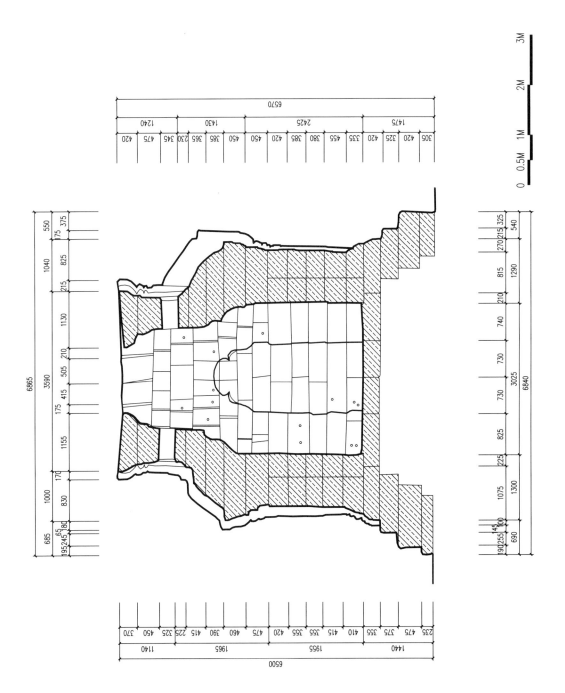

64. 北藏经阁 4-4 剖面图

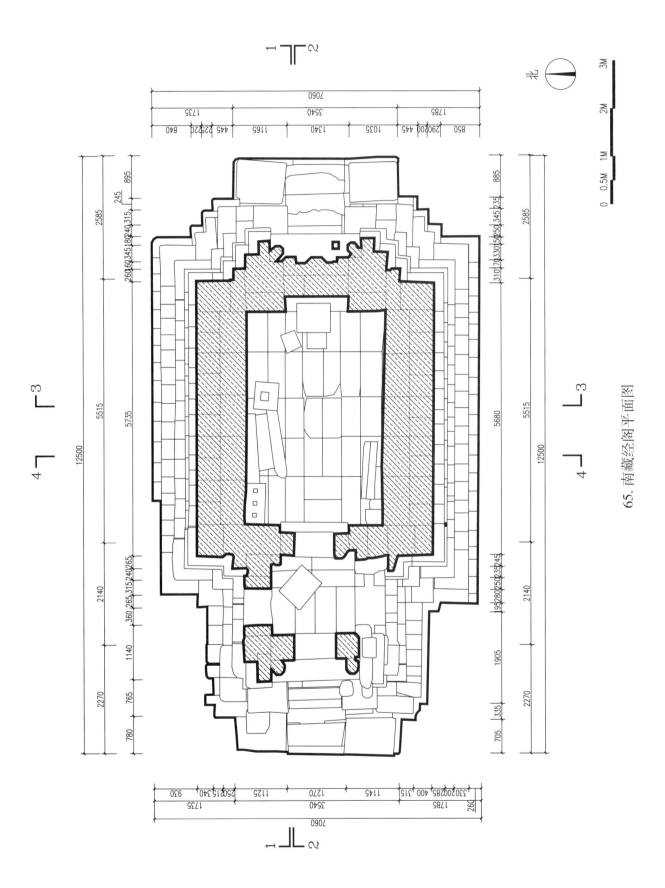

65. 南藏经阁平面图 ——实测图

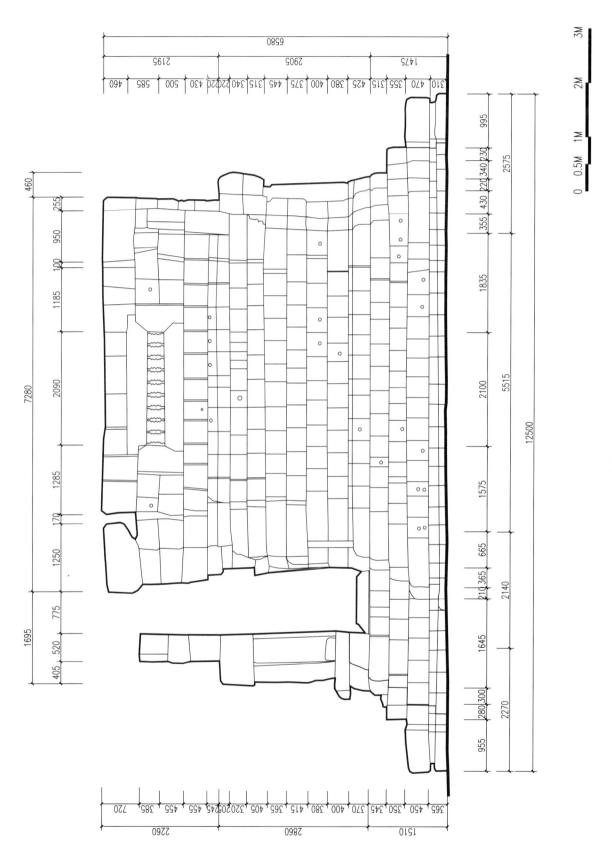

66. 南藏经阁南立面图

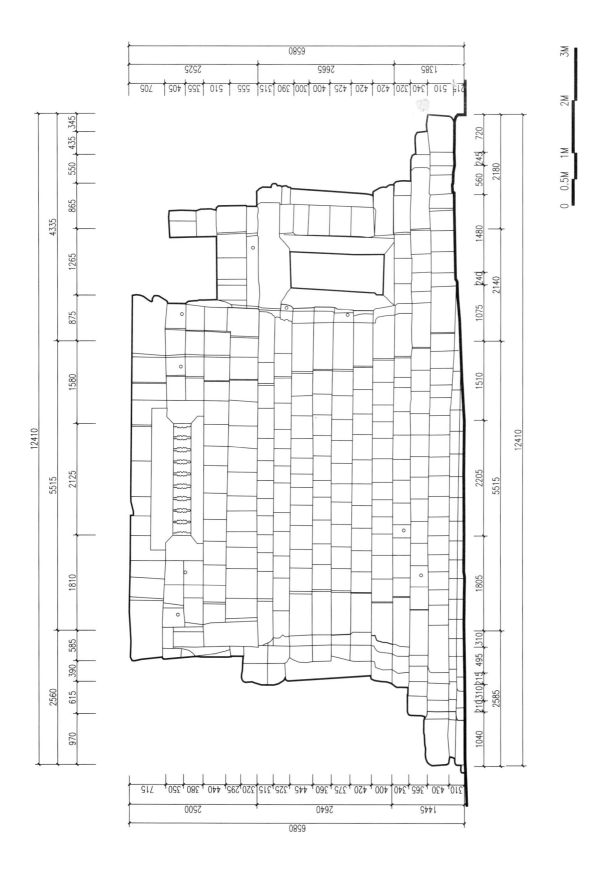

67. 南藏经阁北立面图

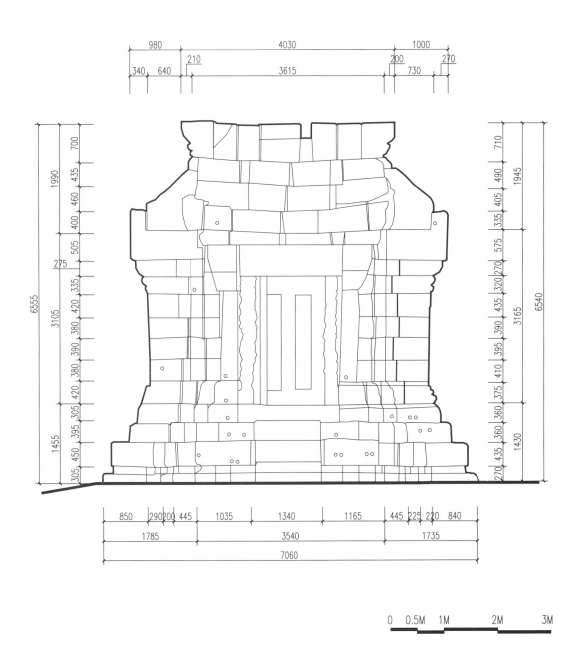

68. 南藏经阁东立面图

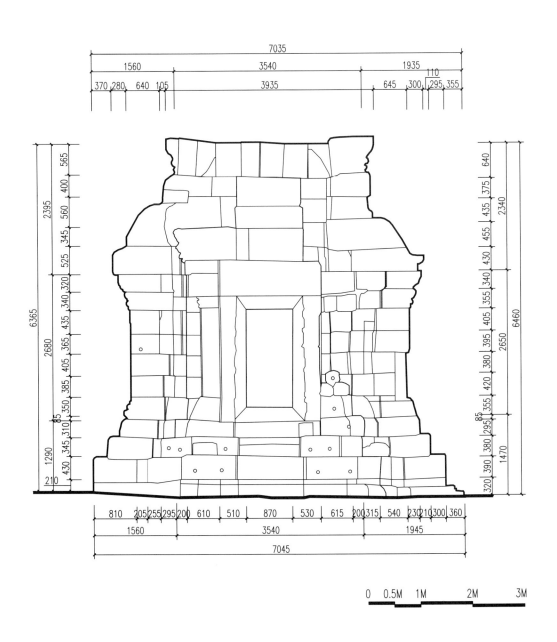

69. 南藏经阁西立面图

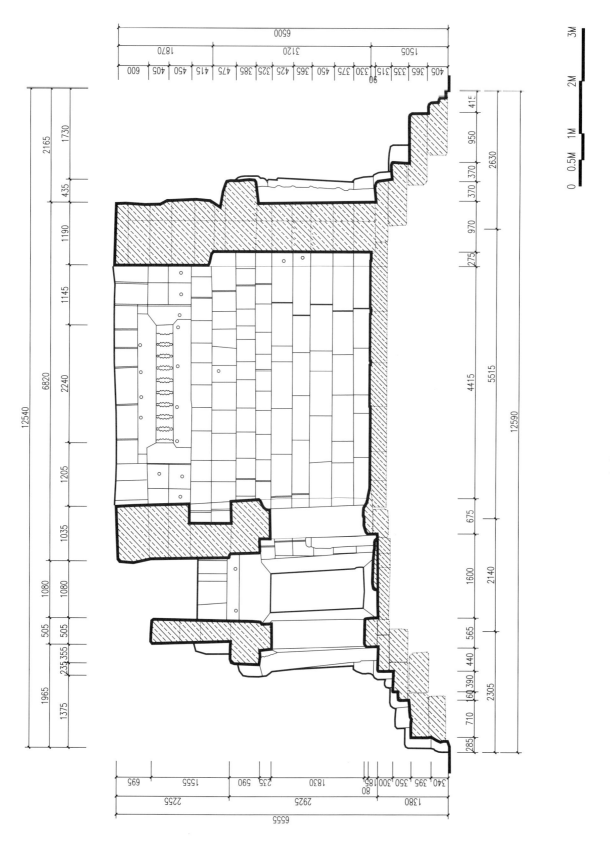

70. 南藏经阁 1-1 剖面图

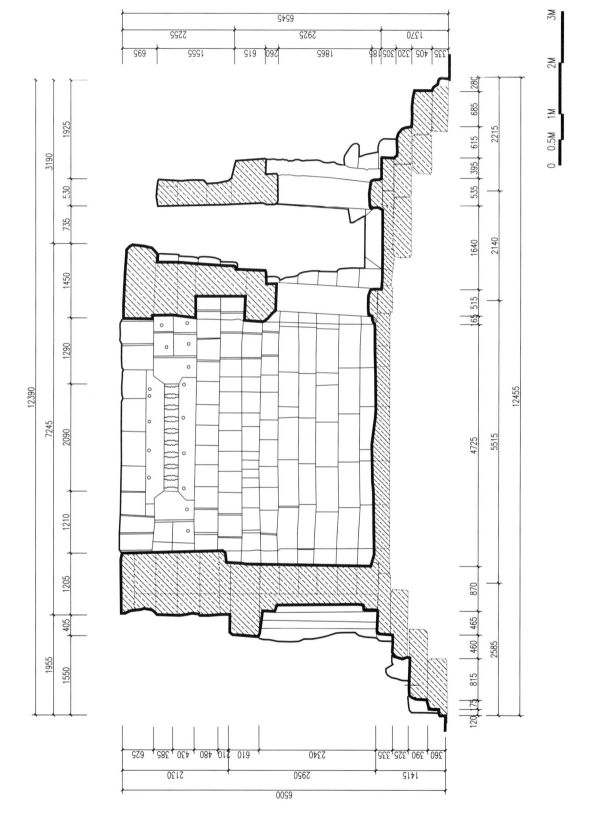

71. 南藏经阁 2-2 剖面图

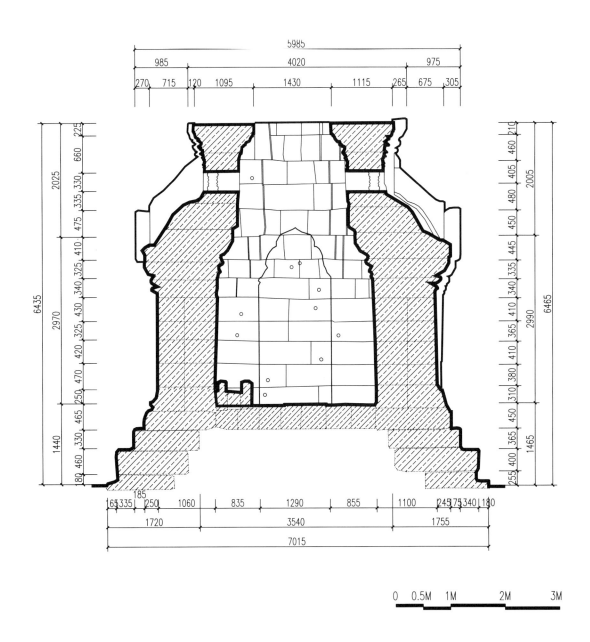

72. 南藏经阁 3-3 剖面图

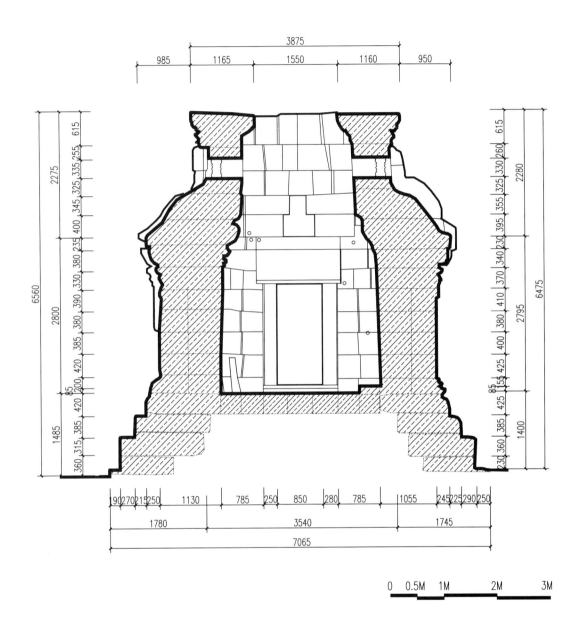

73. 南藏经阁 4-4 剖面图

茶胶寺庙山建筑研究

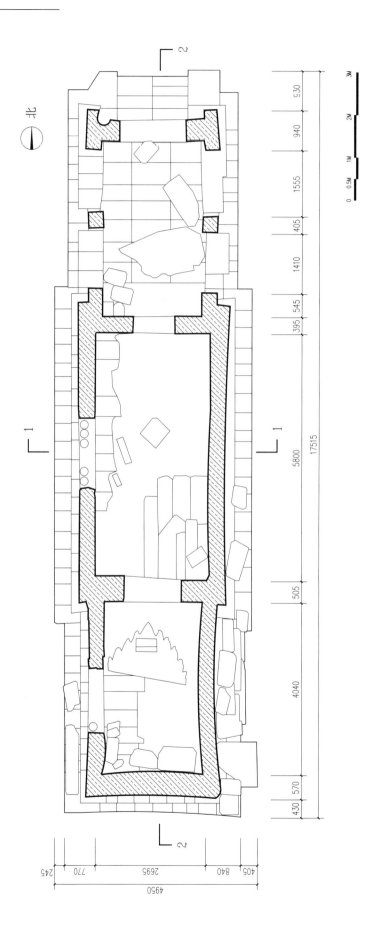

74. 南内长厅平面图

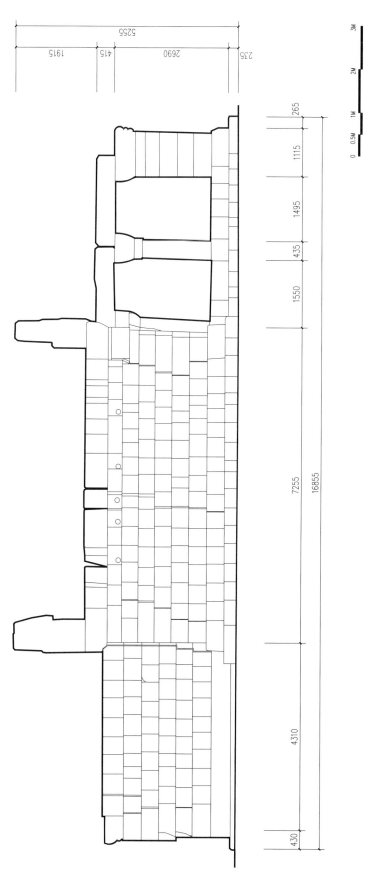

75. 南内长厅东立面图

299

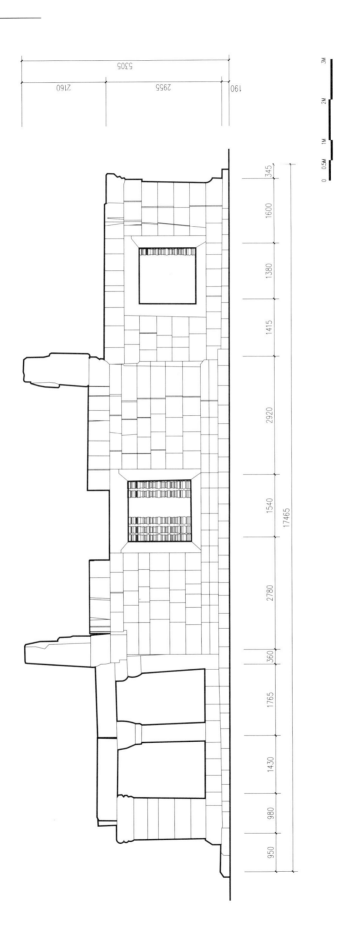

76. 南内长厅西立面图

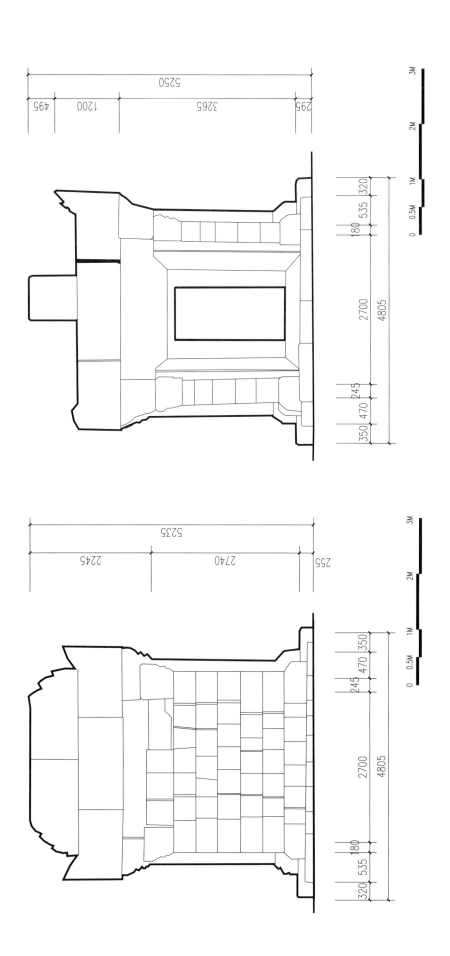

77. 南内长厅南立面图

78. 南内长厅北立面图

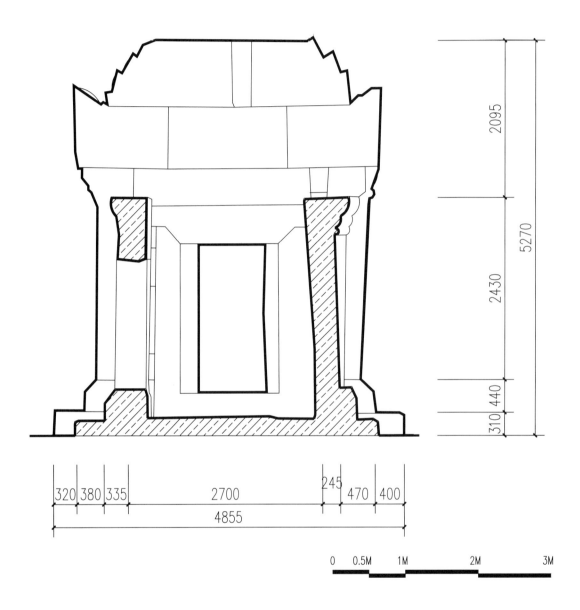

79. 南内长厅 1-1 剖面图

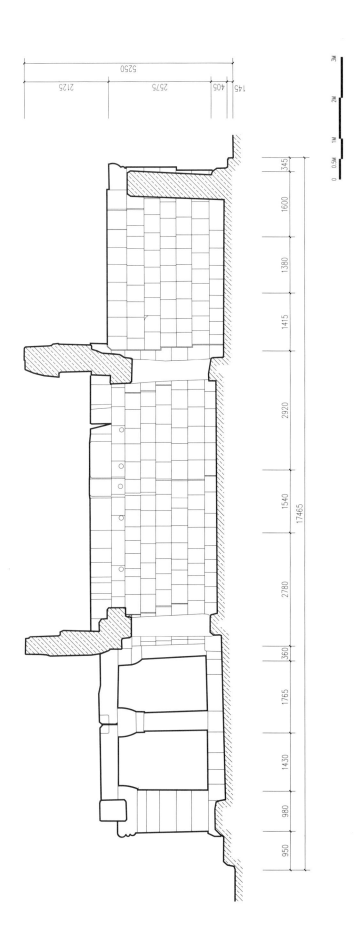

80. 南内长厅 2-2 剖面图

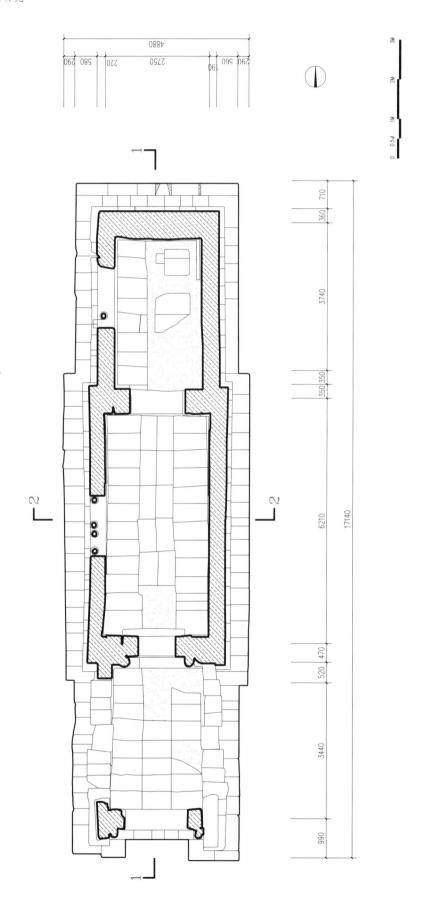

81. 北内长厅平面图

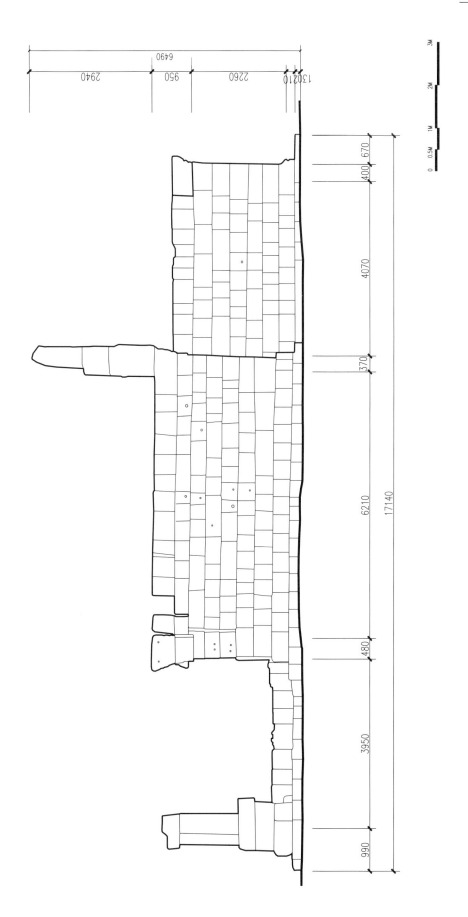

82. 北内长厅东立面图

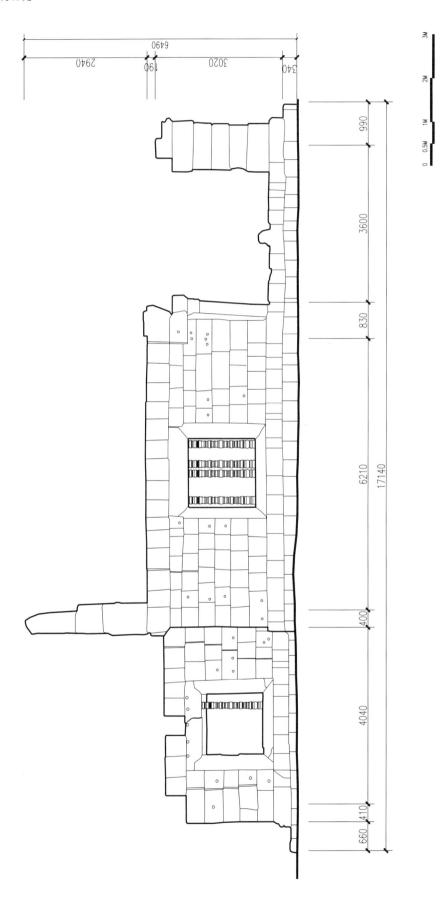

83. 北内长厅西立面图

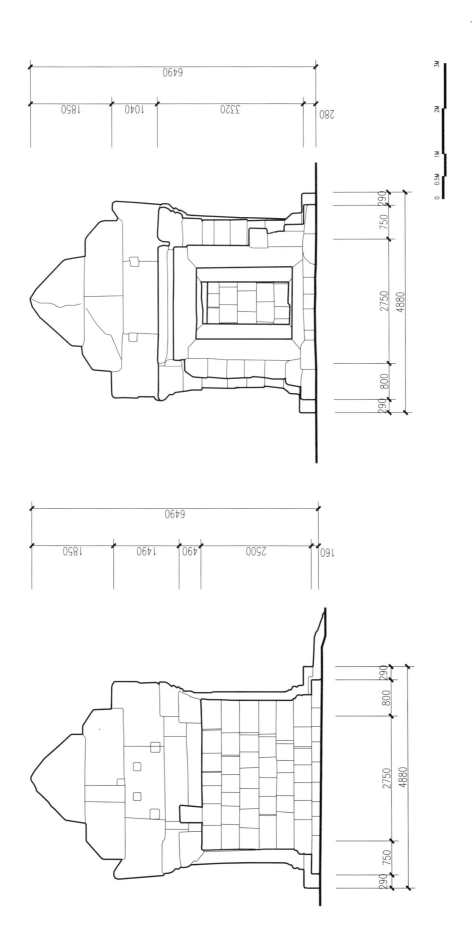

85. 北内长厅南立面图

84. 北内长厅北立面图

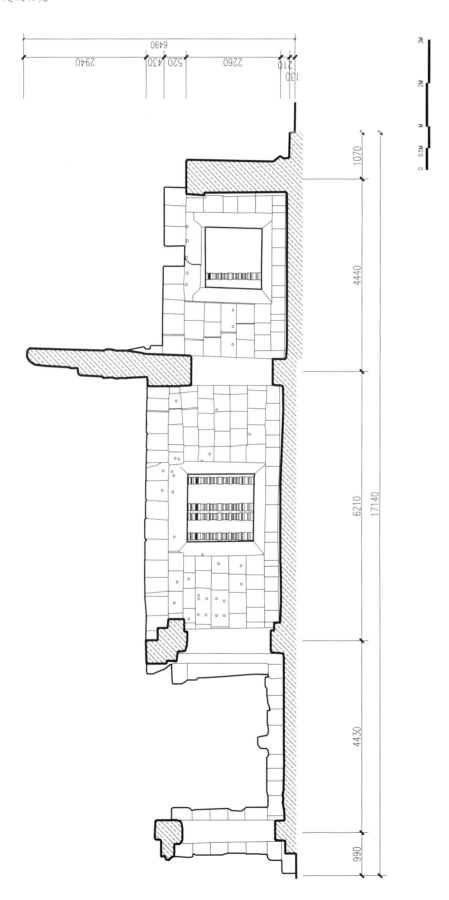

86. 北内长厅 1-1 剖面图

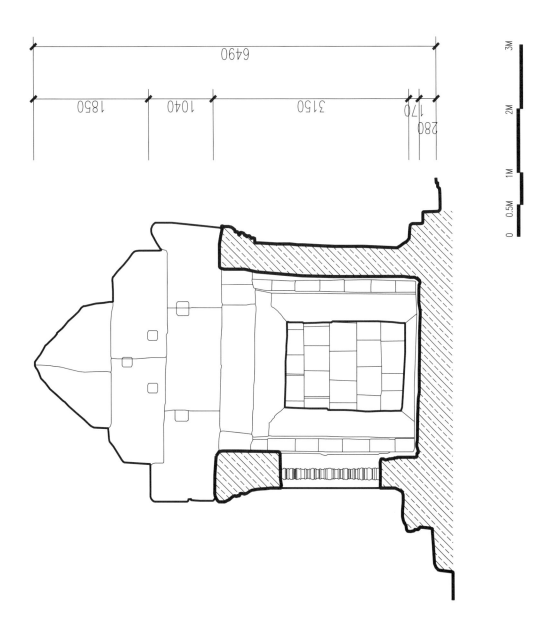

87. 北内长厅 2-2 剖面图

茶胶寺庙山建筑研究

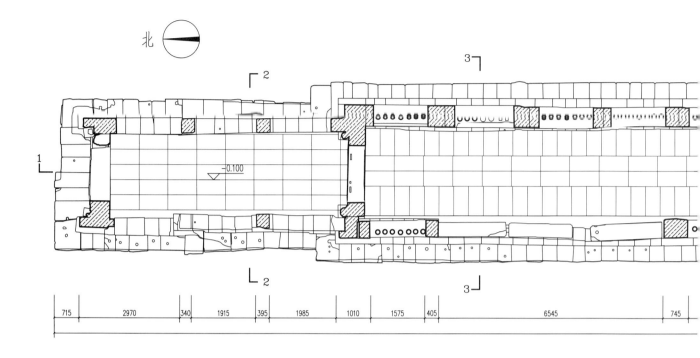

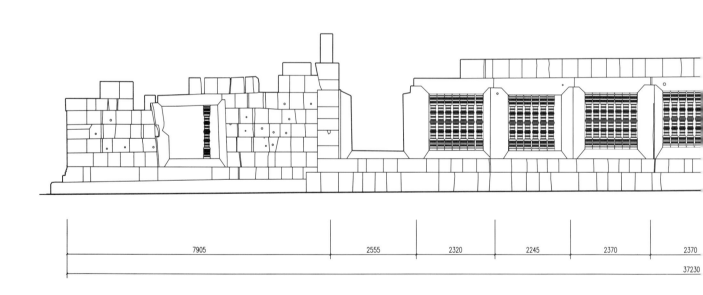

———— 实测图

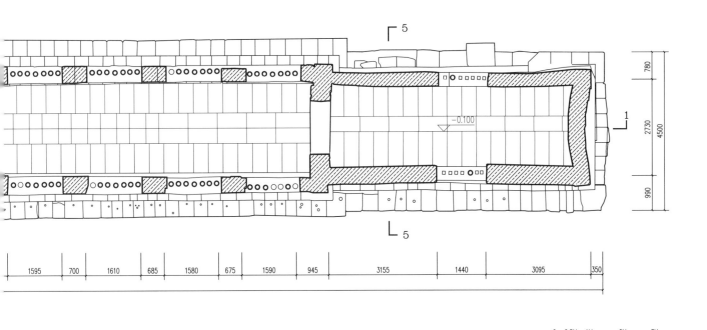

88. 南外长厅平面图

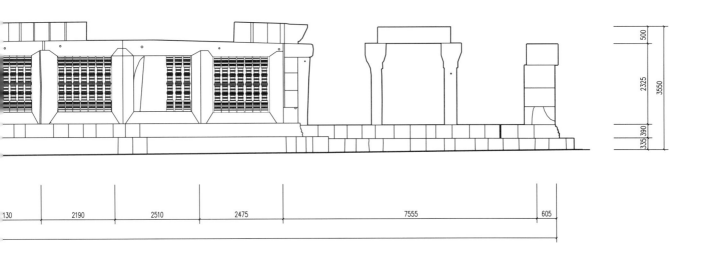

89. 南外长厅东立面图

茶胶寺庙山建筑研究

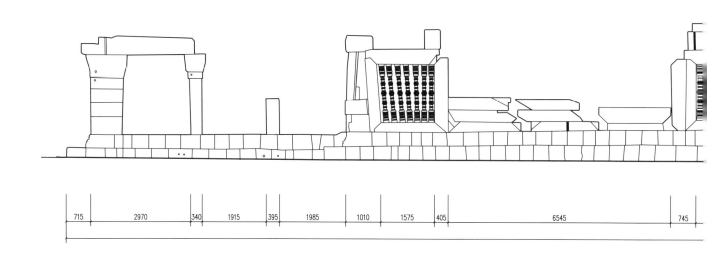

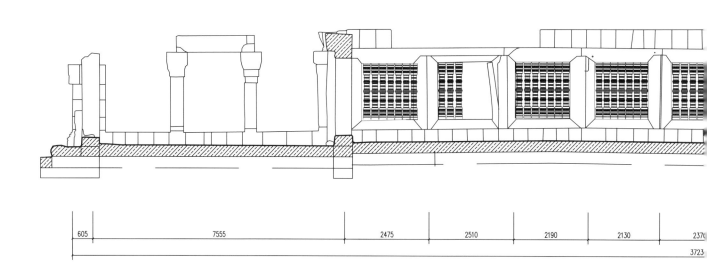

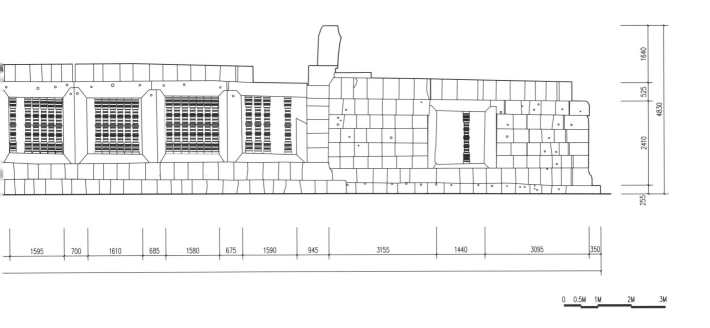

90. 南外长厅西立面图

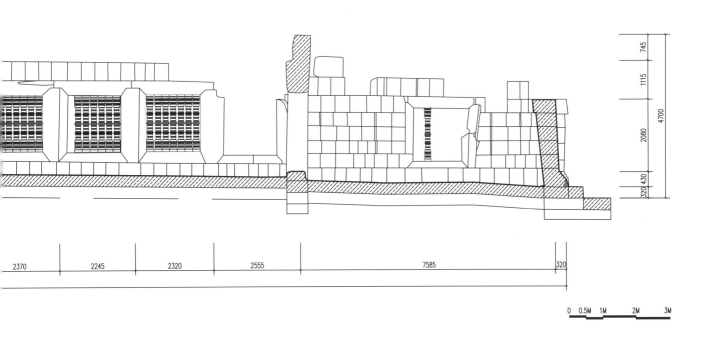

91. 南外长厅 1-1 剖面图

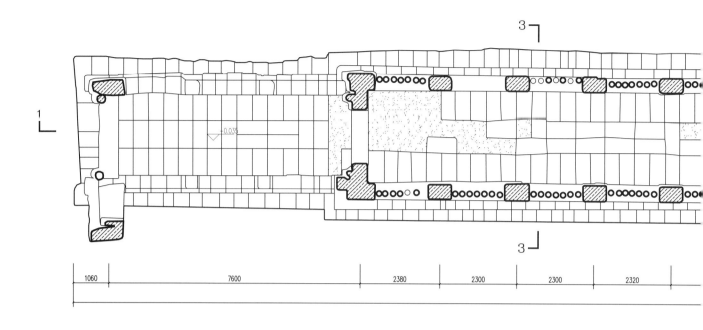

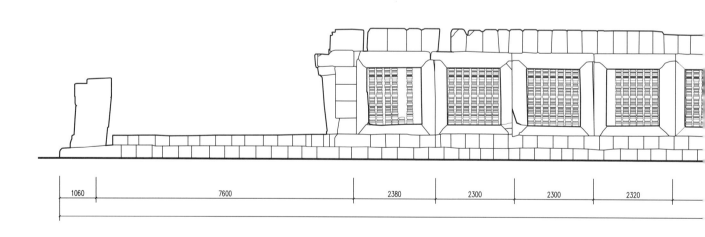

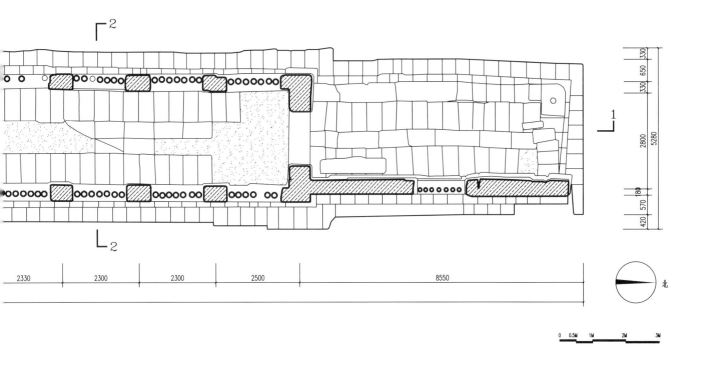

92. 北外长厅平面图

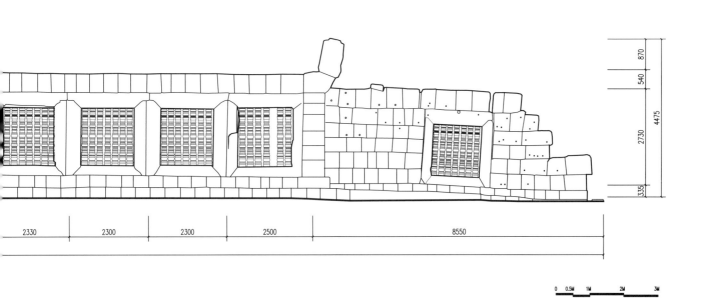

93. 北外长厅东立面图

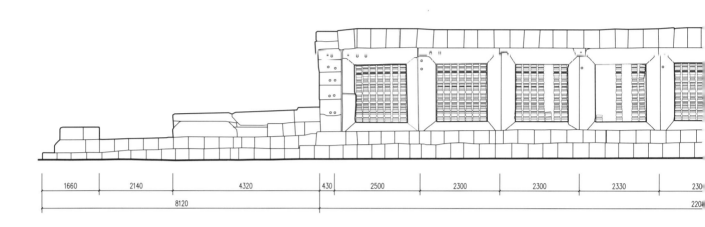

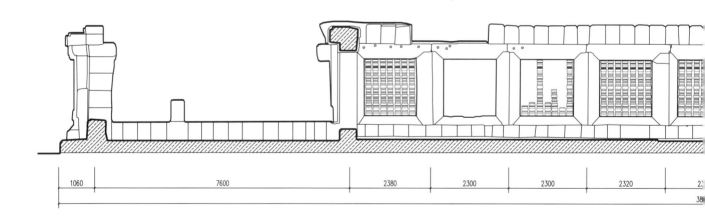

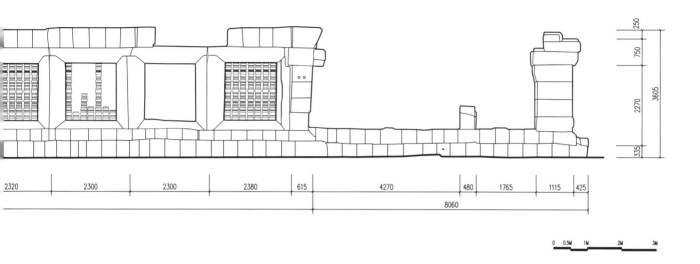

94. 北外长厅西立面图

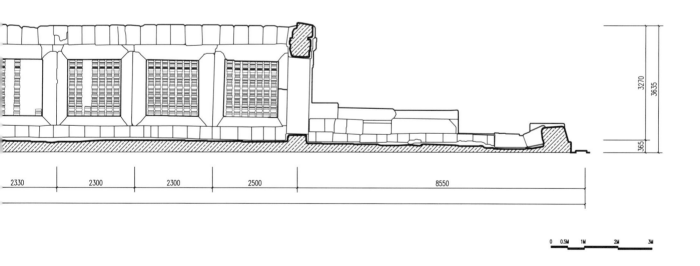

95. 北外长厅 1-1 剖面图

图版

1. 茶胶寺庙山鸟瞰

2. 茶胶寺庙山东侧及神道局部

3. 茶胶寺庙山西南侧全景

4. 茶胶寺庙山东北侧

5. 茶胶寺庙山西北侧

6. 热带暴雨中的茶胶寺庙山东侧

7. 茶胶寺庙山东北侧

8. 茶胶寺庙山西侧

9. 茶胶寺庙山南侧

10. 中央主塔东侧外观

11. 中央主塔中厅仰视

12. 中央主塔西侧过厅内景

13. 中央主塔中厅内景

14. 中央主塔西侧外观

15. 中央主塔顶部假层的砌筑方式

16. 中央主塔东侧抱厦

17. 中央主塔北抱厦的抱框石

18. 中央主塔南侧外观

19. 中央主塔塔顶砌石的构造方式之一

20. 中央主塔塔顶砌石的构造方式之二

21. 中央主塔塔顶俯瞰抱厦顶部

22. 东北角塔东侧外观

23. 东北角塔北侧外观

24. 东北角塔西侧外观

25. 东北角塔南侧外观

26. 东北角塔东侧抱厦内景

27. 东北角塔西侧抱厦内景

28. 东北角塔中厅内景之一

29. 东北角塔中厅内景之二

茶胶寺庙山建筑研究

30. 东北角塔俯瞰

31. 东北角塔中厅内景之三

32. 东南角塔西侧外观 33. 东南角塔东侧外观

34. 东南角塔北侧外观

35. 东南角塔俯瞰

36. 东南角塔中厅内景之一

37. 西北角塔仰视

38. 西北角塔北侧踏道

39. 西北角塔俯瞰

40. 西南角塔东侧外观

41. 西南角塔北侧外观

42. 西南角塔顶部砌石细部之一

43. 西南角塔顶部砌石细部之二

44. 西南角塔俯瞰

45. 须弥台东北角顶部现状

46. 须弥台东侧顶部现状（自北向南拍摄）

47. 须弥台西北侧顶部现状（自东向西拍摄）

48. 须弥台西南侧顶部现状（自北向南拍摄）

49. 须弥台东北角部现状之一

50. 须弥台东北角部现状之二

51. 须弥台之西北侧

52. 须弥台东侧踏道之一

53. 须弥台东侧踏道之二

54. 须弥台东侧踏道局部

55. 须弥台东侧踏道垛台之一

56. 须弥台东侧侧踏道垛台之二

57. 须弥台北侧踏道

58. 须弥台北侧踏道垛台

59. 须弥台西侧踏道

60. 须弥台西侧踏道细部

61. 须弥台东侧的砂岩雕刻遗存

62. 须弥台东侧的砂岩雕刻（仰莲）

63. 须弥台东侧的砂岩雕刻纹样之一

64. 须弥台东侧的砂岩雕刻纹样之二

65. 须弥台东侧的砂岩雕刻纹样之三

66. 须弥台东侧的砂岩雕刻风化现状之一

67. 须弥台东侧的砂岩雕刻风化现状之二

68. 须弥台东侧的砂岩雕刻风化现状之三

69. 须弥台东侧的砂岩雕刻风化现状之四

70. 须弥台东侧的砂岩雕刻风化现状之五

71. 须弥台东侧的砂岩雕刻风化现状之六

72. 东外塔门东侧立面

73. 东外塔门西侧立面

74. 东外塔门北侧室内景

75. 东外塔门主室内景

76. 东外塔门东北侧

77. 东外塔门及东内塔门西侧俯瞰

78. 北外塔门北侧立面

79. 北外塔门南侧俯瞰

80. 北外塔门南侧立面

81. 北外塔门东侧外观

82. 北外塔门西侧俯瞰

83. 西外塔门东侧立面

84. 西外塔门内北侧外观

85. 西外塔门南抱厦

86. 西外塔门西侧

87. 南外塔门西侧

88. 南外塔门东侧

89. 南外塔门东侧室内景

90. 南外塔门北侧

91. 南外塔门西侧室内景

92. 南外塔门南侧

93. 东内塔门东侧立面

94. 东内塔门西侧立面

95. 东内塔门内北侧

96. 东内塔门北侧室砖叠涩屋顶遗迹

97. 东内塔门北侧砖叠涩遗迹

98. 东内塔门外南侧

茶胶寺庙山建筑研究

99. 北内塔门南侧

100. 北内塔门西侧

101. 北内塔门北侧

102. 北内塔门俯瞰

103. 北内塔门主室仰视

104. 西内塔门东侧

105. 西内塔门东侧

106. 西内塔门东侧俯瞰

107. 西内塔门西侧

108. 西内塔门内侧的假层山花

109. 南内塔门内东侧

110. 南内塔门南侧

111. 南内塔门内侧

112. 南内塔门内西侧

———— 图 版

113. 南内塔门内侧俯瞰

114. 南外长厅西侧明窗之局部

115. 南外长厅西北侧

116. 南外长厅西侧的明窗

117. 南外长厅后室的明窗

118. 南外长厅内景

119. 南外长厅北侧抱厦

120. 南外长厅东北侧俯瞰

121. 南外长厅西侧

122. 北外长厅后室明窗细部

123. 北外长厅西南侧俯瞰

124. 北外长厅南侧抱厦遗迹之一

125. 北外长厅抱厦遗迹之二

126. 北外长厅后室西侧

127. 南内长厅西北侧俯瞰

128. 南内长厅西侧窗棂细部

129. 南内长厅北侧抱厦

130. 南内长厅西侧

131. 南内长厅东侧

132. 南内长厅主室内景

133. 南内长厅主室地面铺砌遗迹

134. 北内长厅西南侧俯瞰

135. 北内长厅南侧抱厦入口

136. 北内长厅抱厦西侧

137. 北内长厅西侧

138. 北内长厅主室西侧立面

139. 北内长厅主室内景

140. 北内长厅抱厦内景之一

141. 北内长厅抱厦内景之二

142. 北内长厅山花遗迹

143. 北藏经阁西侧俯瞰

144. 北藏经阁东南侧

145. 北藏经阁东北侧

146. 北藏经阁南侧

147. 北藏经阁西侧抱厦局部

148. 北藏经阁东北侧

149. 北藏经阁西北侧

150. 北藏经阁俯瞰

151. 北藏经阁内景

152. 北藏经阁东侧

153. 北藏经阁西侧抱厦

154. 北藏经阁抱厦门楣

155. 北藏经阁东门头顶部砌石

156. 北藏经阁西南侧之一

157. 北藏经阁西南侧之二

158. 南藏经阁西北侧俯瞰

159. 南藏经阁西南侧

160. 南藏经阁南侧

161. 南藏经阁西南侧俯瞰

162. 南藏经阁东侧

163. 南藏经阁东北侧

164. 南藏经阁西北侧俯瞰

165. 暴雨中的南藏经阁

166. 烈日下的南藏经阁

167. 第二层台回廊西北侧

168. 第二层台回廊西北侧俯瞰

169. 第二层台回廊东南侧俯瞰

170. 第二层台回廊东南角俯瞰

171. 第二层台回廊西北侧俯瞰

172. 庙山内院回廊北侧东段

173. 庙山内院回廊东侧南段

174. 庙山内院回廊西侧局部

175. 庙山内院回廊西侧南段

176. 庙山内院回廊西南角

177. 庙山内院回廊东侧南段

178. 庙山内院回廊

179. 庙山内院回廊西侧南段

180. 庙山内院（二层台）东北角楼俯瞰

181. 庙山内院（二层台）东北角及角楼

182. 庙山内院东南角及角楼内侧

183. 庙山内院（二层台）东南角及角楼

184. 庙山内院（二层台）西南角及角楼俯瞰

185. 庙山内院（二层台）西南角及角楼内侧

186. 庙山内院（二层台）西南角楼俯瞰

187. 庙山内院（二层台）西南角楼内景（自北向南拍摄）

188. 庙山内院东北侧

189. 庙山内院西侧

190. 庙山内院南侧

191. 庙山内院的北藏经阁与北内长厅

192. 庙山内院北侧

193. 庙山内院北侧俯瞰

194. 庙山内院之东南角

195. 庙山内院之东北角

196. 庙山内院东北角及北藏经阁

197. 庙山外院东南侧

198. 庙山外院东北侧

199. 庙山外院北侧俯瞰

200. 庙山外院围墙西侧局部

201. 庙山外院围墙北侧

202. 庙山外院围墙西北角

203. 庙山外院围墙东北角北侧

204. 庙山外院围墙东北侧内部

205. 庙山外院围墙南侧

206. 北环濠西侧（自东向西）

207. 北环濠通道西侧（自西向东）

208. 东门外南池

209. 东门外神道

210. 东门外神道上的建筑基址（自东向西）

211. 西侧环濠（自北向南）

212. 东内塔门编号 K278 古代高棉文碑铭

213. 茶胶寺庙山东内塔门编号 K278 古代高棉文碑铭之局部

214. 茶胶寺庙山东内塔门编号 K275 古代高棉文碑铭之局部

215. 建筑装饰构件之南外长厅明窗窗棂局部

216. 建筑装饰构件之各种类型的窗棂

217. 建筑装饰构件转角部位雕刻的蛇神（NAGA）纹样

218. 建筑装饰构件表面雕刻的植物纹样之一

219. 建筑装饰构件表面雕刻的植物纹样之二

220. 建筑装饰构件之东外塔门主室的山花

221. 建筑装饰构件之花柱

222. 建筑装饰构件之花柱细部

223. 建筑装饰构件之花柱未完成的细部雕饰

224. 建筑装饰构件细部雕刻（摩羯 Makara）之一

225. 建筑装饰构件细部雕刻（摩羯 Makara）之二

226. 建筑装饰构件细部之三

227. 建筑装饰构件细部之四

228. 建筑装饰构件细部之五

229. 施工流程遗迹之一

230. 施工流程遗迹之二

231. 施工流程遗迹之三

232. 施工流程遗迹之四

233. 第二层基台西南角内部角砾岩的砌筑方式

234. 西南角楼砌石间的金属拉结之一

235. 西南角楼砌石间的金属拉结之二

236. 北内长厅山花背面木构架插隼遗迹之一

237. 北内长厅山花背面木构架插隼遗迹之二

238. 南内长厅附近的散落挡头瓦构件

239. 砂岩砌石表面的圆孔及封石（北内塔门）

240. 墙身角部特殊的砌石方式

241. 回廊抱框石的连接方式

242. 塔门窗棱的固定方式

243. 须弥台内部的角砾岩砌石

244. 庙山内院的建筑遗迹之一

245. 庙山内院的建筑遗迹之二

246. 庙山内院的建筑遗迹之三

247. 庙山内院的建筑遗迹之四

248. 庙山内院的建筑遗迹之五

249. 茶胶寺庙山神道全景之一

250. 茶胶寺庙山神道全景之二

251. 位于东池西堤上与茶胶寺庙山相接的建筑基址之东北侧

252. 位于东池西堤上与茶胶寺庙山相接的建筑基址之南侧

253. 位于东池西堤上与茶胶寺庙山相接的建筑基址之西侧

254. 位于东池西堤上与茶胶寺庙山相接的建筑基址之东南侧局部

255. 位于东池西堤上与茶胶寺庙山相接的建筑基址之北侧

256. 位于东池西堤上另外一座与茶胶寺庙山相接的建筑基址之南侧局部

257. 位于东池西堤上另外一座与茶胶寺庙山相接的建筑基址之西侧

258. 位于东池西堤上另外一座与茶胶寺庙山相接的建筑基址东侧

259. 位于东池西堤上另外一座与茶胶寺庙山相接的建筑基址东侧石狮子雕像遗迹

260. 位于东池西堤上另外一座与茶胶寺庙山相接的建筑基址东侧石狮子雕像底部的铭文

261. 石狮子基座底部铭文拓片

262. 位于茶胶寺庙山西侧阇耶跋摩七世时代所建的小型寺庙（HOSPITAL）遗址

263. 从茶胶寺东遗址北至 1 号遗址

264. 神道东端与东池西堤相接的建筑基址北部（自南向北）

265. 茶胶寺庙山东南角的建筑遗址

补 记

2013年5月－7月，中国文化遗产研究院援柬吴哥古迹保护工作队（CACH-CSA）和柬埔寨吴哥古迹保护与发展管理局（APSARA Authority）和金边皇家艺术大学考古系（RUFA）共同组建考古队，联合开展2013年度茶胶寺庙山田野考古发掘工作。这次考古发掘工作选择以茶胶寺庙山外围的南池为中心，采用5×5平方米探方发掘方式，发掘了解东神道、东壕沟南段、南池等相邻遗迹分布范围和结构布局及其建造工艺，取得了重要的发现。另外，此次考古发掘还发现并局部解剖了茶胶寺庙山东侧的一处石构水道遗迹，是探索茶胶寺庙山与吴哥时代城市格局及其水利系统关联情况的重要遗迹。

2013年度茶胶寺庙山田野考古发掘现场

本书著者参加了上述考古发掘工作，鉴于2013年度茶胶寺庙山考古发掘工作尚处于资料整理和报告编写过程中，仅就目前发掘所获，基本可以确定其神道的建筑形制及其砌石构造，并初步探明了神道、南池、环壕等与庙山整体布局的关系，这都在很大程度上加深了我们对茶胶寺庙山整体布局、建筑形制、构造细节等方面的认识，从而也修正了著者从事茶胶寺庙山建筑形制与复原研究过程中形成的某些初步结论。

2013年度茶胶寺庙山田野考古之东环壕局部及其构造
（照片右侧为石构水道遗迹）

本书付梓之际，特此补注说明及最近考古发掘照片如兹，一方面可以提示读者茶胶寺庙山建筑研究取得最新进展的线索；另一方面也表明，正是基于保护修复与考古研究紧密结合的理路，茶胶寺庙山未知领域的隐秘魅力正在渐次展现出来。

2013年度茶胶寺庙山田野考古之神道形制及其构造

后 记

　　2007年岁末，在我毕业分配来中国文化遗产研究院工作的第二年，适逢中国政府援柬一期周萨神庙保护修复工程即将告竣，工程收尾阶段计划进行竣工图的现场实地测绘。根据当时的工作安排，由我代表文研院与我的母校天津大学建筑学院合作承担此项任务。正是这次机缘巧合的安排，使我开始有机会参与到援柬吴哥古迹保护修复工程之中。可是，当自己第一次走进茶胶寺，却是在挥汗如雨和晕头转向间极为潦草地领略了这座吴哥古迹中最为雄伟且具有鲜明特色的庙山遗构。不曾料想，这座伟大的庙山竟会成为自己未来数年间为之好奇、为之困惑、为之着迷、为之探寻、为之付出却始终未能得见其门径的一座壮观无比的 *Hemasringagir*[1]，一座神秘之山，一座信仰之山。

　　如今忆及当时不得其门而入的困惑，甚为感慨怀念自己与已故的罗哲文先生在柬埔寨吴哥一段过从的经历。记得那时我刚刚到吴哥工作不久，罗哲文先生随中国文物学会代表团访问柬埔寨并专程前往吴哥考察。这段记忆的最深处，则是我与罗老从茶胶寺庙山东塔门进入寺内，先行穿过两层高大的基台，最终攀越极其陡峻的须弥台的踏道，罗老几乎是无须搀扶地直至登临茶胶寺庙山四十余米高的中央主塔之上。一行人从茶胶寺返回驻地午餐，罗老意犹未尽，特意将他灌在小二锅头瓶中的好酒拿出，邀我们几个小辈一起品尝。我素不善饮，现在也无从记起当时好酒的滋味，倒是那时的我们皆在不经意间沉入了微醺之中，倏忽间忘却了谁是老人，谁又是少年。如今斯人已逝，忆及自己当年与罗哲文先生踏访吴哥时的印象，似乎都与那个登临茶胶寺庙山中央主塔的正午有关，以及那一壶让众人皆微醺微醉的美酒，却上心头。

　　过去六年的时光之间，自己曾无数次地探寻触摸过这座雄伟壮观的金色神山。从陌生到相识，从相识到熟悉，从熟悉到了解，从了解到爱，继而仍旧回到追寻的起点……转瞬之间，却是轮回；周而复始，甘苦自知。

　　承蒙中国文化遗产研究院刘曙光、柴晓明、马清林、侯卫东、许言、李战崎等院领导一如既往的扶掖与鼓励，特别是副院长、总工程师侯卫东先生，在专项研究和工程设计方面给与我倾力支持和悉心指导，才使本书的编撰得以顺利进行。过去的数年间，仅就援柬茶胶寺保护修复项目而言，我的研究和设计工作始终得到张廷皓、孟宪民、荣大为、余鸣谦、黄克忠、王丹华、姜怀英、付清远、乔梁、刘兰华等文研院专家和领导的提携与解惑；杨新、沈阳、李宏松、王金华、詹长法、丁燕、郑子良、顾军、张宪文、永昕群、颜华、闫明、王林安、于志飞、葛川、张兵峰、刘江、张秋艳、王元林、余建立、胡源、孙延忠、李黎、刘建辉、张念、陈艺文、周西安、陈京南等诸多文研院师友们更是鼎力支持和热心指教，在此谨向他们致以我最诚挚的感谢。

　　另外，还要特别感谢天津大学建筑学院王其亨教授、张玉坤教授、徐苏斌教授、青木信夫教授、

---

[1] 梵文"金角山"之意，是印度教神话中"须弥山"的象征。

吴葱教授；清华大学建筑学院王贵祥教授、吕舟教授、张杰教授；东南大学建筑研究所刘叙杰教授、朱光亚教授、陈薇教授；故宫博物院单霁翔院长；中国社会科学院考古研究所李裕群研究员；北京大学外国语学院段晴教授；《中国建筑文化遗产》杂志社金磊主编等师友们在不同场合曾经给予我的慷慨与帮助，正是聆听他们无私的教诲使我深深懂得了怀揣一颗谦逊和感恩的心对于学术生涯的意义。

作为国际援助柬埔寨吴哥古迹保护国际行动的积极参与者，援柬茶胶寺保护修复项目还得到柬埔寨吴哥古迹保护与发展管理局（APSARA Authority）的 BUN Naith 局长、ROS Borath 副局长、KHUON Khun–Neay 副局长、So Chheng 先生；日本政府援助吴哥古迹保护工作队（JASA）中川武教授、岩崎好规博士、下田一太博士；法国远东学院（EFEO）的 Pascal Royère 博士、Christophe Pottier 博士、Dominique Soutif 博士、Jacques Gaucher 教授、柯兰博士、陆康博士；德国吴哥古迹保护工作队（GACP）的 Hans Leisen 教授、印度吴哥古迹保护工作队（ASI）Sood 先生等国际同行的热心支持和鼓励，谨向他们致以我最诚挚的谢意。

现在看来，家人对我"执迷不悟"的理解和宽容，是何等的珍贵，无以回报，惟有心存感激。万分感激那已永逝的父爱对我的纵容。家国万里的异乡，朗朗的青空在上，您依然无处不在的注视和护佑时常令我感受激心之痛，面对这生命的无常，面对这永不回归的彼岸，我已无法诉说，因为任何企图告慰您安息的文字都是苍白而徒劳的。许久以来，我只有祈求您无声无息的牵挂和宽容再次来临，让漂泊游荡的我渐渐平静下来。

温玉清

2013 年 3 月于京郊百望山寓所